T0324061

Lectures on
$\mathfrak{sl}_2(\mathbb{C})$-modules

Lectures on $\mathfrak{sl}_2(\mathbb{C})$-modules

Volodymyr Mazorchuk
Uppsala University, Sweden

ICP

Imperial College Press

Published by

Imperial College Press
57 Shelton Street
Covent Garden
London WC2H 9HE

Distributed by

World Scientific Publishing Co. Pte. Ltd.
5 Toh Tuck Link, Singapore 596224
USA office: 27 Warren Street, Suite 401-402, Hackensack, NJ 07601
UK office: 57 Shelton Street, Covent Garden, London WC2H 9HE

British Library Cataloguing-in-Publication Data
A catalogue record for this book is available from the British Library.

LECTURES ON SL_2-MODULES

Copyright © 2010 by Imperial College Press

All rights reserved. This book, or parts thereof, may not be reproduced in any form or by any means, electronic or mechanical, including photocopying, recording or any information storage and retrieval system now known or to be invented, without written permission from the Publisher.

For photocopying of material in this volume, please pay a copying fee through the Copyright Clearance Center, Inc., 222 Rosewood Drive, Danvers, MA 01923, USA. In this case permission to photocopy is not required from the publisher.

ISBN-13 978-1-84816-517-5
ISBN-10 1-84816-517-X

Printed in Singapore.

Preface

The representation theory of Lie algebras is an important and intensively studied area of modern mathematics with applications in practically all major areas of mathematics and physics. There are several textbooks which specialize in different aspects of the representation theory of Lie algebras and its applications, but the usual topics covered in such books are finite-dimensional, highest weight or Harish-Chandra modules.

The smallest simple Lie algebra $\mathfrak{sl}_2$ differs in many aspects from all other semi-simple Lie algebras. One could, for example, mention that $\mathfrak{sl}_2$ is the only semi-simple Lie algebra for which *all* simple (not necessarily finite-dimensional) modules are in some sense understood. The algebra $\mathfrak{sl}_2$ is generated by only two elements and hence is an invaluable source of computable examples. Moreover, in many cases the ideas which one gets from working with $\mathfrak{sl}_2$ generalize relatively easily to other Lie algebras with a minimum of extra knowledge required.

The aim of these lecture notes is to give a relatively short introduction to the representation theory of Lie algebras, based on the Lie algebra $\mathfrak{sl}_2$, with a special emphasis on explicit examples. Using this Lie algebra, we can examine and describe many more aspects of the representation theory of Lie algebras than are covered in standard textbooks.

The notes start with two conventional introductory chapters on finite-dimensional modules and the universal enveloping algebra. The third chapter moves on to the study of weight modules, including a complete classification and explicit construction of all weight modules and a description of the category of all weight modules with finite-dimensional weight spaces, via quiver algebras. This is followed by a description and study of the primitive spectrum of the universal enveloping algebra and its primitive quotients. The next step is a relatively complete description of the Bernstein–Gelfand–

Gelfand category $\mathcal{O}$ and its properties. The two last chapters contain a description of all simple $\mathfrak{sl}_2$-modules and various categorifications of simple finite-dimensional modules. The material presented in the last chapter is based on papers which were published in the last two years.

The notes are primarily directed towards postgraduate students interested in learning the basics of the representation theory of Lie algebras. I hope that these notes could serve as a textbook for both lecture courses and reading courses on this subject. Originally, they were written and used for reading courses which I gave in Uppsala in 2008.

The prerequisites for understanding these notes depend on the chapter. For the first two chapters, one needs only some basic knowledge in linear algebra and rings and modules. For the next two chapters, it is assumed that the reader is familiar with the basics of the representation theory of finite-dimensional associative algebras and basic homological algebra. The last three chapters also require some basic experience with category theory.

At the end of each chapter are comments including some historical background, brief descriptions of more advanced results, and references to some original papers. I tried to present these comments to the best of my knowledge and I would like to apologize in advance for any unforeseen errors or omissions.

There are numerous exercises in the main text and at the end of each chapter. The exercises in the main text are usually relatively straightforward and required to understand the material. It is strongly recommended that the reader at least looks through them. Answers and hints are supplied at the end of the notes.

I would like to thank Ekaterina Orekhova and Valentina Chapovalova for their corrections and comments on the earlier version of the manuscript.

Uppsala, August 2009   *Volodymyr Mazorchuk*

Contents

Chapter 1

Finite-dimensional modules

1.1 The Lie algebra $\mathfrak{sl}_2$ and $\mathfrak{sl}_2$-modules

In what follows we will always work over the field $\mathbb{C}$ of complex numbers. Unless stated otherwise, all vector spaces, tensor products and spaces of homomorphisms are taken over $\mathbb{C}$. As usual, we denote by $\mathbb{Z}$, $\mathbb{Q}$ and $\mathbb{R}$ the sets of integer, rational and real numbers, respectively. We also denote by $\mathbb{N}$ the set of all positive integers and by $\mathbb{N}_0$ the set of all non-negative integers.

The Lie algebra $\mathfrak{g} = \mathfrak{sl}_2 = \mathfrak{sl}_2(\mathbb{C})$ consists of the vector space

$$\mathfrak{sl}_2 = \left\{ \begin{pmatrix} a & b \\ c & d \end{pmatrix} : a, b, c, d \in \mathbb{C};\ a + d = 0 \right\}$$

of all complex 2×2 matrices with zero trace and the binary bilinear operation $[X, Y] = XY - YX$ of taking the *commutant* of two matrices on this vector space. Here XY denotes the usual (associative) product of the matrices $X = \begin{pmatrix} x_{11} & x_{12} \\ x_{21} & x_{22} \end{pmatrix}$ and $Y = \begin{pmatrix} y_{11} & y_{12} \\ y_{21} & y_{22} \end{pmatrix}$ given by the following formula:

$$\begin{pmatrix} x_{11} & x_{12} \\ x_{21} & x_{22} \end{pmatrix} \begin{pmatrix} y_{11} & y_{12} \\ y_{21} & y_{22} \end{pmatrix} = \begin{pmatrix} x_{11}y_{11} + x_{12}y_{21} & x_{11}y_{12} + x_{12}y_{22} \\ x_{21}y_{11} + x_{22}y_{21} & x_{21}y_{12} + x_{22}y_{22} \end{pmatrix}.$$

To simplify the notation we will usually denote the Lie algebra $\mathfrak{sl}_2$ simply by $\mathfrak{g}$.

Exercise 1.1. Prove that for any two square complex matrices X and Y of the same size, the matrix $[X, Y]$ has zero trace.

From Exercise 1.1 it follows that the operation $[\cdot, \cdot]$ on $\mathfrak{g}$ is well-defined. The fact that $\mathfrak{g}$ is a *Lie algebra* means that it has the following properties:

Lemma 1.2.

(i) For any $X \in \mathfrak{g}$ we have $[X, X] = 0$.

(ii) For any $X, Y, Z \in \mathfrak{g}$ we have $[X, [Y, Z]] + [Y, [Z, X]] + [Z, [X, Y]] = 0$.

Proof. We have $[X, X] = XX - XX = 0$, proving the statement (i). The statement (ii) is proved by the following computation:

$$[X, [Y, Z]] + [Y, [Z, X]] + [Z, [X, Y]]$$
$$= X(YZ - ZY) - (YZ - ZY)X + Y(ZX - XZ)$$
$$- (ZX - XZ)Y + Z(XY - YX) - (XY - YX)Z$$
$$= XYZ - XZY - YZX + ZYX + YZX - YXZ$$
$$- ZXY + XZY + ZXY - ZYX - XYZ + YXZ$$
$$= 0. \qquad \square$$

Exercise 1.3. Show that the condition in Lemma 1.2(i) is equivalent to the following condition: $[X, Y] = -[Y, X]$ for all $X, Y \in \mathfrak{g}$.

The condition in Lemma 1.2(ii) is called the *Jacobi identity*. The assertion of Exercise 1.3 is true over any field of characteristic different from 2 and basically says that the operation $[\cdot, \cdot]$ is *antisymmetric*.

From the definition we have that elements of the algebra $\mathfrak{g}$ are given by four parameters and one non-trivial linear relation. This means that this algebra has dimension three. We now fix the following *natural* or *standard basis* of $\mathfrak{g}$:

$$\mathbf{e} = \begin{pmatrix} 0 & 1 \\ 0 & 0 \end{pmatrix}, \quad \mathbf{f} = \begin{pmatrix} 0 & 0 \\ 1 & 0 \end{pmatrix}, \quad \mathbf{h} = \begin{pmatrix} 1 & 0 \\ 0 & -1 \end{pmatrix}.$$

By a direct calculation one gets the following *Cayley table* for the operation $[\cdot, \cdot]$ in the standard basis:

$[\cdot, \cdot]$	e	f	h
e	0	h	-2e
f	$-$h	0	2f
h	2e	-2f	0

Another way to notate the essential information from the above Cayley table (the diagonal of the table is fairly obvious and given by Lemma 1.2(i)) is the following:

$$\begin{aligned} [\mathbf{e}, \mathbf{f}] &= \mathbf{ef} - \mathbf{fe} = \mathbf{h}, \\ [\mathbf{h}, \mathbf{e}] &= \mathbf{he} - \mathbf{eh} = 2\mathbf{e}, \\ [\mathbf{h}, \mathbf{f}] &= \mathbf{hf} - \mathbf{fh} = -2\mathbf{f}. \end{aligned} \qquad (1.1)$$

A *module over* $\mathfrak{g}$ (or, simply, a $\mathfrak{g}$-*module*) is a vector space V together with three fixed linear operators $E = E_V$, $F = F_V$ and $H = H_V$ on V, which satisfy the right-hand side equalities in (1.1), that is

$$EF - FE = H, \quad HE - EH = 2E, \quad HF - FH = -2F. \qquad (1.2)$$

It is worth noting that the last two relations can be rewritten as follows:

$$HE = E(H + 2), \quad HF = F(H - 2). \qquad (1.3)$$

Example 1.4. Let $V = \mathbb{C}$ and $E = F = H = 0$. Then all equalities in (1.2) obviously hold and we get the *trivial* $\mathfrak{g}$-module.

Example 1.5. Let $V = \mathbb{C}^2$. In the usual way we identify the set of all linear operators on V with the set of all complex 2×2 matrices. Set $E = \mathbf{e}$, $F = \mathbf{f}$ and $H = \mathbf{h}$. All equalities in (1.2) hold because of (1.1) and we get the *natural* $\mathfrak{g}$-module.

Example 1.6. Take now $V = \mathfrak{g}$. Let E denote the linear operator on V given by $[\mathbf{e}, \_]$ (that is the linear operator of taking the commutator with $\mathbf{e}$, the latter standing on the left). Analogously define F as $[\mathbf{f}, \_]$ and H as $[\mathbf{h}, \_]$. From Lemma 1.7 below we have that these linear operators satisfy (1.2) and we get the *adjoint* $\mathfrak{g}$-module.

Lemma 1.7. *For any $X \in \mathfrak{g}$ we have*

$$[\mathbf{e}, [\mathbf{f}, X]] - [\mathbf{f}, [\mathbf{e}, X]] = [\mathbf{h}, X],$$
$$[\mathbf{h}, [\mathbf{e}, X]] - [\mathbf{e}, [\mathbf{h}, X]] = [2\mathbf{e}, X],$$
$$[\mathbf{h}, [\mathbf{f}, X]] - [\mathbf{f}, [\mathbf{h}, X]] = [-2\mathbf{f}, X].$$

Proof. The equality $[\mathbf{e}, [\mathbf{f}, X]] - [\mathbf{f}, [\mathbf{e}, X]] = [\mathbf{h}, X]$ can be rewritten as follows:

$$[\mathbf{e}, [\mathbf{f}, X]] - [\mathbf{f}, [\mathbf{e}, X]] - [\mathbf{h}, X] = 0. \qquad (1.4)$$

Recall that $\mathbf{h} = [\mathbf{e}, \mathbf{f}]$. Applying now Exercise 1.3 to the inner bracket of the second summand and the outer bracket of the third summand reduces the equality (1.4) to the Jacobi identity. Hence the first equality from the formulation follows from Lemma 1.2(ii). The rest is proved similarly. $\square$

Given two $\mathfrak{g}$-modules V and W, a *homomorphism* from V to W (or a $\mathfrak{g}$-*homomorphism*, or, simply, a *morphism*) is a linear map $\Phi : V \to W$ that makes the following diagram commutative for all $X \in \{E, F, H\}$:

$$
\begin{array}{ccc}
V & \xrightarrow{\ X_V\ } & V \\
\Phi \downarrow & & \downarrow \Phi \\
W & \xrightarrow{\ X_W\ } & W
\end{array}
$$

In other words, the linear map Φ intertwines the actions of $\mathbf{e}$, $\mathbf{f}$ and $\mathbf{h}$ on V and W in the following sense:

$$\Phi E_V = E_W \Phi, \quad \Phi F_V = F_W \Phi, \quad \Phi H_V = H_W \Phi. \tag{1.5}$$

The set of all homomorphisms from V to W is denoted by $\mathrm{Hom}_\mathfrak{g}(V, W)$.

Example 1.8. For any two $\mathfrak{g}$-modules V and W, the zero linear map from V to W obviously satisfies (1.5). This is the so-called *zero homomorphism*.

From Example 1.8 it follows that the set $\mathrm{Hom}_\mathfrak{g}(V, W)$ is always non-empty.

Exercise 1.9. Show that $\mathrm{Hom}_\mathfrak{g}(V, W)$ is closed with respect to the usual addition of linear maps and multiplication of linear maps by complex numbers. In particular, show that the set $\mathrm{Hom}_\mathfrak{g}(V, W)$ is a vector space.

Example 1.10. For any $\mathfrak{g}$-module V the identity map id_V on V obviously satisfies (1.5) (where $V = W$). This is the so-called *identity homomorphism*.

An injective homomorphism is called a *monomorphism*, a surjective homomorphism is called an *epimorphism* and a bijective homomorphism is called an *isomorphism*. As usual, it only makes sense to study $\mathfrak{g}$-modules up to isomorphism. The fact that two modules V and W are isomorphic is usually denoted by $V \cong W$.

Let V be a $\mathfrak{g}$-module. A subspace $W \subset V$ is called an *submodule* (or a $\mathfrak{g}$-*submodule*) of V provided that it is invariant with respect to the action of E_V, F_V and H_V, that is

$$E_V W \subset W, \quad F_V W \subset W, \quad H_V W \subset W. \tag{1.6}$$

For example, the module V always has two obvious submodules; namely, the zero subspace and the whole space V. Any submodule, different from these two is called a *proper* submodule. A module which does not have any proper submodules is called *simple*. For example, any module of dimension one is simple.

Exercise 1.11. Show that all $\mathfrak{g}$-modules from Examples 1.4, 1.5 and 1.6 are simple.

Exercise 1.12. Let $V = \mathbb{C}^2$ and $E = F = H = 0$. Show that this defines on V the structure of a $\mathfrak{g}$-module that is not simple.

Exercise 1.13. Let V be a $\mathfrak{g}$-module and W a submodule of V. Show that the quotient space V/W carries the natural structure of a $\mathfrak{g}$-module given by $E(v+W) = E(v)+W$, $F(v+W) = F(v)+W$ and $H(v+W) = H(v)+W$. The module V/W is called the *quotient* or the *factor* of V by W.

Lemma 1.14. *Let V and W be two $\mathfrak{g}$-modules and $\Phi \in \mathrm{Hom}_\mathfrak{g}(V,W)$. Then*

(i) The kernel $\mathrm{Ker}(\Phi)$ of Φ is a submodule of V.
(ii) The image $\mathrm{Im}(\Phi)$ of Φ is a submodule of W.

Proof. Let $v \in \mathrm{Ker}(\Phi)$ and $X \in \{E, F, H\}$. This gives us

$$\Phi(X_V(v)) = \Phi X_V(v) \overset{(1.5)}{=} X_W \Phi(v) = 0,$$

implying $X_V(v) \in \mathrm{Ker}(\Phi)$. This proves (i). To prove (ii) is left as an exercise to the reader. □

1.2 Classification of simple finite-dimensional modules

This section contains perhaps the most classical part of the $\mathfrak{sl}_2$-representation theory; namely, a classification of all simple *finite-dimensional* $\mathfrak{sl}_2$-modules. As we will see later, such modules form only a very small family of simple $\mathfrak{sl}_2$-modules. A description of *all* simple $\mathfrak{sl}_2$-modules is an ultimate goal of this book, but that will require much more theory and effort. The beauty of finite-dimensional modules is in the fact that their description is absolutely elementary.

Let $V \neq 0$ be a finite-dimensional $\mathfrak{g}$-module. For $\lambda \in \mathbb{C}$ set

$$V(\lambda) = \{v \in V : (H - \lambda)^k v = 0 \text{ for some } k \in \mathbb{N}\},$$
$$V_\lambda = \{v \in V : Hv = \lambda v\}$$

(here, as usual, we identify $\mathbb{C}$ with multiples of id_V).

As we are working over the algebraically closed field of complex numbers, from the Jordan Decomposition Theorem we have that

$$V \cong \bigoplus_{\lambda \in \mathbb{C}} V(\lambda). \tag{1.7}$$

Set $W = \bigoplus_{\lambda \in \mathbb{C}} V_\lambda \subset V$ and note that $W \neq 0$ as H must have at least one eigenvalue and hence at least one non-zero eigenvector in V.

Lemma 1.15. *Let $\lambda \in \mathbb{C}$.*

(i) $EV(\lambda) \subset V(\lambda + 2)$ and $EV_\lambda \subset V_{\lambda+2}$.
(ii) $FV(\lambda) \subset V(\lambda - 2)$ and $FV_\lambda \subset V_{\lambda-2}$.
(iii) $HV(\lambda) \subset V(\lambda)$ and $HV_\lambda \subset V_\lambda$.

Proof.   For $v \in V_\lambda$ we have

$$H(E(v)) = HEv \overset{(1.2)}{=} EHv + 2Ev = \lambda Ev + 2Ev = (\lambda + 2)E(v),$$

which implies the second part of the statement (i). Similarly, for $v \in V(\lambda)$ let $k \in \mathbb{N}_0$ be such that $(H - \lambda)^k v = 0$. Then we have

$$(H - (\lambda + 2))^k (E(v)) = (H - (\lambda + 2))^k Ev \overset{(1.3)}{=}$$
$$= E(H + 2 - (\lambda + 2))^k v = E(H - \lambda)^k v = 0,$$

which implies the first part of the statement (i).

The statement (ii) is proved similarly. The statement (iii) is obvious from the definitions. $\square$

Exercise 1.16. Generalizing the arguments from the proof of Lemma 1.15, show that for any $f(x) \in \mathbb{C}[x]$ one has the equalities $f(H)E = Ef(H + 2)$ and $f(H)F = Ff(H - 2)$.

From Lemma 1.15 we immediately obtain:

Corollary 1.17. *The space W is a submodule of V, in particular, we have $W = V$ if the module V is simple.*

If the module V is simple, we can use Corollary 1.17 to improve the decomposition given by (1.7) as follows:

$$V \cong \bigoplus_{\lambda \in \mathbb{C}} V_\lambda. \tag{1.8}$$

Since V is finite-dimensional, the decomposition (1.8) must be finite in the sense that only finitely many summands are non-zero. Thus we can fix some $\mu \in \mathbb{C}$ such that $V_\mu \neq 0$ and $V_{\mu+2k} = 0$ for all $k \in \mathbb{N}$. Let $v \in V_\mu$ be some non-zero element. As $V_{\mu-2k}$ must be zero for some $k \in \mathbb{N}$, from Lemma 1.15(ii) it follows that there exists a minimal $n \in \mathbb{N}$ such that $F^n v = 0$. For $i \in \{1, 2, \ldots, n-1\}$ set $v_i = F^i v$, and also set $v_0 = v$. From (1.8) it follows that the v_i's are linearly independent. From Lemma 1.15(ii) we have $Hv_i = (\mu - 2i)v_i$ for all i. Let N denote the linear span of all v_i's.

Lemma 1.18. *We have $Ev_0 = 0$ and $Ev_i = i(\mu - i + 1)v_{i-1}$ for all $i \in \{1, 2, \ldots, n-1\}$.*

Proof. That $Ev_0 = 0$ is obvious. To prove the rest we proceed by induction on i. For $i = 1$ we have

$$Ev_1 = EFv_0 \overset{(1.2)}{=} FEv_0 + Hv_0 = 0 + \mu v_0 = \mu v_0,$$

which agrees with our formula. When $i > 1$ for the induction step we have:

$$\begin{aligned} Ev_i &= EFv_{i-1} \\ &\overset{(1.2)}{=} FEv_{i-1} + Hv_{i-1} \\ \text{(inductive assumption)} \quad &= (i-1)(\mu - i + 2)Fv_{i-2} + (\mu - 2(i-1))v_{i-1} \\ &= i(\mu - i + 1)v_{i-1}. \end{aligned}$$

This completes the proof. $\qquad\square$

Corollary 1.19. *N is a submodule of V, in particular, $N = V$ provided that V is simple.*

Proof. That N is invariant with respect to the action of H and F is obvious. By Lemma 1.18 it is also invariant with respect to the action of E. The claim follows. $\qquad\square$

Lemma 1.20. *$\mu = n - 1$.*

Proof. From the inductive argument used in the proof of Lemma 1.18 we get $EFv_{n-1} = n(\mu - n + 1)v_{n-1}$. However, $Fv_{n-1} = 0$ by our assumptions, hence $n(\mu - n + 1) = 0$ implying $\mu = n - 1$. $\qquad\square$

Assuming that V is simple, let us sum up the information which we now have about this module. It has the basis $\{v_0, v_1, \ldots, v_{n-1}\}$, in which the action of the operators E, F and H can be depicted as follows:

$$(1.9)$$

Here $a_i = i(n-i)$. The double arrow represents the action of F, the regular arrow represents the action of E and the dotted arrow represents the action of H. The numbers over arrows are coefficients.

Exercise 1.21. Check that for any $n \in \mathbb{N}$ the picture (1.9) defines on the formal linear span of $\{v_0, \ldots, v_{n-1}\}$ the structure of a $\mathfrak{g}$-module. We will denote this module by $\mathbf{V}^{(n)}$.

Now we are ready to formulate the main result of this section.

Theorem 1.22 (Classification of simple finite-dimensional modules).

(i) *For every $n \in \mathbb{N}$ the module $\mathbf{V}^{(n)}$ is a simple $\mathfrak{g}$-module of dimension n.*

(ii) *For any $n, m \in \mathbb{N}$ we have $\mathbf{V}^{(n)} \cong \mathbf{V}^{(m)}$ if and only if $n = m$.*

(iii) *Let V be a simple finite-dimensional $\mathfrak{g}$-module of dimension n. Then $V \cong \mathbf{V}^{(n)}$.*

Proof. That $\mathbf{V}^{(n)}$ is a module follows from Exercise 1.21. Let $M \subset \mathbf{V}^{(n)}$ be a non-zero submodule and $v \in M$, $v \neq 0$. From (1.9) we have that $E^n v = 0$, in particular, $E^n M = 0$ and hence M must have a non-trivial intersection with the kernel of E. Again, from (1.9) it follows that the kernel of E is just the linear span of v_0 and is, in particular, one-dimensional. Hence M contains v_0. Applying to v_0 the operator F inductively we get that M contains all the v_i's. Hence $M = \mathbf{V}^{(n)}$. This proves the statement (i).

As $\dim \mathbf{V}^{(n)} = n$, the statement (ii) is obvious. The statement (iii) follows from the analysis leading to the picture (1.9). $\qquad\square$

Exercise 1.23. Show that after rescaling the basis $\{v_i\}$ in the following way: $w_i = \frac{1}{i!} v_i$ the picture (1.9) transforms into the following symmetric form:

$$(1.10)$$

Exercise 1.24. Show that one can rescale the basis $\{v_i\}$ so that in the new basis $\{\hat{w}_i\}$ the picture (1.9) transforms into the following symmetric form:

$$(1.11)$$

Exercise 1.25. Let V be a simple finite-dimensional $\mathfrak{g}$-module which contains a non-zero vector v such that $E(v) = 0$ and $H(v) = (n-1)v$. Show that $V \cong \mathbf{V}^{(n)}$.

In the basis $\{w_0, w_1, \ldots, w_{n-1}\}$ from Exercise 1.23 the linear operators E, F and H are given by the following matrices:

$$E = \begin{pmatrix} 0 & n\text{-}1 & 0 & \ldots & 0 & 0 & 0 \\ 0 & 0 & n\text{-}2 & \ldots & 0 & 0 & 0 \\ 0 & 0 & 0 & \ldots & 0 & 0 & 0 \\ \vdots & \vdots & \vdots & \vdots & \vdots & \vdots & \vdots \\ 0 & 0 & 0 & \ldots & 0 & 2 & 0 \\ 0 & 0 & 0 & \ldots & 0 & 0 & 1 \\ 0 & 0 & 0 & \ldots & 0 & 0 & 0 \end{pmatrix} \quad F = \begin{pmatrix} 0 & 0 & 0 & \ldots & 0 & 0 & 0 \\ 1 & 0 & 0 & \ldots & 0 & 0 & 0 \\ 0 & 2 & 0 & \ldots & 0 & 0 & 0 \\ \vdots & \vdots & \vdots & \vdots & \vdots & \vdots & \vdots \\ 0 & 0 & 0 & \ldots & 0 & 0 & 0 \\ 0 & 0 & 0 & \ldots & n\text{-}2 & 0 & 0 \\ 0 & 0 & 0 & \ldots & 0 & n\text{-}1 & 0 \end{pmatrix}$$

$$H = \begin{pmatrix} n\text{-}1 & 0 & 0 & \ldots & 0 & 0 & 0 \\ 0 & n\text{-}3 & 0 & \ldots & 0 & 0 & 0 \\ 0 & 0 & n\text{-}5 & \ldots & 0 & 0 & 0 \\ \vdots & \vdots & \vdots & \vdots & \vdots & \vdots & \vdots \\ 0 & 0 & 0 & \ldots & 5\text{-}n & 0 & 0 \\ 0 & 0 & 0 & \ldots & 0 & 3\text{-}n & 0 \\ 0 & 0 & 0 & \ldots & 0 & 0 & 1\text{-}n \end{pmatrix}$$

We complete this section with a description of homomorphisms between simple modules:

Theorem 1.26 (Schur's lemma).

(i) Any non-zero homomorphism between two simple $\mathfrak{g}$-modules is an isomorphism.

(ii) For any two simple finite-dimensional $\mathfrak{g}$-modules V and W we have

$$\mathrm{Hom}_{\mathfrak{g}}(V, W) \cong \begin{cases} \mathbb{C}, & V \cong W; \\ 0, & \text{otherwise.} \end{cases}$$

Proof. Let $\Phi \in \mathrm{Hom}_{\mathfrak{g}}(V, W)$ be some non-zero homomorphism. Applying Lemma 1.14(i) we have that the kernel of Φ is a submodule of V. As V is simple and $\Phi \neq 0$, we thus get that $\mathrm{Ker}(\Phi) = 0$ and Φ is injective. By Lemma 1.14(ii) the image of Φ is a submodule of W. As W is simple and $\Phi \neq 0$, we thus get that $\mathrm{Im}(\Phi) = W$ and Φ is surjective. Therefore any non-zero element of $\mathrm{Hom}_{\mathfrak{g}}(V, W)$ is an isomorphism, proving the statement (i). In particular, $\mathrm{Hom}_{\mathfrak{g}}(V, W) = 0$ if $V \not\cong W$.

Assume now that $V \cong W$ and $0 \neq \Psi \in \mathrm{Hom}_{\mathfrak{g}}(V, W)$. We then have an obvious isomorphism from $\mathrm{Hom}_{\mathfrak{g}}(V, V)$ to $\mathrm{Hom}_{\mathfrak{g}}(V, W)$ given by $\Phi \mapsto \Psi \circ \Phi$. Let us show that

$$\mathrm{Hom}_{\mathfrak{g}}(V, V) = \mathbb{C}\,\mathrm{id}_V \cong \mathbb{C}.$$

If $\Phi \in \mathrm{Hom}_{\mathfrak{g}}(V, V)$ is non-zero, it must have a non-zero eigenvalue, $\lambda \in \mathbb{C}$ say (as $\mathbb{C}$ is algebraically closed). Then $\Phi - \lambda \cdot \mathrm{id}_V \in \mathrm{Hom}_{\mathfrak{g}}(V, V)$. However, any eigenvector of Φ with eigenvalue λ belongs to the kernel of $\Phi - \lambda \cdot \mathrm{id}_V$. Thus $\Phi - \lambda \cdot \mathrm{id}_V$ is not an isomorphism, and hence it must be zero by the previous paragraph. This yields $\Phi = \lambda \cdot \mathrm{id}_V$ and completes the proof. $\square$

1.3 Semi-simplicity of finite-dimensional modules

Given two $\mathfrak{g}$-modules V and W the vector space $V \oplus W$ can be endowed with the structure of a $\mathfrak{g}$-module as follows:

$$\begin{aligned} E(v \oplus w) &= E(v) \oplus E(w), \\ F(v \oplus w) &= F(v) \oplus F(w), \\ H(v \oplus w) &= H(v) \oplus H(w). \end{aligned} \tag{1.12}$$

The module $V \oplus W$ is called the *direct sum* of V and W. For $n \in \mathbb{N}$ and any $\mathfrak{g}$-module V we denote by nV the $\mathfrak{g}$-module

$$\underbrace{V \oplus V \oplus \cdots \oplus V}_{n \text{ summands}}.$$

Exercise 1.27. Check that the formulae (1.12) indeed define on $V \oplus W$ the structure of a $\mathfrak{g}$-module.

A $\mathfrak{g}$-module V is called *decomposable* provided that $V \cong V_1 \oplus V_2$ for some non-zero $\mathfrak{g}$-modules V_1 and V_2. Those $\mathfrak{g}$-modules which are not decomposable are called *indecomposable*. A module, which is isomorphic to a direct sum of (possibly finitely many) simple modules is called *semi-simple*.

Example 1.28. Every simple module is indecomposable; in particular, every one-dimensional module is indecomposable. Indeed, if $V \cong V_1 \oplus V_2$ and both V_1 and V_2 are non-zero, then V_1 is a proper submodule of V and hence V cannot be simple.

In the general situation (for example if one considers all $\mathfrak{g}$-modules) there exist many indecomposable modules which are not simple. We will see many examples later on. However, the case of finite-dimensional modules turns out to be very special. The main aim of the present section is to prove the following statement:

Theorem 1.29 (Weyl's Theorem). *Every indecomposable finite-dimensional $\mathfrak{g}$-module is simple. Equivalently, every finite-dimensional $\mathfrak{g}$-module is semi-simple.*

To prove this theorem we will need some preparation. From now and until the end of the proof we assume that V is a finite-dimensional $\mathfrak{g}$-module. Consider the *Casimir operator* $C = C_V$ on V, defined as follows:

$$C = (H+1)^2 + 4FE.$$

Lemma 1.30.

(i) $C = (H-1)^2 + 4EF = H^2 + 1 + 2EF + 2FE.$
(ii) $HC = CH,\ EC = CE,\ FC = CF.$

Proof. The statement (i) follows from the definition of C and the equality $EF = FE + H$. To prove the statement (ii) we use (i) and Exercise 1.16 as follows:

$$\begin{aligned}
HC &= H((H+1)^2 + 4FE) \\
&= H(H+1)^2 + 4HFE \\
\text{(by Exercise 1.16)} &= (H+1)^2 H + 4F(H-2)E \\
\text{(by Exercise 1.16)} &= (H+1)^2 H + 4FEH \\
&= ((H+1)^2 + 4FE)H \\
&= CH;
\end{aligned}$$

$$EC = E((H+1)^2 + 4FE)$$
$$= E(H+1)^2 + 4EFE$$
$$\text{(by Exercise 1.16)} = (H-1)^2 E + 4EFE$$
$$= ((H-1)^2 + 4EF)E$$
$$\text{by (i)} = CE.$$

The equality $FC = CF$ is checked in the same way. $\qquad\square$

To proceed, we need to recall the following result from linear algebra:

Exercise 1.31. Let W be a vector space, A and B two linear commuting operators on the space W and $\lambda \in \mathbb{C}$. Show that both the subspace

$$\{w \in W \ : \ Aw = \lambda w\}$$

and the subspace

$$\{w \in W \ : \ (A - \lambda)^k w = 0 \text{ for some } k \in \mathbb{N}\}$$

are invariant with respect to B.

Applying the Jordan Decomposition Theorem to the linear operator C on V we find that

$$V \cong \bigoplus_{\tau \in \mathbb{C}} V(C, \tau),$$

where

$$V(C, \tau) = \{v \in V \ : \ (C - \tau)^k v = 0 \text{ for some } k \in \mathbb{N}\}.$$

Lemma 1.32. *For any $\tau \in \mathbb{C}$ the subspace $V(C, \tau)$ is a $\mathfrak{g}$-submodule of V. In particular, if V is indecomposable, then $V = V(C, \tau)$ for some $\tau \in \mathbb{C}$.*

Proof. By Lemma 1.30(ii), the operator C commutes with the operators E, F and H. Hence all these operators preserve $V(C, \tau)$ by Exercise 1.31. The claim follows. $\qquad\square$

Exercise 1.33. Check that $C_{V^{(n)}} = n^2 \cdot \mathrm{id}_{V^{(n)}}$ for all $n \in \mathbb{N}$.

Now we are ready to prove Theorem 1.29:

Proof.   Let V be a non-zero indecomposable finite-dimensional $\mathfrak{g}$-module. This means that it has a non-trivial simple submodule and hence from Lemma 1.32 and Exercise 1.33 we obtain that $V = V(C, n^2)$ for some $n \in \mathbb{N}$.

Consider the decomposition (1.7). First, we claim that E acts injectively on any $V(\lambda)$, $\lambda \neq n-1, -n-1$. Indeed, for any $v \in V(\lambda) \cap \mathrm{Ker}(E)$ we have

$$E(H(v)) = EHv \stackrel{(1.2)}{=} HEv - 2Ev = 0,$$

and hence $V(\lambda) \cap \mathrm{Ker}(E)$ is invariant under the action of H. If we have the inequality $V(\lambda) \cap \mathrm{Ker}(E) \neq 0$, then $V_\lambda \cap \mathrm{Ker}(E) \neq 0$ and for any $v \in V_\lambda \cap \mathrm{Ker}(E)$ we have

$$Cv = ((H+1)^2 + 4FE)v = (H+1)^2 v + 4FEv = (\lambda+1)^2 v.$$

At the same time $Cv = n^2 v$ as $V = V(C, n^2)$, which implies $\lambda = n - 1$ or $\lambda = -n-1$. Analogously, one can show that F acts injectively on any $V(\lambda)$ such that $\lambda \neq 1 - n, n + 1$.

Since V is finite-dimensional, the previous paragraph implies that the inequality $V(\lambda) \neq 0$ is possible only if $\lambda \in \{-n+1, -n+3, \ldots, n-1\}$ and that $\mathrm{Ker}(E) = V(n-1)$, $\mathrm{Ker}(F) = V(1-n)$. In particular, $\dim V(\lambda) = \dim V(\mu)$ for any $\lambda, \mu \in \{-n+1, -n+3, \ldots, n-1\}$. Furthermore, for any $i \in \{1, 2, \ldots, n-1\}$ the restriction A_i of the operator F^i to the subspace $V(n-1)$ gives a linear isomorphism from $V(n-1)$ to $V(n-1-2i)$. Hence we can identify $V(n-1)$ and $V(n-1-2i)$ as vector spaces via the action of A_i. Set $A = A_{n-1}$.

As C commutes with H, all $V(\lambda)$'s are invariant with respect to C. Denote by C_1 and H_1 the restrictions of C and H to $V(n-1)$, respectively. Denote by C_2 and H_2 the restrictions of C and H to $V(1-n)$ respectively. Restricting $CF^{n-1} = F^{n-1}C$ to $V(n-1)$ we get

$$AC_1 = C_2 A. \tag{1.13}$$

Analogously, using $FH = (H+2)F$ (see Exercise 1.16) we get

$$AH_1 = (H_2 + 2(n-1))A. \tag{1.14}$$

As $\mathrm{Ker}(E) = V(n-1)$ and $C = (H+1)^2 + 4FE$, we have

$$C_1 = (H_1 + 1)^2. \tag{1.15}$$

As $\mathrm{Ker}(F) = V(1-n)$ and $C = (H-1)^2 + 4EF$, we have

$$C_2 = (H_2 - 1)^2. \tag{1.16}$$

Thus we have:

$$(H_1 + 1)^2 \overset{(1.15)}{=} C_1$$
$$= A^{-1} A C_1$$
$$(\text{by } (1.13)) = A^{-1} C_2 A$$
$$(\text{by } (1.16)) = A^{-1}(H_2 - 1)^2 A$$
$$(\text{by } (1.14)) = A^{-1} A (H_1 - 1 - 2(n-1))^2$$
$$= (H_1 - 1 - 2(n-1))^2.$$

Hence $(H_1 + 1)^2 = (H_1 - 1 - 2(n-1))^2$, which reduces to $H_1 = n - 1$. This means that $V(n-1) = V_{n-1}$. Since $A_i H = (H + 2i) A_i$ and A_i identifies the space $V(n-1)$ with the space $V(n-1-2i)$ for all i, we get $V(\lambda) = V_\lambda$ for all $\lambda \in \{-n+1, -n+3, \ldots, n-1\}$.

Let $\{v_1, \ldots, v_k\}$ be a basis of V_{n-1}. For $i \in \{1, \ldots, k\}$ denote by W_i the linear span of $\{v_i, F v_i, \ldots, F^{n-1} v_i\}$. From the above we have

$$V \cong W_1 \oplus W_2 \oplus \cdots \oplus W_k$$

and, by Corollary 1.19, each W_i is a submodule of V. Since V is indecomposable by our assumptions, we get $k = 1$ and thus $\dim V_{n-1} = 1$. In this case Corollary 1.19 and (1.9) imply that $V \cong \mathbf{V}^{(n)}$, which completes the proof. $\qquad\square$

Corollary 1.34. *Let V be a finite-dimensional $\mathfrak{g}$-module. Then*

$$V \cong \bigoplus_{n \in \mathbb{N}} m_n \mathbf{V}^{(n)},$$

where

$$m_n = \dim \operatorname{Hom}_{\mathfrak{g}}(\mathbf{V}^{(n)}, V) = \dim \operatorname{Hom}_{\mathfrak{g}}(V, \mathbf{V}^{(n)}).$$

Proof. From Theorem 1.29 it follows that we can decompose V into a direct sum of simple modules, say $V \cong V_1 \oplus \cdots \oplus V_k$, where all the V_i's are simple. Now

$$\dim \operatorname{Hom}_{\mathfrak{g}}(\mathbf{V}^{(n)}, V) = \dim \operatorname{Hom}_{\mathfrak{g}}(\mathbf{V}^{(n)}, \oplus_i V_i)$$
$$= \sum_i \dim \operatorname{Hom}_{\mathfrak{g}}(\mathbf{V}^{(n)}, V_i)$$
$$(\text{by Schur's lemma}) = |\{i \,:\, \mathbf{V}^{(n)} \cong V_i\}|.$$

This proves the first equality for m_n and the second equality is proved similarly. $\qquad\square$

1.4 Tensor products of finite-dimensional modules

Given two $\mathfrak{g}$-modules V and W the vector space $V \otimes W$ can be endowed with the structure of a $\mathfrak{g}$-module as follows:

$$
\begin{aligned}
E(v \otimes w) &= E(v) \otimes w + v \otimes E(w), \\
F(v \otimes w) &= F(v) \otimes w + v \otimes F(w), \\
H(v \otimes w) &= H(v) \otimes w + v \otimes H(w).
\end{aligned}
\tag{1.17}
$$

The module $V \otimes W$ is called the *tensor product* of V and W. For $n \in \mathbb{N}$ and any $\mathfrak{g}$-module V we denote by $V^{\otimes n}$ the $\mathfrak{g}$-module

$$
\underbrace{V \otimes V \otimes \cdots \otimes V}_{n \text{ factors}}.
$$

Exercise 1.35. Check that the formulae (1.17) do indeed define on $V \otimes W$ the structure of a $\mathfrak{g}$-module.

Exercise 1.36. Let V and W be two $\mathfrak{g}$-modules. Check that the map $v \otimes w \mapsto w \otimes v$ induces an isomorphism between $V \otimes W$ and $W \otimes V$.

Exercise 1.37. Let V_1, V_2 and W be $\mathfrak{g}$-modules. Prove that

$$
(V_1 \oplus V_2) \otimes W \cong (V_1 \otimes W) \oplus (V_2 \otimes W).
$$

Exercise 1.38. Let V, W and U be $\mathfrak{g}$-modules. Prove that

$$
V \otimes (W \otimes U) \cong (V \otimes W) \otimes U.
$$

If both V and W are finite-dimensional, the module $V \otimes W$ is finite-dimensional as well. Due to Corollary 1.34, it is natural to ask how $V \otimes W$ decomposes into a direct sum of simple modules (depending on V and W). Exercises 1.36 and 1.37 mean that to answer this question it is sufficient to consider the case when both V and W are simple modules. This is what we will do in this section. Our main result is the following:

Theorem 1.39. *Let $m, n \in \mathbb{N}$ be such that $m \le n$. Then*

$$
\mathbf{V}^{(n)} \otimes \mathbf{V}^{(m)} \cong \mathbf{V}^{(n-m+1)} \oplus \mathbf{V}^{(n-m+3)} \oplus \cdots \oplus \mathbf{V}^{(n+m-3)} \oplus \mathbf{V}^{(n+m-1)}.
\tag{1.18}
$$

Proof. We proceed by induction on m. If $m = 1$, the module $\mathbf{V}^{(1)} \cong \mathbb{C}$ is the trivial $\mathfrak{g}$-module and hence $\mathbf{V}^{(n)} \otimes \mathbf{V}^{(1)}$ is isomorphic to $\mathbf{V}^{(n)}$, for example via the isomorphism $v \otimes 1 \mapsto v$.

Let $m = 2$. Then $\mathbf{V}^{(2)}$ is the natural $\mathfrak{g}$-module. Let $\{e_1, e_2\}$ be the natural basis of $\mathbf{V}^{(2)}$. Then the action of E, F and H in this basis is given by the following picture (see (1.9)):

$$
\tag{1.19}
$$

Assume that $\mathbf{V}^{(n)}$ is given by (1.9). Then from the formulae (1.17) we obtain that the vector $v_0 \otimes e_1 \in \mathbf{V}^{(n)} \otimes \mathbf{V}^{(2)}$ satisfies $E(v_0 \otimes e_1) = 0$ and $H(v_0 \otimes e_1) = n(v_0 \otimes e_1)$. The only $\mathbf{V}^{(i)}$ which contains a non-zero vector with such properties is $\mathbf{V}^{(n+1)}$ (see Exercise 1.25). Hence $\mathbf{V}^{(n+1)}$ is a direct summand of $\mathbf{V}^{(n)} \otimes \mathbf{V}^{(2)}$.

Let $w = v_1 \otimes e_1 - (n - 1)v_0 \otimes e_2 \neq 0$. Using the definitions, one can easily check that $E(w) = 0$ and $H(w) = (n - 2)w$. The only $\mathbf{V}^{(i)}$ that contains a non-zero vector with such properties is $\mathbf{V}^{(n-1)}$ (Exercise 1.25). Hence $\mathbf{V}^{(n-1)}$ is a direct summand of $\mathbf{V}^{(n)} \otimes \mathbf{V}^{(2)}$. But

$$
\begin{aligned}
\dim \mathbf{V}^{(n)} \otimes \mathbf{V}^{(2)} &= \dim \mathbf{V}^{(n)} \times \dim \mathbf{V}^{(2)} \\
&= 2n \\
&= (n - 1) + (n + 1) \\
&= \dim \mathbf{V}^{(n-1)} + \dim \mathbf{V}^{(n+1)}.
\end{aligned}
$$

This implies that

$$
\mathbf{V}^{(n)} \otimes \mathbf{V}^{(2)} \cong \mathbf{V}^{(n-1)} \oplus \mathbf{V}^{(n+1)}. \tag{1.20}
$$

Now let us prove the induction step. We assume that $k > 2$ and that (1.18) is true for all $m = 1, \ldots, k - 1$. Let us compute $\mathbf{V}^{(n)} \otimes \mathbf{V}^{(k-1)} \otimes \mathbf{V}^{(2)}$ in two different ways. On the one hand we have

$$
\begin{aligned}
\mathbf{V}^{(n)} \otimes \mathbf{V}^{(k-1)} \otimes \mathbf{V}^{(2)} &\overset{(1.20)}{\cong} \mathbf{V}^{(n)} \otimes (\mathbf{V}^{(k)} \oplus \mathbf{V}^{(k-2)}) \\
\text{(by Exercise 1.37)} \quad &\cong \mathbf{V}^{(n)} \otimes \mathbf{V}^{(k)} \oplus \mathbf{V}^{(n)} \otimes \mathbf{V}^{(k-2)} \\
\text{(by inductive assumption)} \quad &\cong \mathbf{V}^{(n)} \otimes \mathbf{V}^{(k)} \oplus \mathbf{V}^{(n-k+3)} \oplus \ldots \\
&\quad \ldots \oplus \mathbf{V}^{(n+k-5)} \oplus \mathbf{V}^{(n+k-3)}.
\end{aligned}
\tag{1.21}
$$

On the other hand, using the inductive assumption we have

$$\mathbf{V}^{(n)} \otimes \mathbf{V}^{(k-1)} \otimes \mathbf{V}^{(2)} \cong \left(\bigoplus_{i=0}^{k-2} \mathbf{V}^{(n-k+2+2i)} \right) \otimes \mathbf{V}^{(2)}$$

$$(\text{by Exercise 1.37}) \cong \bigoplus_{i=0}^{k-2} \mathbf{V}^{(n-k+2+2i)} \otimes \mathbf{V}^{(2)}$$

$$(\text{by (1.20)}) \cong \bigoplus_{i=0}^{k-2} \left(\mathbf{V}^{(n-k+3+2i)} \oplus \mathbf{V}^{(n-k+1+2i)} \right)$$

$$\cong \mathbf{V}^{(n-k+1)} \oplus \mathbf{V}^{(n-k+3)} \oplus \cdots \oplus \mathbf{V}^{(n+k-1)} \oplus$$
$$\oplus \mathbf{V}^{(n-k+3)} \oplus \mathbf{V}^{(n-k+5)} \oplus \cdots \oplus \mathbf{V}^{(n+k-3)}.$$
$$(1.22)$$

The statement of the theorem now follows, comparing (1.21) with (1.22) and using the uniqueness of the decomposition of $\mathbf{V}^{(n)} \otimes \mathbf{V}^{(k-1)} \otimes \mathbf{V}^{(2)}$ into a direct sum of irreducible modules (Corollary 1.34). □

1.5 Unitarizability of finite-dimensional modules

The correspondence

$$\mathbf{e}^\star = \mathbf{f}, \quad \mathbf{f}^\star = \mathbf{e}, \quad \mathbf{h}^\star = \mathbf{h}$$

uniquely extends to a *skew-linear* involution $\star$ on the vector space $\mathfrak{g}$ in the sense that $(\lambda x)^\star = \bar{\lambda} x^\star$ for all $x \in \mathfrak{g}$ and $\lambda \in \mathbb{C}$, where ¯ denotes the complex conjugation. This involution satisfies

$$[x^\star, y^\star] = [y, x]^\star$$

for all $x, y \in \mathfrak{g}$ and hence is a (skew) *anti-involution* of the Lie algebra $\mathfrak{g}$. The involution $\star$ induces an involution on the set $\{E, F, H\}$, which we will denote by the same symbol.

A $\mathfrak{g}$-module V is called *unitarizable* with respect to the involution $\star$ provided that there exists a (positive definite) Hermitian inner product $(\cdot, \cdot)$ on V such that

$$(X(v), w) = (v, X^\star(w)) \qquad (1.23)$$

for all $v, w \in V$ and $X \in \{E, F, H\}$. The aim of this section is to prove the following result:

Theorem 1.40. *Every finite-dimensional $\mathfrak{g}$-module is unitarizable.*

Exercise 1.41. Show that a direct sum $V \oplus W$ of two $\mathfrak{g}$-modules V and W is unitarizable if, and only if, each summand is unitarizable.

Proof. By Corollary 1.34, every finite-dimensional $\mathfrak{g}$-module decomposes into a direct sum of modules $\mathbf{V}^{(n)}$, $n \in \mathbb{N}$. Hence Exercise 1.41 implies that it is enough to prove the statement of the theorem for the modules $\mathbf{V}^{(n)}$, $n \in \mathbb{N}$.

Assume that $n \in \mathbb{N}$ and the module $\mathbf{V}^{(n)}$ is given by (1.9). Note that all $a_i > 0$ and define

$$c_0 = 1, \qquad c_i = \frac{c_{i-1}}{\sqrt{a_i}}, \; i = 1, \ldots, n-1.$$

Then $u_i = c_i v_i$ defines a diagonal change of basis in $\mathbf{V}^{(n)}$. In the basis $\{u_i\}$ the action of E, F and H is given by:

$$(1.24)$$

Let $(\cdot, \cdot)$ be the inner product on $\mathbf{V}^{(n)}$ with respect to which the basis $\{u_0, \ldots, u_{n-1}\}$ is orthonormal. From (1.24) it follows by a direct calculation that in this basis the linear operators E, F and H satisfy (1.23). This proves that $\mathbf{V}^{(n)}$ is unitarizable. As mentioned above, the general statement follows. $\square$

The anti-involution $\star$ is not the only anti-involution on $\mathfrak{g}$. The correspondence

$$\mathbf{e}^\diamond = \mathbf{e}, \quad \mathbf{f}^\diamond = \mathbf{f}, \quad \mathbf{h}^\diamond = -\mathbf{h}$$

uniquely extends to a linear involution $\diamond$ on $\mathfrak{g}$. This involution satisfies

$$[x^\diamond, y^\diamond] = [y, x]^\diamond$$

for all $x, y \in \mathfrak{g}$ and hence is an anti-involution of the Lie algebra $\mathfrak{g}$. The involution $\diamond$ induces an involution on the set $\{E, F, \pm H\}$, which we will denote by the same symbol. A $\mathfrak{g}$-module V is called a $\diamond$-*module* provided that there exists a non-degenerate symmetric bilinear form $(\cdot, \cdot)$ on V such that

$$(X(v), w) = (v, X^\diamond(w)) \tag{1.25}$$

for all $v, w \in V$ and $X \in \{E, F, H\}$.

Exercise 1.42. Let V be a non-trivial simple finite-dimensional $\diamond$-module with the corresponding symmetric bilinear form $(\cdot, \cdot)$. Show that $(\cdot, \cdot)$ is neither positive nor negative definite.

Exercise 1.43. Show that a direct sum $V \oplus W$ of two $\diamond$-modules is a $\diamond$-module.

Theorem 1.44. *Every finite-dimensional $\mathfrak{g}$-module is a $\diamond$-module.*

Proof. By Corollary 1.34, every finite-dimensional $\mathfrak{g}$-module decomposes into a direct sum of modules $\mathbf{V}^{(n)}$, $n \in \mathbb{N}$. Hence Exercise 1.43 implies that it is enough to prove the statement of the theorem for the modules $\mathbf{V}^{(n)}$, $n \in \mathbb{N}$.

Assume that $n \in \mathbb{N}$ and the module $\mathbf{V}^{(n)}$ is given by (1.9). Let $(\cdot, \cdot)$ be the symmetric bilinear form on $\mathbf{V}^{(n)}$ which is given by the matrix

$$\begin{pmatrix} 0 & 0 & 0 & \dots & 0 & 0 & 1 \\ 0 & 0 & 0 & \dots & 0 & 1 & 0 \\ 0 & 0 & 0 & \dots & 1 & 0 & 0 \\ \vdots & \vdots & \vdots & \dots & \vdots & \vdots & \vdots \\ 0 & 0 & 1 & \dots & 0 & 0 & 0 \\ 0 & 1 & 0 & \dots & 0 & 0 & 0 \\ 1 & 0 & 0 & \dots & 0 & 0 & 0 \end{pmatrix}$$

in the basis $\{v_0, v_1, \dots, v_{n-1}\}$. From (1.9) it follows by a direct calculation that in this basis the linear operators E, F and H satisfy (1.25). This proves that $\mathbf{V}^{(n)}$ is a $\diamond$-module and completes the proof. $\qquad \square$

Note that the proof of Theorem 1.44 can be seen as a kind of justification of the basis $\{v_0, v_1, \dots, v_{n-1}\}$ of the module $\mathbf{V}^{(n)}$.

After the above results, given some $\mathfrak{g}$-module V it is natural to ask how many different forms $(\cdot, \cdot)$ on V the module V is unitarizable for (or a $\diamond$-module). The answer turns out to be easy for simple finite-dimensional modules.

Proposition 1.45. *Let V be a simple finite-dimensional $\mathfrak{g}$-module. Then the Hermitian inner product with respect to which the module V is unitarizable is unique up to a positive real scalar.*

Proof. Let $V \cong \mathbf{V}^{(n)}$, $n \in \mathbf{N}$, and $(\cdot, \cdot)$ be an Hermitian inner product on V with respect to which V is unitarizable. Consider the basis $\{u_0, \dots, u_{n-1}\}$ from the proof of Theorem 1.40. The vectors in this basis are eigenvectors of the self-adjoint linear operator H corresponding to pairwise different eigenvalues. Hence $(u_i, u_j) = 0$ for all $i \neq j$. Let $(u_0, u_0) = c$. Then c is a positive real number. Let us prove that $(u_i, u_i) = c$ for all

$i \in \{0, \ldots, n-1\}$ by induction on i. The basis of the induction is trivial.
For all $i \in \{1, \ldots, n-1\}$ we have

$$(u_i, u_i) \overset{(1.24)}{=} \left(\frac{1}{\sqrt{a_i}} F(u_{i-1}), \frac{1}{\sqrt{a_i}} F(u_{i-1}) \right)$$

$$\text{(by (1.23))} \quad = \quad \frac{1}{a_i} (u_{i-1}, E(F(u_{i-1})))$$

$$\text{(by (1.24))} \quad = \quad \frac{1}{a_i} (u_{i-1}, a_i u_{i-1})$$

$$= \quad (u_{i-1}, u_{i-1})$$

$$\text{(by induction)} \quad = \quad c.$$

The claim follows. $\qquad\qquad\qquad\qquad\qquad\qquad\qquad\qquad\qquad\qquad\qquad$ $\square$

Proposition 1.46. *Let V be a simple finite-dimensional $\mathfrak{g}$-module. Then the non-degenerate symmetric bilinear form with respect to which the module V is a $\diamond$-module is unique up to a non-zero complex scalar.*

Proof. Let $V \cong \mathbf{V}^{(n)}$, $n \in \mathbf{N}$, and $(\cdot, \cdot)$ be a non-degenerate symmetric bilinear form on V with respect to which V is a $\diamond$-module. Consider the basis $\{v_0, v_1, \ldots, v_{n-1}\}$ from (1.9). For $i, j \in \{0, \ldots, n-1\}$ by (1.25) we have $(H(v_i), v_j) = -(v_i, H(v_j))$. As all elements of our basis are eigenvectors to H with pairwise different eigenvalues, it follows that $(v_i, v_j) \neq 0$ implies that the eigenvalues λ_i and λ_j of v_i and v_j, respectively, satisfy $\lambda_i = -\lambda_j$. Hence $(v_i, v_j) \neq 0$ implies $i = n - 1 - j$. Let $(v_0, v_{n-1}) = c$. As $(\cdot, \cdot)$ is non-degenerate, we have $c \neq 0$. Let us show by induction on i that $(v_i, v_{n-1-i}) = c$ for all $i \in \{0, 1, \ldots, n-1\}$. For all such $i > 0$ we have

$$(v_i, v_{n-1-i}) \overset{(1.9)}{=} \left(F(v_{i-1}), \frac{1}{a_{n-i}} E(v_{n-i}) \right)$$

$$\text{(by (1.25))} \quad = \quad \frac{1}{a_{n-i}} (v_{i-1}, F(E(v_{n-i})))$$

$$\text{(by (1.9))} \quad = \quad \frac{1}{a_{n-i}} (v_{i-1}, a_{n-i} v_{n-i})$$

$$= \quad (v_{i-1}, v_{n-i})$$

$$\text{(by induction)} \quad = \quad c.$$

The claim follows. $\qquad\qquad\qquad\qquad\qquad\qquad\qquad\qquad\qquad\qquad\qquad$ $\square$

For a direct sum of simple modules, the description of bilinear forms analogous to Propositions 1.45 and 1.46 will be more complicated. In particular, as an obvious observation one could point out that it is possible to independently rescale the restrictions of the bilinear form to pairwise orthogonal direct summands.

1.6 Bilinear forms on tensor products

Let V and W be two vector spaces and $(\cdot,\cdot)_1$ and $(\cdot,\cdot)_2$ be bilinear forms on V and W respectively. Then the assignment

$$(v \otimes w, v' \otimes w') = (v, v')_1 \cdot (w, w')_2 \tag{1.26}$$

extends to a bilinear form on the tensor product $V \otimes W$.

Exercise 1.47. Check that the form $(\cdot,\cdot)$ is symmetric provided that both $(\cdot,\cdot)_1$ and $(\cdot,\cdot)_2$ are symmetric; that the form $(\cdot,\cdot)$ is non-degenerate provided that both $(\cdot,\cdot)_1$ and $(\cdot,\cdot)_2$ are non-degenerate; and that the form $(\cdot,\cdot)$ is Hermitian provided that both $(\cdot,\cdot)_1$ and $(\cdot,\cdot)_2$ are Hermitian.

Proposition 1.48. *Assume that V and W are unitarizable modules (resp. $\diamond$-modules) for the forms $(\cdot,\cdot)_1$ and $(\cdot,\cdot)_2$ respectively. Then $V \otimes W$ is unitarizable (resp. a $\diamond$-module) for $(\cdot,\cdot)$.*

Proof. We prove the statement for unitarizable modules. For $\diamond$-modules the proof is similar. Due to Exercise 1.47 it is sufficient to check (1.23) for $X \in \{E, F, H\}$. For $v, v' \in V$ and $w, w' \in W$ we have

$$\begin{aligned}
(X(v \otimes w), v' \otimes w') &\overset{(1.17)}{=} (X(v) \otimes w + v \otimes X(w), v' \otimes w') \\
\text{(by linearity)} \quad &= (X(v) \otimes w, v' \otimes w') + (v \otimes X(w), v' \otimes w') \\
\text{(by (1.26))} \quad &= (X(v), v')_1 \cdot (w, w')_2 + (v, v')_1 (X(w), w')_2 \\
\text{(by (1.23))} \quad &= (v, X^\star(v'))_1 \cdot (w, w')_2 + (v, v')_1 (w, X^\star(w'))_2 \\
\text{(by (1.26))} \quad &= (v \otimes w, X^\star(v') \otimes w') + (v \otimes w, v' \otimes X^\star(w')) \\
\text{(by linearity)} \quad &= (v \otimes w, X^\star(v') \otimes w' + v' \otimes X^\star(w')) \\
\text{(by (1.17))} \quad &= (v \otimes w, X^\star(v' \otimes w')).
\end{aligned}$$

The claim follows. $\qquad\qquad\qquad\qquad\qquad\qquad\qquad\qquad\qquad\qquad\quad\square$

We know that the tensor product of two simple finite-dimensional $\mathfrak{g}$-modules is not simple in general (see Theorem 1.39). Hence the bilinear form making this tensor product module unitarizable or a ◇-module is usually not unique (even up to some scalar). However, we would like to finish this section with a description of one invariant, which turns out in the real case.

Exercise 1.49. Consider the real Lie algebra $\mathfrak{sl}_2(\mathbb{R})$. Show that (1.9) still defines on the real span $\mathbf{V}_{\mathbb{R}}^{(n)}$ of $\{v_0, \ldots, v_{n-1}\}$ the structure of a simple $\mathfrak{sl}_2(\mathbb{R})$-module. Check that the analogues of Theorem 1.39 and all the above results from Sections 1.5 and 1.6 are true for $\mathfrak{sl}_2(\mathbb{R})$ with the same proofs.

After Exercise 1.49 one could point out one striking difference between the real versions of Proposition 1.45 and Proposition 1.46. It is the possibility of the sign change in the assertion of Proposition 1.46 (note that two forms which differ by a sign change cannot be obtained from each other by a base change in the original module). Let us call the form on $\mathbf{V}^{(n)}$, described in the proof of Proposition 1.46, *standard*, and the form, obtained from the standard form by multiplying with -1, *non-standard*. Our main result in this section is the following:

Theorem 1.50. *Let $m, n \in \mathbb{N}$, $m \leq n$; $(\cdot, \cdot)_1$ and $(\cdot, \cdot)_2$ be standard forms on $\mathbf{V}_{\mathbb{R}}^{(m)}$ and $\mathbf{V}_{\mathbb{R}}^{(n)}$ respectively; and $(\cdot, \cdot)$ be the form on $\mathbf{V}_{\mathbb{R}}^{(n)} \otimes \mathbf{V}_{\mathbb{R}}^{(m)}$ given by (1.26). Then, up to multiplication with a positive real number, for $i = 0, 1, \ldots, m$ the restriction of $(\cdot, \cdot)$ to the direct summand $\mathbf{V}_{\mathbb{R}}^{(n+m-1-2i)}$ of $\mathbf{V}_{\mathbb{R}}^{(n)} \otimes \mathbf{V}_{\mathbb{R}}^{(m)}$ is standard for all even i and non-standard for all odd i.*

Proof. As in the proof of Theorem 1.39 we use induction on m. For $m = 1$ the statement is obvious. To proceed we will need the following lemma:

Lemma 1.51. *Assume that the form $(\cdot, \cdot)'$ on $\mathbf{V}_{\mathbb{R}}^{(n)}$ makes $\mathbf{V}_{\mathbb{R}}^{(n)}$ into a ◇-module. Then $(\cdot, \cdot)'$ is standard if, and only if, $(v_0, F^{n-1}(v_0))' > 0$ and is non-standard if, and only if, $(v_0, F^{n-1}(v_0))' < 0$.*

Proof. From the definition we have that the form $(\cdot, \cdot)'$ is standard if, and only if, $(v_0, v_{n-1})' > 0$. From (1.9) we have $F^{n-1}(v_0) = v_{n-1}$. The claim follows. □

Let $m = 2$, $n \geq 2$ and assume that $\mathbf{V}_{\mathbb{R}}^{(n)}$ is given by (1.9) and $\mathbf{V}_{\mathbb{R}}^{(2)}$ is given by (1.19). As all coefficients in (1.17) are positive, we get that $F^n(v_0 \otimes e_1) = c\, v_{n-1} \otimes e_2$, where $c > 0$. As

$$(v_0 \otimes e_1, v_{n-1} \otimes e_2) = (v_0, v_{n-1})_1 (e_1, e_2)_2 = 1 > 0$$

(here we used that both $(\cdot,\cdot)_1$ and $(\cdot,\cdot)_2$ are standard), from Lemma 1.51 we obtain that the restriction of $(\cdot,\cdot)$ to the direct summand $\mathbf{V}_{\mathbb{R}}^{(n+1)}$ of $\mathbf{V}_{\mathbb{R}}^{(n)} \otimes \mathbf{V}_{\mathbb{R}}^{(2)}$ is standard.

For the element $w = v_1 \otimes e_1 - (n-1)v_0 \otimes e_2 \neq 0$ we have $E(w) = 0$, so w generates the direct summand $\mathbf{V}_{\mathbb{R}}^{(n-1)}$ of $\mathbf{V}_{\mathbb{R}}^{(n)} \otimes \mathbf{V}_{\mathbb{R}}^{(2)}$. A direct computation shows that

$$F^{n-2}(w) = v_{n-1} \otimes e_1 - v_{n-2} \otimes e_2.$$

Another direct computation then shows that

$$(v_1 \otimes e_1 - (n-1)v_0 \otimes e_2, v_{n-1} \otimes e_1 - v_{n-2} \otimes e_2) = -n < 0.$$

Hence from Lemma 1.51 we obtain that the restriction of $(\cdot,\cdot)$ to the direct summand $\mathbf{V}_{\mathbb{R}}^{(n+1)}$ of $\mathbf{V}_{\mathbb{R}}^{(n)} \otimes \mathbf{V}_{\mathbb{R}}^{(2)}$ is non-standard. This completes the proof of the theorem in the case $m = 2$.

For $k \in \mathbb{N}$ let us denote $\mathbf{V}_{\mathbb{R}}^{(k,+)}$ and $\mathbf{V}_{\mathbb{R}}^{(k,-)}$ the module $\mathbf{V}_{\mathbb{R}}^{(k)}$ endowed with a standard and non-standard (up to a positive real scalar) form, respectively. Then we have just proved that

$$\mathbf{V}_{\mathbb{R}}^{(n,+)} \otimes \mathbf{V}_{\mathbb{R}}^{(2,+)} \cong \mathbf{V}_{\mathbb{R}}^{(n+1,+)} \oplus \mathbf{V}_{\mathbb{R}}^{(n-1,-)}. \tag{1.27}$$

Note that we obviously have

$$\mathbf{V}_{\mathbb{R}}^{(n,+)} \otimes \mathbf{V}_{\mathbb{R}}^{(1,+)} \cong \mathbf{V}_{\mathbb{R}}^{(n,+)}, \quad \mathbf{V}_{\mathbb{R}}^{(n,+)} \otimes \mathbf{V}_{\mathbb{R}}^{(1,-)} \cong \mathbf{V}_{\mathbb{R}}^{(n,-)}. \tag{1.28}$$

In this notation the statement of our theorem can be written as follows:

$$\mathbf{V}_{\mathbb{R}}^{(n,+)} \otimes \mathbf{V}_{\mathbb{R}}^{(m,+)} \cong \mathbf{V}_{\mathbb{R}}^{(n+m-1,+)} \oplus \mathbf{V}_{\mathbb{R}}^{(n+m-3,-)} \oplus \mathbf{V}_{\mathbb{R}}^{(n+m-5,+)} \oplus \ldots.$$

The induction step now follows using (1.27) and (1.28) and, rewriting in this new notation, the calculations in (1.21) and (1.22). We leave the details to the reader. $\qquad\square$

1.7 Addenda and comments

1.7.1

Alternative expositions for the material presented in Sections 1.1–1.4 can be found in a large number of books and articles, see for example [37, 46, 49, 57, 106]. Many of the results are true or have analogs in much more general contexts (which also can be found in the books listed above). In particular, simple finite-dimensional modules are classified (see Theorem 1.22) and Weyl's Theorem (Theorem 1.29) is true for all simple finite-dimensional complex Lie algebras. For all such algebras there is also an analog of Theorem 1.39, however its formulation is more complicated, as higher multiplicities appear on the right hand side.

1.7.2

If A is an associative algebra with associative multiplication $\cdot$, then one can define on A the structure of a Lie algebra using the operation of taking the commutator with respect to $\cdot$: $[a, b] = a \cdot b - b \cdot a$. The Lie algebra $(A, [\cdot, \cdot])$ is called the Lie algebra *underlying* the associative algebra $(A, \cdot)$ and is often denoted by $A^{(-)}$. In particular, if V is a vector space, one can consider the associative algebra $\mathcal{L}(V)$ of all linear operators on V and the underlying Lie algebra $\mathcal{L}(V)^{(-)}$.

An $\mathfrak{sl}_2$-module is then given by a Lie algebra *homomorphism* from $\mathfrak{sl}_2$ to $\mathcal{L}(V)^{(-)}$, that is a linear map $\varphi : \mathfrak{sl}_2 \to \mathcal{L}(V)$, which satisfies

$$\varphi([x, y]) = [\varphi(x), \varphi(y)] \tag{1.29}$$

for all $x, y \in \mathfrak{sl}_2$. For such φ in the notation of Section 1.1 we simply have $H = \varphi(\mathbf{h})$, $F = \varphi(\mathbf{f})$ and $E = \varphi(\mathbf{e})$.

Substituting $\mathfrak{sl}_2$ with an arbitrary Lie algebra, one obtains the notion of a *module* over any Lie algebra. The homomorphism φ is usually called a *representation* of the Lie algebra. Hence the notions of module and representation are equivalent, differing only in their emphasis on the underlying vector space V (for modules) or the homomorphism φ (for representations). Sometimes one can also say that a representation defines an *action* of the Lie algebra on the underlying vector space V.

1.7.3

The Lie algebra $\mathfrak{sl}_2$ is a subalgebra of the Lie algebra $\mathfrak{gl}_2$, the latter being the underlying Lie algebra of the associative algebra of all complex 2×2 matrices. Moreover, the algebra $\mathfrak{gl}_2$ is a direct sum of $\mathfrak{sl}_2$ and the commutative Lie subalgebra of all scalar matrices.

1.7.4

Weyl's Theorem can be proved using the notion of unitarizability of finite-dimensional modules. Let V be an arbitrary finite-dimensional $\mathfrak{g}$-module. Using the exponential map one first could lift the $\mathfrak{g}$-action on V to the action of the group SL(2) and further SU(2). In fact one can show that there is a natural bijection between finite-dimensional $\mathfrak{g}$-modules, finite-dimensional SL(2)-modules and finite-dimensional SU(2)-modules. As SU(2) is compact, all finite-dimensional SU(2)-modules and hence all finite-dimensional g-module are completely reducible. We refer the reader to Chapter III, §6 of [106] for details.

1.7.5

So far, all modules which we considered were *left* modules. There is also the natural notion of a *right* module. The triple E', F' and H' of linear operators on a vector space V defines on V the structure of a *right* g-module provided that the operators E', F' and H' satisfy

$$F'E' - E'F' = H', \quad E'H' - H'E' = 2E', \quad F'H' - H'F' = -2F'.$$

This corresponds to an *antihomomorphism* from g to $\mathcal{L}(V)^{(-)}$, that is to a linear map $\varphi : \mathfrak{g} \to \mathcal{L}(V)$, which satisfies

$$\varphi([x, y]) = [\varphi(y), \varphi(x)] \tag{1.30}$$

for all $x, y \in \mathfrak{sl}_2$. In what follows, by "module" we will always mean a left module.

1.7.6

If A is an algebra (associative or Lie), then for any left A-module V the dual space $V^* = \text{Hom}(V, \mathbb{C})$ carries the natural structure of a right A-module given by $(a(f))(v) = f(a(v))$ for $a \in A$, $v \in V$ and $f \in V^*$.

However, if A is an algebra (associative or Lie) with a fixed anti-involution $\natural$, then for any left A-module V the space V^* carries the natural structure of a left A-module given by $(a(f))(v) = f(a^\natural(v))$ for $a \in A$, $v \in V$ and $f \in V^*$.

Any element $\varphi \in \text{Hom}_A(V, V^*)$ defines a bilinear form on V as follows: $(v, w)_\varphi = \varphi(v)(w)$. This form obviously satisfies

$$(a(v), w) = (v, a^\natural(w)). \tag{1.31}$$

If V is finite-dimensional then every bilinear form on V satisfying (1.31) has the form $(v, w)_\varphi$ for some $\varphi \in \text{Hom}_A(V, V^*)$. The form $(v, w)_\varphi$ is non-degenerate if, and only if, φ is an isomorphism. These general arguments give an alternative proof of Propositions 1.45 and 1.46. More details and some further related results can be found in [93].

1.7.7

Theorem 1.50 appears in [6] (in the form presented later on in Exercise 1.70) in connection to the study of Hodge–Riemann relations for polytopes. Our proof follows the general idea of [6]. There exists an alternative "brute force" argument for Theorem 1.50 worked out in [71]. Here is its outline:

Let $m, n \in \mathbb{N}$ and $n \geq m$. Assume that $\mathbf{V}^{(n)}$ is given by (1.9) and $\mathbf{V}^{(m)}$ is similarly given by (1.9) in the basis $\{w_0, w_1, \ldots, w_{m-1}\}$. For $k = 0, 1, \ldots, m-1$ set $l_k = m + n - 2 - 2k$. A direct calculation shows that the element

$$u_k = \sum_{i=0}^{k} (-1)^i v_{n-1+i} \otimes w_{m-1+k-i}$$

satisfies

$$H(u_k) = -l_k u_k, \quad F(u_k) = 0.$$

In the same way as in Lemma 1.51, one shows that the form $(\cdot, \cdot)'$ on $\mathbf{V}_{\mathbb{R}}^{(n)}$, making $\mathbf{V}_{\mathbb{R}}^{(n)}$ into a $\diamond$-module, is standard if, and only if, $(v_{n-1}, E^{n-1}(v_{n-1}))' > 0$ and non-standard if, and only if, $(v_{n-1}, E^{n-1}(v_{n-1}))' < 0$. Hence to prove Theorem 1.50 one simply has to show that for any $k = 0, 1, \ldots, m-1$ the sign of the number $(u_k, E^{l_k}(u_k))$ alternates with k. This reduces to the computation of $E^{l_k}(v_k)$, which is not entirely straightforward. However, using the following *Karlsson–Minton identity* for the hypergeometric series $_3F_2$: for all integers a, b, c, d such that $0 \leq b \leq a \leq \min(c, d)$ we have

$$\sum_{i=\max(0, 2a-b-d)}^{\min(a, c-b)} (-1)^i \frac{(c-i)!(d-a+i)!}{i!(a-i)!(c-b-i)!(d+b-2a+i)!} = (-1)^{a+b},$$

one proves by a direct calculation that

$$E^{l_k}(u_k) = l_k! \sum_{i=0}^{k} (-1)^{k+i} \frac{(m-1-i)!\,(n-1-k+i)!}{i!\,(k-i)!} v_{k-i} \otimes w_i.$$

Using the last formula, the computation of $(u_k, E^{n+m-2-2k}(u_k))$ is fairly straightforward and yields the necessary result. We refer the reader to [71] for details.

1.8   Additional exercises

Exercise 1.52. Let $\mathfrak{a}$ denote the vector space with the basis $\{e_{-1}, e_0, e_1\}$. Define the bilinear operation $[\cdot, \cdot]$ on $\mathfrak{a}$ via

$$[e_i, e_j] = \begin{cases} (j-i)e_{i+j}, & i+j \in \{-1, 0, 1\}; \\ 0, & \text{otherwise.} \end{cases}$$

Show that this makes $\mathfrak{a}$ into a Lie algebra. Show further that $\mathfrak{a}$ is isomorphic to $\mathfrak{sl}_2$.

Exercise 1.53. Let $\mathfrak{b}$ denote the vector space with the basis $\{a, b, c\}$. Define the antisymmetric bilinear operation $[\cdot, \cdot]$ on $\mathfrak{b}$ via

$$[a, b] = c, \quad [b, c] = a, \quad [c, a] = b.$$

Show that this makes $\mathfrak{b}$ into a Lie algebra. Show further that $\mathfrak{b}$ is isomorphic to $\mathfrak{sl}_2$.

Exercise 1.54. Consider the vector space $V = \mathbb{C}[x, y]$ and the linear operators

$$E = x \cdot \frac{\partial}{\partial y}, \quad F = y \cdot \frac{\partial}{\partial x}, \quad H = x \cdot \frac{\partial}{\partial x} - y \cdot \frac{\partial}{\partial y}$$

on V.

(a) Show that the operators E, F and H make V into a $\mathfrak{g}$-module.
(b) Show that for every $n \in \mathbb{N}_0$ the linear span of all homogeneous polynomials of degree n is a submodule of V, isomorphic to $\mathbf{V}^{(n+1)}$.

Exercise 1.55. Consider the vector space $V = \mathrm{Mat}_{3\times 3}(\mathbb{C})$ of all complex 3×3 matrices and the matrices

$$X = \begin{pmatrix} 0 & 0 & 1 \\ 0 & 0 & 0 \\ 0 & 0 & 0 \end{pmatrix}, \quad Y = \begin{pmatrix} 0 & 0 & 0 \\ 0 & 0 & 0 \\ 1 & 0 & 0 \end{pmatrix}, \quad Z = \begin{pmatrix} 1 & 0 & 0 \\ 0 & 0 & 0 \\ 0 & 0 & -1 \end{pmatrix}.$$

Show that the linear operators

$$E = [X, \_], \quad F = [Y, \_], \quad H = [Z, \_]$$

(here $E(A) = [X, A] = XA - AX$ for all $A \in \mathrm{Mat}_{3\times 3}(\mathbb{C})$ and similarly for F and H) define on V the structure of a $\mathfrak{g}$-module and determine its decomposition into a direct sum of $\mathbf{V}^{(n)}$'s.

Exercise 1.56. Let $n \in \mathbb{N}$. For every $\lambda \in \{-n+1, -n+3, \ldots, n-3, n-1\}$ fix some non-zero element $x_\lambda \in \mathbf{V}_\lambda^{(n)}$ (see (1.8)). Show that the set

$$\mathbf{x} = \{x_\lambda : \lambda \in \{-n+1, -n+3, \ldots, n-3, n-1\}\}$$

is a basis of $\mathbf{V}^{(n)}$.

Exercise 1.57 (Gelfand–Zetlin model, [52]). Let $c \in \mathbb{C}$ be a fixed complex number. For $n \in \mathbb{N}$ set $c' = c - n$ and consider the set $\mathbf{T}_{c,n}$ consisting of all tableaux

$$t_c(a) = \begin{array}{|c|c|} \hline c & c' \\ \hline \end{array} \atop \begin{array}{|c|} \hline a \\ \hline \end{array}$$

where $a \in \{c - i \ : \ i = 0, 1, \ldots, n - 1\}$. Let $V = V_{c,n}$ denote the linear span of all elements from $\mathbf{T}_{c,n}$. Define the linear operators E, F and H on V as follows:

$$F(t_c(a)) = \begin{cases} t_c(a - 1), & t_c(a - 1) \in \mathbf{T}_{c,n}; \\ 0, & \text{otherwise.} \end{cases}$$

$$E(t_c(a)) = \begin{cases} -(c - a)(c' - a)t_c(a + 1), & t_c(a + 1) \in \mathbf{T}_{c,n}; \\ 0, & \text{otherwise.} \end{cases}$$

$$H(t_c(a)) = (2a - c - c' - 1) \cdot t_c(a).$$

Show that this turns V into a $\mathfrak{g}$-module, which is isomorphic to $\mathbf{V}^{(n)}$.

Exercise 1.58. Write a Cayley table of the Lie algebra $\mathfrak{gl}_2$ in the standard basis $\{e_{11}, e_{12}, e_{21}, e_{22}\}$ consisting of matrix units.

Exercise 1.59 (Gelfand–Zetlin model for $\mathfrak{gl}_2$, [52]).
(a) Show that the $\mathfrak{sl}_2$-module structure on the module $V_{c,n}$ from Exercise 1.57 can be extended to a $\mathfrak{gl}_2$-module structure in the following way:

$$e_{12}(t_c(a)) = E(t_c(a)),$$
$$e_{21}(t_c(a)) = F(t_c(a)),$$
$$e_{11}(t_c(a)) = a \cdot t_c(a),$$
$$e_{22}(t_c(a)) = (c + c' - a + 1) \cdot t_c(a).$$

(b) Show that the module $V_{c,n}$ is a simple $\mathfrak{gl}_2$-module.
(c) Show that $V_{c,n} \cong V_{d,m}$ if, and only if, $c = d$ and $n = m$.
(d) Show that every simple $\mathfrak{gl}_2$-module is isomorphic to $V_{c,n}$ for some $c \in \mathbb{C}$ and $n \in \mathbb{N}$.

Exercise 1.60. Construct a counterexample which shows that Weyl's theorem fails for finite-dimensional $\mathfrak{gl}_2$-modules.

Exercise 1.61. Let A and B be two linear operators on some finite-dimensional vector space V.

(a) Prove that $[A, B] = \lambda \cdot A$ for some $\lambda \in \mathbb{C}$, $\lambda \neq 0$, implies that the operator A is nilpotent.
(b) Prove that $[A, B] = A^2$ implies that the operator A is nilpotent.
(c) Prove that $[A, [A, B]] = 0$ implies that the operator $[A, B]$ is nilpotent.

Exercise 1.62. Let V be a $\mathfrak{g}$-module and $n \in \mathbb{N}$. Consider the n-th tensor power $V^{\otimes n}$ of V. Let $\mathbf{S}_n$ denote the symmetric group on $\{1, 2, \ldots, n\}$. Show that the linear span of all vectors of the form

$$\sum_{\sigma \in S_n} v_{\sigma(1)} \otimes v_{\sigma(2)} \otimes \cdots \otimes v_{\sigma(n-1)} \otimes v_{\sigma(n)},$$

where $v_1, v_2, \ldots, v_n \in V$ forms a $\mathfrak{g}$-submodule of $V^{\otimes n}$. This submodule is called the n-th *symmetric power* of V and is denoted by $\mathrm{Sym}^n(V)$.

Exercise 1.63. Prove that $\mathrm{Sym}^n(\mathbf{V}^{(2)}) \cong \mathbf{V}^{(n+1)}$.

Exercise 1.64. Let V and $n \in \mathbb{N}$ be as in Exercise 1.62. Show that the linear span of all vectors of the form

$$\sum_{\sigma \in S_n} \mathrm{sign}(\sigma) \cdot v_{\sigma(1)} \otimes v_{\sigma(2)} \otimes \cdots \otimes v_{\sigma(n-1)} \otimes v_{\sigma(n)},$$

where $v_1, v_2, \ldots, v_n \in V$ forms a $\mathfrak{g}$-submodule of V. This submodule is called the n-th *exterior power* of V and is denoted by $\bigwedge^n(V)$.

Exercise 1.65. Prove that $\bigwedge^n(\mathbf{V}^{(n)}) \cong \mathbf{V}^{(1)}$.

Exercise 1.66. Prove that $\bigwedge^n(V) \cong 0$ provided that $n > \dim V$.

Exercise 1.67.

(a) Show that the correspondence

$$\mathbf{e}^\natural = -\mathbf{e}, \quad \mathbf{f}^\natural = -\mathbf{f}, \quad \mathbf{h}^\natural = -\mathbf{h}$$

uniquely extends to an anti-involution $\natural$ on $\mathfrak{g}$.
(b) Prove that for every $n \in \mathbb{N}$ there exists a unique (up to a non-zero scalar) non-degenerate bilinear form $(\cdot, \cdot)_n$ on $\mathbf{V}^{(n)}$ such that

$$(X(v), w)_n = (v, X^\natural(w))_n$$

for all $X \in \{\pm E, \pm F, \pm H\}$ and all $v, w \in \mathbf{V}^{(n)}$.
(c) Prove that the form $(\cdot, \cdot)_n$ is symmetric for odd n and antisymmetric for even n.

Exercise 1.68. Formulate and prove an analogue of Proposition 1.48 for the anti-involution $\natural$ from Exercise 1.67.

Exercise 1.69. Formulate and prove an analog of Theorem 1.50 for the form $(\cdot, \cdot)_n$ from Exercise 1.67.

Exercise 1.70 ([6]). For $n \in \mathbb{N}$ set

$$\tilde{\mathbf{V}}^{(n)} = \begin{cases} \mathbf{V}^{(n,+)}, & n \text{ is odd}; \\ \mathbf{V}^{(n,-)}, & n \text{ is even} \end{cases}$$

(see notation of Theorem 1.50). Show that for any $k, m \in \mathbb{N}$ the tensor product $\tilde{\mathbf{V}}^{(k)} \otimes \tilde{\mathbf{V}}^{(m)}$ decomposes into a direct sum of modules of the form $\tilde{\mathbf{V}}^{(n)}$, $n \in \mathbb{N}$.

Exercise 1.71. Fix $n \in \mathbb{N}$. For $i = 1, \ldots, n-1$ let $\mathbf{s}_i$ denote the involution $(i, i+1)$ of the symmetric group $\mathbf{S}_n$. The group $\mathbf{S}_n$ is a Coxeter group in the natural way. Denote by R the set of all reflections in $\mathbf{S}_n$. In particular, the $\mathbf{s}_i$'s are simple reflections. Denote by $\mathbf{e}$ the identity element of $\mathbf{S}_n$. For $i = 0, \ldots, n-1$ and $j = 0, \ldots, i$ set

$$x_{i,j} = \begin{cases} \mathbf{e}, & j = 0; \\ \mathbf{s}_i \mathbf{s}_{i-1} \cdots \mathbf{s}_{i-j+1}, & j > 0; \end{cases}$$

and define $X_i = \{x_{i,0}, x_{i,1}, \ldots, x_{i,i-1}\}$. Let $\mathbf{X}_i$ denote the formal linear span of the elements from X_i. Define on $\mathbf{X}_i$ the structure of a $\mathfrak{g}$-module via (1.9) using the convention $v_j = x_{i,i-1-j}$.

(a) Show that every $\alpha \in \mathbf{S}_n$ admits a unique decomposition of the form $\alpha = \alpha_0 \alpha_1 \ldots \alpha_{n-1}$, where $\alpha_i \in X_i$ for all i.

(b) Show that the underlying space of the tensor product $\mathbf{X}_0 \otimes \mathbf{X}_1 \otimes \cdots \otimes \mathbf{X}_{n-1}$ can be canonically identified with $\mathbb{C}[\mathbf{S}_n]$ via the map

$$x_{0,j_0} \otimes x_{1,j_1} \otimes \cdots \otimes x_{n-1,j_{n-1}} \mapsto x_{0,j_0} x_{1,j_1} \cdots x_{n-1,j_{n-1}}.$$

This equips $\mathbb{C}[\mathbf{S}_n]$ with the structure of a $\mathfrak{g}$-module. (This structure comes from the Hard Lefschetz theorem (see [53]) applied to the cohomology algebra of the flag variety, which can be naturally identified with $\mathbb{C}[\mathbf{S}_n]$ as a vector space, see [56].)

(c) Show that every $\alpha \in \mathbf{S}_n$ is a eigenvector for $\mathbf{h}$ (with respect to the $\mathfrak{g}$-module structure described in (b)) with the eigenvalue

$$|\{r \in R : r\alpha < \alpha\}| - |\{r \in R : r\alpha > \alpha\}|,$$

where $<$ denotes the Bruhat order on $\mathbf{S}_n$. Show that the same number equals $-\frac{n(n-1)}{2} + 2l(\alpha)$, where $l(\alpha)$ is the length of α with respect to the set $\{\mathbf{s}_1, \mathbf{s}_2, \ldots \mathbf{s}_{n-1}\}$ of simple reflections.

Exercise 1.72. Generalize Exercise 1.71 to other Coxeter groups.

Exercise 1.73. For every finite-dimensional $\mathfrak{g}$-module V we define the function $\mathrm{ch}_V : \mathbb{Z} \to \mathbb{N}_0$ as follows:

$$\mathrm{ch}_V(\lambda) = \dim V_\lambda, \quad \lambda \in \mathbb{Z}.$$

(a) Show that $\mathrm{ch}_V(\lambda) = 0$ for all $\lambda \in \mathbb{Z}$ such that $|\lambda|$ is big enough.
(b) Show that $\mathrm{ch}_V(\lambda) = \mathrm{ch}_V(-\lambda)$ for all $\lambda \in \mathbb{Z}$.
(c) Show that $\mathrm{ch}_V(\lambda) \geq \mathrm{ch}_V(\mu)$ for all elements $\lambda, \mu \in \mathbb{Z}$ of the same parity such that $0 \leq |\lambda| \leq |\mu|$.
(d) Show that $\mathrm{ch}_V = \mathrm{ch}_W$ if and only if $V \cong W$.
(e) Show that for any function $\mathrm{ch} : \mathbb{Z} \to \mathbb{N}_0$, which has the properties, described in (a)–(c) above, there exists a unique (up to isomorphism) $\mathfrak{g}$-module V such that $\mathrm{ch} = \mathrm{ch}_V$.

Exercise 1.74 ([60]). Show that the elements $\mathbf{x} = \mathbf{h}$, $\mathbf{y} = 2\mathbf{e} - \mathbf{h}$ and $\mathbf{z} = -2\mathbf{f} - \mathbf{h}$ form a basis of $\mathfrak{g}$ and that we have

$$[\mathbf{x}, \mathbf{y}] = 2\mathbf{x} + 2\mathbf{y}, \quad [\mathbf{y}, \mathbf{z}] = 2\mathbf{y} + 2\mathbf{z}, \quad [\mathbf{z}, \mathbf{x}] = 2\mathbf{z} + 2\mathbf{x}$$

The basis $\{\mathbf{x}, \mathbf{y}, \mathbf{z}\}$ is called the *equitable* basis of $\mathfrak{g}$.

Chapter 2

The universal enveloping algebra

2.1 Construction and the universal property

As we saw in 1.7.2, a $\mathfrak{g}$-module corresponds to a Lie algebra homomorphism $\varphi : \mathfrak{g} \to \mathcal{L}(V)^{(-)}$. In all nontrivial cases, φ has one very annoying property: the image of φ is *not* closed with respect to the composition of linear operators, but is only closed with respect to taking the commutator of linear operators (since the latter is the Lie algebra structure on $\mathcal{L}(V)^{(-)}$). This heavily restricts our possibilities to analyze the structure of V as we are forced to look for some *external* objects which initially seem unrelated to the algebra $\mathfrak{g}$. A good example is the Casimir operator C_V from Section 1.3. This operator does not belong to the image of the homomorphism defining the module structure on V in general, so its appearance should be a total mystery at the moment. At the same time, it played a crucial role in the proof of Weyl's theorem, so it is clear that this operator is very important.

The aim of the present chapter is to clarify the situation in the following way: We will define a certain associative algebra $U(\mathfrak{g})$, called the *universal enveloping algebra* of $\mathfrak{g}$ and show that it has the following properties:

- The Lie algebra $\mathfrak{g}$ is a canonical subalgebra of the underlying Lie algebra $U(\mathfrak{g})^{(-)}$.
- Any $\mathfrak{g}$-action on any vector space (that is, any $\mathfrak{g}$-module) canonically extends to a $U(\mathfrak{g})$-action on the same vector space (that is, to a module over the associative algebra $U(\mathfrak{g})$).
- This extension, and the restriction from $U(\mathfrak{g})$ to $\mathfrak{g}$, are mutually inverse isomorphisms between the categories of all $\mathfrak{g}$-modules and all $U(\mathfrak{g})$-modules.

Basically, this says that the study of $\mathfrak{g}$-modules is the same as the study of

modules over the associative algebra $U(\mathfrak{g})$. Any $U(\mathfrak{g})$-module corresponds
to a homomorphism $\psi : U(\mathfrak{g}) \to \mathcal{L}(V)$ and the image of this homomorphism
is always closed with respect to the composition of operators. Hence the
study of $U(\mathfrak{g})$ mostly reduces to the study of the internal structure of $U(\mathfrak{g})$.
In particular, the Casimir operator C_V from Section 1.3 will turn out to be
the image of a special central element from $U(\mathfrak{g})$. The only disadvantage
of $U(\mathfrak{g})$ is that this algebra is infinite-dimensional, while the Lie algebra $\mathfrak{g}$
has finite dimension. But the above benefits mean that this is a reasonable
price to pay.

Consider the free associative algebra R with generators e, f and h and
denote by $U(\mathfrak{g})$ the quotient of R modulo the ideal generated by the following relations:

$$ef - fe = h, \quad he - eh = 2e, \quad hf - fh = -2f. \tag{2.1}$$

The algebra $U(\mathfrak{g})$ is called the *universal enveloping algebra* of $\mathfrak{g}$. Abusing
notation we will usually identify the elements of R with their images in the
quotient algebra $U(\mathfrak{g})$.

Exercise 2.1. Prove that the image of the element $(h + 1)^2 + 4fe$ from R
in the algebra $U(\mathfrak{g})$ belongs to the center of $U(\mathfrak{g})$.

Exercise 2.2. Prove that the elements $(h + 1)^2 + 4fe$ and $(h - 1)^2 + 4ef$
are different elements in R, while their images in $U(\mathfrak{g})$ coincide.

Exercise 2.3. Show that there exists a unique anti-involution σ on $U(\mathfrak{g})$
such that $\sigma(f) = e$ and $\sigma(h) = h$.

Lemma 2.4.

(i) *There is a unique linear map* $\varepsilon : \mathfrak{g} \to U(\mathfrak{g})$ *satisfying*

$$\varepsilon(\mathbf{e}) = e, \quad \varepsilon(\mathbf{f}) = f, \quad \varepsilon(\mathbf{h}) = h.$$

(ii) *The map* ε *is a Lie algebra homomorphism from* $\mathfrak{g}$ *to* $U(\mathfrak{g})^{(-)}$.

Proof.   The statement (i) follows from the fact that $\{\mathbf{e}, \mathbf{f}, \mathbf{h}\}$ is a basis
of $\mathfrak{g}$. The statement (ii) follows from (1.1), the definition of $U(\mathfrak{g})$ and the
definition of the underlying Lie algebra. $\square$

The map ε is called the *canonical embedding* of $\mathfrak{g}$ into $U(\mathfrak{g})^{(-)}$. However,
it is not obvious that this map is injective. We will prove this in the
next section. For the moment we would like to present the following very
important result:

Theorem 2.5 (Universal property of $U(\mathfrak{g})$). *Let A be any associative algebra and $\varphi : \mathfrak{g} \to A^{(-)}$ be any homomorphism of Lie algebras. Then there exists a unique homomorphism $\overline{\varphi} : U(\mathfrak{g}) \to A$ of associative algebras such that $\varphi = \overline{\varphi} \circ \varepsilon$. In other words, the following diagram commutes:*

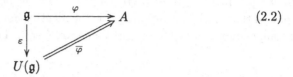

$$(2.2)$$

Proof. We start with the existence of $\overline{\varphi}$. For the free algebra R with generators e, f and h we have a unique homomorphism $\psi : R \to A$ of associative algebras such that

$$\psi(e) = \varphi(\mathbf{e}), \quad \psi(f) = \varphi(\mathbf{f}), \quad \psi(h) = \varphi(\mathbf{h}). \qquad (2.3)$$

Let K be the kernel of the natural projection $R \twoheadrightarrow U(\mathfrak{g})$. We have

$$\psi(ef - fe) = \psi(e)\psi(f) - \psi(f)\psi(e)$$
$$\text{(by definition)} = \varphi(\mathbf{e})\varphi(\mathbf{f}) - \varphi(\mathbf{f})\varphi(\mathbf{e})$$
$$= [\varphi(\mathbf{e}), \varphi(\mathbf{f})]$$
$$\text{(as } \varphi \text{ is a Lie alg. homom.)} = \varphi([\mathbf{e}, \mathbf{f}])$$
$$\text{(by (1.1))} = \varphi(\mathbf{h})$$
$$\text{(by definition)} = \psi(h).$$

This yields $\psi(ef - fe - h) = 0$. Similarly one shows that $\psi(he - eh - 2e) = 0$ and $\psi(hf - fh + 2f) = 0$. This means that $\psi(K) = 0$ and hence ψ factors through $R/K \cong U(\mathfrak{g})$. Denote by $\overline{\varphi}$ the induced homomorphism from $U(\mathfrak{g})$ to A. Then $\varphi = \overline{\varphi} \circ \varepsilon$ follows from the definitions.

The uniqueness of $\overline{\varphi}$ follows from the uniqueness of ψ as the equality $\varphi = \overline{\varphi} \circ \varepsilon$ forces formulae (2.3). $\qquad \square$

As usual, the universal property from Theorem 2.5 guarantees that the universal enveloping algebra is defined uniquely up to isomorphism:

Proposition 2.6 (Uniqueness of $U(\mathfrak{g})$). *Let $U(\mathfrak{g})'$ be another associative algebra such that there exists a fixed homomorphism $\varepsilon' : \mathfrak{g} \to (U(\mathfrak{g})')^{(-)}$ of Lie algebras with the universal property as described in Theorem 2.5. Then there exists a unique isomorphism $\overline{\varepsilon} : U(\mathfrak{g})' \to U(\mathfrak{g})$ such that $\overline{\varepsilon} \circ \varepsilon' = \varepsilon$; in particular, $U(\mathfrak{g})'$ and $U(\mathfrak{g})$ are canonically isomorphic.*

Proof. First we note that for $A = U(\mathfrak{g})$ and $\varphi = \varepsilon$ in (2.2) we have $\overline{\varphi} = \mathrm{id}_{U(\mathfrak{g})}$ (because $\mathrm{id}_{U(\mathfrak{g})}$ works and is unique by the universal property).

Now take $A = U(\mathfrak{g})'$ and $\varphi = \varepsilon'$. The universal property for $U(\mathfrak{g})$ gives a homomorphism $\overline{\varepsilon'} : U(\mathfrak{g}) \to U(\mathfrak{g})'$. The universal property for $U(\mathfrak{g})'$ gives a homomorphism $\overline{\varepsilon} : U(\mathfrak{g})' \to U(\mathfrak{g})$ (see (2.4)). From the previous paragraph we obtain that the composition $\overline{\varepsilon} \circ \overline{\varepsilon'}$ is the identity map on $U(\mathfrak{g})$. Similarly, the composition $\overline{\varepsilon'} \circ \overline{\varepsilon}$ is the identity map on $U(\mathfrak{g})'$. The claim follows.

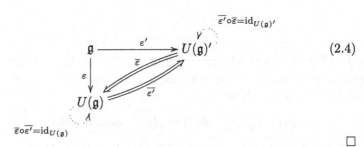

$$\tag{2.4}$$

$\square$

Having the universal property it is now easy to relate $\mathfrak{g}$-modules and $U(\mathfrak{g})$-modules.

Proposition 2.7.

(i) *Let V be a $\mathfrak{g}$-module given by a homomorphism $\varphi : \mathfrak{g} \to \mathcal{L}(V)^{(-)}$ of Lie algebras. Then the homomorphism $\overline{\varphi} : U(\mathfrak{g}) \to \mathcal{L}(V)$, given by the universal property, endows V with the canonical structure of a $U(\mathfrak{g})$-module.*

(ii) *Let V be a $U(\mathfrak{g})$-module given by the homomorphism $\psi : U(\mathfrak{g}) \to \mathcal{L}(V)$. Then the composition $\psi \circ \varepsilon$ is a Lie algebra homomorphism from $\mathfrak{g}$ to $\mathcal{L}(V)^{(-)}$, which endows V with the canonical structure of a $\mathfrak{g}$-module.*

(iii) *Let V and W be two $\mathfrak{g}$-modules with the induced structures of $U(\mathfrak{g})$-modules given by (i). Then $\mathrm{Hom}_{\mathfrak{g}}(V, W) = \mathrm{Hom}_{U(\mathfrak{g})}(V, W)$.*

(iv) *Let V and W be two $U(\mathfrak{g})$-modules with the induced structures of $\mathfrak{g}$-modules given by (ii). Then $\mathrm{Hom}_{U(\mathfrak{g})}(V, W) = \mathrm{Hom}_{\mathfrak{g}}(V, W)$.*

(v) *The operations described in (i) and (ii) are mutually inverse to one another.*

Exercise 2.8. Let A and B be two associative algebras and $\psi : A \to B$ be a homomorphism of algebras. Check that $\psi : A^{(-)} \to B^{(-)}$ is also a homomorphism of Lie algebras.

Proof. The statement (i) is obvious, as well as a large part of the statement (ii). The only thing in (ii) that one must check is that $\psi \circ \varepsilon$ is a Lie algebra homomorphism. However, this follows from Exercise 2.8. The statements (iii) and (iv) follow from the fact that for any $\mathfrak{g}$-module and the associated $U(\mathfrak{g})$-module V, the image of $\mathfrak{g}$ in $\mathcal{L}(V)^{(-)}$ is generated by the same elements as the image of $U(\mathfrak{g})$ in $\mathcal{L}(V)$. The statement (v) follows from the definitions and the uniqueness of ε (Lemma 2.4). □

Denote by $\mathfrak{g}$-mod the category of all (left) $\mathfrak{g}$-modules, and by $U(\mathfrak{g})$-mod the category of all (left) $U(\mathfrak{g})$-modules. From Proposition 2.7 we immediately obtain:

Corollary 2.9. *Operations described in Proposition 2.7(i) and Proposition 2.7(ii) give mutually inverse isomorphisms between the categories $\mathfrak{g}$-mod and $U(\mathfrak{g})$-mod.*

Remark 2.10. From now on we will call every $\mathfrak{g}$-module, if necessary, a $U(\mathfrak{g})$-module, and vice versa. Moreover, if V is a $\mathfrak{g}$-module, $v \in V$ and $u \in U(\mathfrak{g})$, we will usually denote the action of u on v by $u(v)$ or, simply, uv. In particular, we always have $e(v) = E(v)$, $f(v) = F(v)$ and $h(v) = H(v)$.

Exercise 2.11. Let V be a finite-dimensional $\mathfrak{g}$-module. Then V is also a $U(\mathfrak{g})$-module via the construction from Proposition 2.7(i). Denote by $\psi_V : U(\mathfrak{g}) \to \mathcal{L}(V)$ the corresponding homomorphism. Show that the Casimir element C_V, considered in Section 1.3, is the image of the central element $(h + 1)^2 + 4fe \in U(\mathfrak{g})$ under ψ_V.

Exercise 2.12. Let $\mathfrak{a}$ be a one-dimensional Lie algebra with basis $\mathbf{x}$. Show that the polynomial algebra $\mathbb{C}[x]$ is the universal enveloping algebra of $\mathfrak{a}$ in the sense that $\mathbb{C}[x]$ together with the homomorphism $\epsilon : \mathfrak{a} \to \mathbb{C}[x]^{(-)}$, defined via $\epsilon(\mathbf{x}) = x$, have the universal property, analogous to that from Theorem 2.5. Prove also analogs of Proposition 2.6 and Proposition 2.7 in this situation.

2.2 Poincaré–Birkhoff–Witt Theorem

Corollary 2.9 says that the universal enveloping algebra $U(\mathfrak{g})$, defined in the previous section, is extremely important for the study of $\mathfrak{g}$-modules. However, the definition of $U(\mathfrak{g})$ does not give us much information about this algebra. It is not at all clear, for example, whether $U(\mathfrak{g})$ is finite-dimensional or infinite-dimensional (or even whether it is non-trivial). Nor is it clear at this stage, as mentioned in the previous section, whether the canonical map ε is injective. In this section we will clarify all this by giving a very explicit basis of $U(\mathfrak{g})$.

Theorem 2.13 (Poincaré–Birkhoff–Witt). *The set*

$$\{f^i h^j e^k : i, j, k \in \mathbb{N}_0\}$$

is a basis of $U(\mathfrak{g})$.

Theorem 2.13 is usually called the *PBW Theorem*. The monomials of the form $f^i h^j e^k$ are usually called *standard monomials*. Note that standard monomials also form a basis of the polynomial algebra $\mathbb{C}[f, h, e]$. Hence the PBW Theorem says that the non-commutative algebra $U(\mathfrak{g})$ is "of the same size" as the commutative algebra $\mathbb{C}[f, h, e]$. To prove the PBW Theorem we will need some preparation.

Lemma 2.14. *Standard monomials generate* $U(\mathfrak{g})$.

Proof. The standard basis of the free algebra R with generators e, f and h is given by arbitrary monomials $x_1 x_2 \cdots x_k$, where $k \in \mathbb{N}_0$ and $x_i \in \{e, f, h\}$ for all $i = 1, \ldots, k$. Hence to prove the statement of our lemma we just have to check that each $x_1 x_2 \cdots x_k$ can be written as a linear combination of standard monomials.

We proceed by induction on k. For $k = 1$ the statement is obvious. Let us now prove the induction step. For $k > 1$ consider some $x_1 x_2 \cdots x_k$ as above. A pair (i, j), $1 \leq i < j \leq k$, will be called an *inversion* provided that one of the following holds:

$$x_i = h \text{ and } x_j = f; \qquad x_i = e \text{ and } x_j = f; \qquad x_i = e \text{ and } x_j = h.$$

We proceed by induction on the number of inversions in $x_1 x_2 \cdots x_k$. If there are no inversions, the monomial $x_1 x_2 \cdots x_k$ is standard and we have nothing to prove. Otherwise, we can fix some inversion $(i, i + 1)$. We have

$$x_1 \cdots x_{i-1} x_i x_{i+1} x_{i+2} \cdots x_k$$
$$\overset{(2.1)}{=} x_1 \cdots x_{i-1} x_{i+1} x_i x_{i+2} \cdots x_k + x_1 \cdots x_{i-1} [x_i, x_{i+1}] x_{i+2} \cdots x_k.$$

As $[x_i, x_{i+1}] \in \{\pm h, \pm 2e, \pm 2f\}$, the second summand is a linear combination of monomials of degree $k - 1$ and hence is dealt with by induction on k. The first summand, in turn, has one inversion less than $x_1 x_2 \cdots x_k$ and hence is dealt with by induction on the number of inversions. Thus these two inductions complete the proof. □

Consider the vector space $V = \mathbb{C}[a, b, c]$. Define, using the induction on the degree of a monomial, the following linear operators F, H and E on V:

$$F(a^i b^j c^k) = a^{i+1} b^j c^k; \tag{2.5}$$

$$H(a^i b^j c^k) = \begin{cases} b^{j+1} c^k, & i = 0, \\ F(H(a^{i-1} b^j c^k)) - 2a^i b^j c^k, & i \neq 0; \end{cases} \tag{2.6}$$

$$E(a^i b^j c^k) = \begin{cases} c^{k+1}, & i, j = 0, \\ H(E(b^{j-1} c^k)) - 2E(b^{j-1} c^k), & i = 0, j \neq 0, \\ F(E(a^{i-1} b^j c^k)) + H(a^{i-1} b^j c^k), & i \neq 0. \end{cases} \tag{2.7}$$

where $i, j, k \in \mathbb{N}_0$.

Exercise 2.15. Check that the formulae (2.5)–(2.7) do give well-defined linear operators on $\mathbb{C}[a, b, c]$.

Exercise 2.16. Check that the formulae (2.5)–(2.7) can be rewritten as follows:

$$F(a^i b^j c^k) = a^{i+1} b^j c^k;$$

$$H(a^i b^j c^k) = \begin{cases} b^{j+1} c^k, & i = 0, \\ F(H(a^{i-1} b^j c^k)) + [H, F] a^{i-1} b^j c^k, & i \neq 0; \end{cases}$$

$$E(a^i b^j c^k) = \begin{cases} c^{k+1}, & i, j = 0, \\ H(E(b^{j-1} c^k)) + [E, H](b^{j-1} c^k), & i = 0, j \neq 0, \\ F(E(a^{i-1} b^j c^k)) + [E, F](a^{i-1} b^j c^k), & i \neq 0. \end{cases}$$

Lemma 2.17. *The formulae (2.5)–(2.7) define on V the structure of a $\mathfrak{g}$-module.*

Proof. We have to check the three relations from (1.2). Let us start with the relation $[H, F] = -2F$. For $i, j, k \in \mathbb{N}_0$ we have

$$H(F(a^i b^j c^k)) \overset{(2.5)}{=} H(a^{i+1} b^j c^k)$$
$$\text{(by (2.6))} \quad = \quad F(H(a^i b^j c^k)) - 2a^{i+1} b^j c^k$$
$$\text{(by (2.5))} \quad = \quad F(H(a^i b^j c^k)) - 2F(a^i b^j c^k)$$

and the relation $[H, F] = -2F$ is proved.

The relation $[E, F] = H$ is proved using the following computation: for $i, j, k \in \mathbb{N}_0$ we have

$$E(F(a^i b^j c^k)) \overset{(2.5)}{=} E(a^{i+1} b^j c^k)$$
$$\text{(by (2.7))} \quad = \quad F(E(a^i b^j c^k)) + H(a^i b^j c^k).$$

Finally, let us prove the relation $[H, E] = 2E$, which we first write in the form $EH - HE = -2E$. For any $j, k \in \mathbb{N}_0$ we have

$$E(H(b^j c^k)) \overset{(2.6)}{=} E(b^{j+1} c^k)$$
$$\text{(by (2.7))} = H(E(b^j c^k)) - 2E(b^j c^k)$$

and the relation $[H, E] = 2E$ is proved on monomials of the form $b^j c^k$. The really tricky thing is to prove this relation on monomials $a^i b^j c^k$, where $i \in \mathbb{N}$ and $j, k \in \mathbb{N}_0$. We do this by induction on i. The case $i = 0$ is already established, so we prove the induction step.

We rewrite $[H, E] = 2E$ as $HE - EH - 2E = 0$. Applying $HE - EH - 2E$ to $a^i b^j c^k$, where $i \in \mathbb{N}$ and $j, k \in \mathbb{N}_0$, and using Exercise 2.16 we obtain

$$(HE - EH - 2E)(a^i b^j c^k)$$
$$= (HFE + H[E,F] - EFH - E[H,F] - 2FE - 2[E,F])(a^{i-1} b^j c^k).$$
$$(2.8)$$

By induction we have $-2FE = F[E, H]$. Using the definition of the commutator and the relation $[H, F] = -2F$, which we proved above, we also have

$$H[E,F] = HEF - HFE,$$
$$E[H,F] = EHF - EFH,$$
$$-2[E,F] = [E, [H,F]].$$

This reduces the equality (2.8) to

$$(HE - EH - 2E)(a^i b^j c^k) = ([F, [E, H]] + [E, [H, F]])(a^{i-1} b^j c^k). \quad (2.9)$$

As we have already proved that $[E, F] = H$, we can add the following zero term:

$$0 = [H, H] = -[H, H] = [H, [F, E]]$$

to the equality (2.9) and obtain

$$(HE - EH - 2E)(a^i b^j c^k)$$
$$= ([F, [E, H]] + [E, [H, F]] + [H, [F, E]])(a^{i-1} b^j c^k).$$

The right-hand side of the latter is equal to zero because of the Jacobi identity for $\mathcal{L}(V)^{(-)}$. This completes the proof. $\qquad \square$

Now we are ready to prove the PBW Theorem 2.13.

Proof. To prove that standard monomials form a basis in $U(\mathfrak{g})$, we have to check that they generate $U(\mathfrak{g})$ and that they are linearly independent. The fact that they generate $U(\mathfrak{g})$ was proved in Lemma 2.14.

To prove that standard monomials are linearly independent, consider the $U(\mathfrak{g})$-module V from Lemma 2.17. Note that for all $i, j, k \in \mathbb{N}_0$ for the constant polynomial $1 \in V$ we have

$$F^i H^j E^k (1) = a^i b^j c^k.$$

Now it is left to observe that the elements $a^i b^j c^k \in V$ are linearly independent. Hence the linear operators $F^i H^j E^k$ are also linearly independent. Since these linear operators are exactly the images of standard monomials under the homomorphism defining the $U(\mathfrak{g})$-module structure on V, we conclude that standard monomials are linearly independent as well. This completes the proof. □

Exercise 2.18. Let x_1, x_2 and x_3 be the elements e, f and h written in some order. Show that the standard monomials $x_1^i x_2^j x_3^k$, $i, j, k \in \mathbb{N}_0$, also form a basis of $U(\mathfrak{g})$.

Corollary 2.19. *The canonical embedding ε of $\mathfrak{g}$ into $U(\mathfrak{g})^{(-)}$ is injective.*

Proof. This follows from the fact that the elements $\mathbf{e}$, $\mathbf{f}$ and $\mathbf{h}$ form a basis of $\mathfrak{g}$ and the fact that the elements $\varepsilon(\mathbf{e}) = e$, $\varepsilon(\mathbf{f}) = f$ and $\varepsilon(\mathbf{h}) = h$ are linearly independent in $U(\mathfrak{g})$ by Theorem 2.13. □

After Corollary 2.19 it is natural to identify $\mathfrak{g}$ with $\varepsilon(\mathfrak{g})$.

Remark 2.20. There exists an alternative (and somewhat easier) argument for Corollary 2.19. The elements $\mathbf{e}$, $\mathbf{f}$ and $\mathbf{h}$, which form a basis of $\mathfrak{g}$, act linearly independently on the natural module (since it is given by the identity map). From Proposition 2.7 we have that this action coincides with the action of $\varepsilon(\mathbf{e}) = e$, $\varepsilon(\mathbf{f}) = f$ and $\varepsilon(\mathbf{h}) = h$. Hence the latter elements must be linearly independent in $U(\mathfrak{g})$ and thus the map ε must be injective.

2.3　Filtration on $U(\mathfrak{g})$ and the associated graded algebra

As usual, for a monomial $x_1 x_2 \cdots x_k \in U(\mathfrak{g})$ (where $x_i \in \{f, h, e\}$ for all i) the number k is called the *degree* of the monomial. The degree of the monomial u is usually denoted by $\deg(u)$. For $i \in \mathbb{N}_0$ denote by $U(\mathfrak{g})^{(i)}$ the linear span of all monomials of degree at most i (we also set $U(\mathfrak{g})^{(-1)} = 0$). This gives us the following filtration on $U(\mathfrak{g})$:

$$U(\mathfrak{g}) = \bigcup_{i \in \mathbb{N}_0} U(\mathfrak{g})^{(i)}.$$

Note that $U(\mathfrak{g})^{(0)} = \mathbb{C}$. We obviously have

$$U(\mathfrak{g})^{(i)} U(\mathfrak{g})^{(j)} \subset U(\mathfrak{g})^{(i+j)}, \qquad (2.10)$$

which means that $U(\mathfrak{g})$ is a *filtered algebra*.

Exercise 2.21. Let $k \in \mathbb{N}_0$. Show that standard monomials of degree at most k form a basis of $U(\mathfrak{g})^{(k)}$.

Lemma 2.22. *Let* $i, j \in \mathbb{N}_0$, $u \in U(\mathfrak{g})^{(i)}$ *and* $v \in U(\mathfrak{g})^{(j)}$. *Then* $[u, v] \in U(\mathfrak{g})^{(i+j-1)}$.

Proof. By Exercise 2.21 the space $U(\mathfrak{g})^{(k)}$, $k \in \mathbb{N}_0$, has a basis consisting of standard monomials. Hence, by the linearity of $[\_, \_]$, it is enough to prove the statement in the case when both u and v are standard monomials. We prove the statement by induction on $\deg(u) + \deg(v)$. If $\deg(u) + \deg(v) \leq 1$, then at least one of u and v must be a scalar and the statement becomes obvious. If both $u, v \in \mathfrak{g}$, then $[u, v] \in \mathfrak{g}$ as well and the statement is true again. Now, to prove the induction step we may assume that $u \notin U(\mathfrak{g})^{(1)}$ and write $u = xu'$, where $x \in \{f, h, e\}$ and u' is a standard monomial. Using the definition of the commutator we have

$$\begin{aligned}
[u, v] &= uv - vu \\
&= xu'v - vxu' \\
&= xvu' + x[u', v] - vxu' \\
&= vxu' + [x, v]u' + x[u', v] - vxu' \\
&= [x, v]u' + x[u', v].
\end{aligned}$$

By induction we have $[u', v] \in U(\mathfrak{g})^{(i+j-2)}$ and hence $x[u', v] \in U(\mathfrak{g})^{(i+j-1)}$. As $u \notin U(\mathfrak{g})^{(1)}$ by our assumption, u' is not a scalar monomial and thus $x \in U(\mathfrak{g})^{(1)} \subset U(\mathfrak{g})^{(i-1)}$. Hence, applying the inductive assumption to $[x, v]$ we get $[x, v]u' \in U(\mathfrak{g})^{(j+1-1)}u' \subset U(\mathfrak{g})^{(i+j-1)}$ and the claim follows. $\qquad \square$

For $i \in \mathbb{N}_0$ put

$$G(\mathfrak{g})_i = U(\mathfrak{g})^{(i)} / U(\mathfrak{g})^{(i-1)}.$$

Consider the vector space

$$G(\mathfrak{g}) = \bigoplus_{i \in \mathbb{N}_0} G(\mathfrak{g})_i.$$

For $u \in U(\mathfrak{g})^{(i)}/U(\mathfrak{g})^{(i-1)}$ and $v \in U(\mathfrak{g})^{(j)}/U(\mathfrak{g})^{(j-1)}$ consider some representatives $\overline{u} \in U(\mathfrak{g})^{(i)}$ and $\overline{v} \in U(\mathfrak{g})^{(j)}$. In this situation, we have $\overline{u}\overline{v} \in U(\mathfrak{g})^{(i+j)}$ by (2.10) and hence we can define uv to be the image of $\overline{u}\overline{v}$ in $U(\mathfrak{g})^{(i+j)}/U(\mathfrak{g})^{(i+j-1)}$.

Exercise 2.23. Show that the product uv is well-defined in the sense that it does not depend on the choice of the representatives $\overline{u}$ and $\overline{v}$. Show further that the binary operation $(u, v) \mapsto uv$ uniquely extends to an associative multiplication on $G(\mathfrak{g})$.

From the above definition we have that

$$G(\mathfrak{g})_i \, G(\mathfrak{g})_j \subset G(\mathfrak{g})_{i+j}$$

for all $i, j \in \mathbb{N}_0$. This means that the associative algebra $G(\mathfrak{g})$ is a *graded algebra*. The algebra $G(\mathfrak{g})$ is called the graded algebra *associated* with the filtered algebra $U(\mathfrak{g})$. Note that $G(\mathfrak{g})_0 = \mathbb{C}$ and that $G(\mathfrak{g})_1$ can be canonically identified with $\mathfrak{g}$ because of Corollary 2.19.

Exercise 2.24. Show that the images in $G(\mathfrak{g})$ of standard monomials from $U(\mathfrak{g})$ form a basis of $G(\mathfrak{g})$.

Proposition 2.25. *The algebra $G(\mathfrak{g})$ is canonically isomorphic to the polynomial algebra $\mathbb{C}[a, b, c]$.*

Proof. From Lemma 2.22 and the definitions it follows that the algebra $G(\mathfrak{g})$ is commutative. The images of standard monomials in $G(\mathfrak{g})$ form a basis of $G(\mathfrak{g})$ by Exercise 2.24. Since any monomial is a product of monomials of degree one, it follows that $G(\mathfrak{g})$ is generated by monomials of degree one. In particular, the map

$$a \mapsto f, \quad b \mapsto h, \quad c \mapsto e$$

uniquely extends to an epimorphism from $\mathbb{C}[a, b, c]$ to $G(\mathfrak{g})$. This epimorphism is injective since standard monomials in a, b and c, which form a basis in $\mathbb{C}[a, b, c]$, are mapped to linearly independent images in $G(\mathfrak{g})$ of standard monomials from $U(\mathfrak{g})$ (again Exercise 2.24). This completes the proof. $\square$

Corollary 2.26. *The algebra $U(\mathfrak{g})$ is a* domain, *that is for arbitrary nonzero elements $u, v \in U(\mathfrak{g})$, we have $uv \neq 0$.*

Proof. The algebra $G(\mathfrak{g})$ is a polynomial algebra and hence is a domain. Let i and j be minimal such that $u \in U(\mathfrak{g})^{(i)}$ and $v \in U(\mathfrak{g})^{(j)}$ respectively. For the corresponding images $u' \in G(\mathfrak{g})_i$ and $v' \in G(\mathfrak{g})_j$ we then have $u', v' \neq 0$ and hence $u'v' \neq 0$. This implies $uv \neq 0$, as required. $\qquad\square$

2.4   Centralizer of h and center of $U(\mathfrak{sl}_2)$

The one-dimensional Lie subalgebra $\mathfrak{h}$ of $\mathfrak{g}$ with basis $\{\mathbf{h}\}$ is called the *Cartan subalgebra*. The very special property of $\mathfrak{h}$ is that the action of every element from $\mathfrak{h}$ on the adjoint module $\mathfrak{g}$ (see Example 1.6) is diagonalizable. Indeed, directly from the definition it follows that the elements $\mathbf{f}$, $\mathbf{h}$ and $\mathbf{e}$ are eigenvectors for the operator $[\mathbf{h}, \_]$ with eigenvalues -2, 0 and 2, respectively.

Exercise 2.27. Find some $x \in \mathfrak{g} \setminus \mathfrak{h}$, whose action on the the adjoint module is diagonalizable.

Exercise 2.28. For $u \in U(\mathfrak{g})$ set

$$E(u) = [e, u], \quad F(u) = [f, u], \quad H(u) = [h, u].$$

Show that this defines on $U(\mathfrak{g})$ the structure of a $\mathfrak{g}$-module, called the *adjoint* module.

The property of $\mathfrak{h}$ to act diagonalizably on the adjoint module $\mathfrak{g}$ lifts up to the analogous property for the adjoint module $U(\mathfrak{g})$ (where the adjoint module $\mathfrak{g}$ is a submodule).

Lemma 2.29. *For $i, j, k \in \mathbb{N}_0$ the standard monomial $f^i h^j e^k$ is an eigenvector for the operator $[h, \_]$ with eigenvalue $2(k - i)$. In particular, the adjoint action of $\mathfrak{h}$ on $U(\mathfrak{g})$ is diagonalizable.*

Proof. We have

$$\begin{aligned}
[h, f^i h^j e^k] &= h f^i h^j e^k - f^i h^j e^k h \\
\text{(by (2.1))} &= f^i (h - 2i) h^j e^k - f^i h^j e^k h \\
&= f^i h^j (h - 2i) e^k - f^i h^j e^k h \\
\text{(by (2.1))} &= f^i h^j e^k (h - 2i + 2k) - f^i h^j e^k h \\
&= 2(k - i) f^i h^j e^k,
\end{aligned}$$

which completes the proof. $\qquad\square$

For $s \in \mathbb{Z}$ denote by $U(\mathfrak{g})_{2s}$ the subspace of $U(\mathfrak{g})$ with the basis $f^i h^j e^k$, $i, j, k \in \mathbb{Z}$, such that $k - i = s$. Then, by the PBW Theorem, we have the decomposition

$$U(\mathfrak{g}) = \bigoplus_{s \in \mathbb{Z}} U(\mathfrak{g})_{2s}, \qquad (2.11)$$

where every summand on the right-hand side is a submodule with respect to the adjoint action of $\mathfrak{h}$ (and, actually, consists of all eigenvectors of **h** with some fixed eigenvalue). In particular, the summand $U(\mathfrak{g})_0$ consists of all elements of $U(\mathfrak{g})$, which commute with $\mathfrak{h}$. Hence $U(\mathfrak{g})_0$ is none other than the *centralizer* of the Cartan subalgebra $\mathfrak{h}$ in $U(\mathfrak{g})$. In particular, $U(\mathfrak{g})_0$ is a subalgebra of $U(\mathfrak{g})$. The algebra $U(\mathfrak{g})_0$ obviously contains h and the central *Casimir element* $c = (h + 1)^2 + 4fe$ (see Exercise 2.1).

Proposition 2.30.

(i) The elements h and c generate $U(\mathfrak{g})_0$.

(ii) The elements h and c are algebraically independent in $U(\mathfrak{g})_0$.

(iii) The algebra $U(\mathfrak{g})_0$ coincides with the polynomial ring $\mathbb{C}[h, c]$.

To prove Proposition 2.30 we will need one abstract definition. We say that the monomial $h^x c^y \in \mathbb{C}[h, c]$ is smaller than the monomial $h^u c^v \in \mathbb{C}[h, c]$ with respect to the *inverse lexicographic order*, provided that we have $y < v$ or $y = v$ and $x < u$. For $g \in \mathbb{C}[h, c]$ we can define the *lexicographic degree* of g as (i, j) provided that $h^i c^j$ is the maximal (with respect to the inverse lexicographic order) monomial which occurs in g with a non-zero coefficient. Similarly, we say that the standard monomial $f^y h^x e^y \in U(\mathfrak{g})_0$ is smaller than the standard monomial $f^v h^u e^v \in U(\mathfrak{g})_0$ with respect to the *lexicographic order* provided that $y < v$ or $y = v$ and $x < u$, and can define the lexicographic degree for elements of $U(\mathfrak{g})_0$ in the corresponding way. Now we are ready to prove Proposition 2.30.

Proof. Denote by A the (unital) subalgebra of $U(\mathfrak{g})_0$ generated by h and c. From Lemma 2.29 we know that the standard monomials $f^i h^j e^i$, $i, j \in \mathbb{N}_0$, form a basis of $U(\mathfrak{g})_0$. Hence to prove the statement (i) it is enough to show that every such $f^i h^j e^i$ belongs to A. Using (2.1) we have $f^i h^j e^i = (h + 2i)^j f^i e^i$ and hence it is enough to show that $f^i e^i \in A$ for all $i \in \mathbb{N}_0$. We prove this by induction on i. For $i = 0$ the statement is obvious. F or $i = 1$ we have

$$fe = \frac{1}{4}(c - (h + 1)^2) \in A$$

by definitions. To prove the induction step we write
$$f^i e^i = fef^{i-1}e^{i-1} + f[f^{i-1}, e]e^{i-1}.$$
The summand $fef^{i-1}e^{i-1}$ belongs to A as both fe and $f^{i-1}e^{i-1}$ belong to A by induction. By Lemma 2.22, the summand $f[f^{i-1}, e]e^{i-1}$ is a linear combination of monomials of degree at most $2i-1$. Hence $f[f^{i-1}, e]e^{i-1} \in A$ by induction. This completes the proof of the statement (i).

The statement (i) says that the canonical homomorphism φ from the polynomial algebra $\mathbb{C}[h, c]$ to $U(\mathfrak{g})_0$ given by $\varphi(h) = h$ and $\varphi(c) = c$ is surjective. An important property of φ is the following:

Lemma 2.31. *Let* $i, j \in \mathbb{N}_0$. *Then*
$$\varphi(h^i c^j) = 4^j f^j h^i e^j + \text{terms of smaller lexicographic degree.}$$

Proof. We have $\varphi(h^i c^j) = h^i((h+1)^2 + 4fe)^j$ and the statement follows by induction on j using Lemma 2.22. We leave the details to the reader.□

Take any $g(h, c) \in \mathbb{C}[h, c]$, $g(h, c) \neq 0$, and write it as follows:
$$g(h, c) = \alpha h^i c^j + \text{terms of smaller lexicographic degree}$$
for some $i, j \in \mathbb{N}_0$ and $\alpha \in \mathbb{C}$, $\alpha \neq 0$. Write $\varphi(g(h, c))$ in the basis consisting of standard monomials. By Lemma 2.31, the term $\varphi(\alpha h^i c^j)$ will contribute with the coefficient $4^j \alpha \neq 0$ to the standard monomial $f^j h^i e^j$, while all other terms will not contribute to $f^j h^i e^j$ at all. Since the standard monomials are linearly independent, we get $\varphi(g(h, c)) \neq 0$. This means that φ is injective and completes the proof of both (ii) and (iii).   □

Denote by $Z(\mathfrak{g})$ the *center* of the algebra $U(\mathfrak{g})$. Since $U(\mathfrak{g})$ is generated by e, f and h, we have
$$Z(\mathfrak{g}) = \{u \in U(\mathfrak{g}) : [e, u] = [f, u] = [h, u] = 0\},$$
in particular, $Z(\mathfrak{g}) \subset U(\mathfrak{g})_0$. Obviously, $U(\mathfrak{g})^{(0)} = \mathbb{C} \subset Z(\mathfrak{g})$. From Exercise 2.1 we have that $c \in Z(\mathfrak{g})$. A complete description of $Z(\mathfrak{g})$ is given by the following:

Theorem 2.32. $Z(\mathfrak{g}) = \mathbb{C}[c] \subset U(\mathfrak{g})_0$, *in particular,* $Z(\mathfrak{g})$ *is a polynomial algebra in one variable.*

Proof. We know that $c \in Z(\mathfrak{g})$ by Exercise 2.1 and hence $\mathbb{C}[c] \subset Z(\mathfrak{g})$ since $\mathbb{C}[c]$ is a subalgebra of $U(\mathfrak{g})$ by Proposition 2.30. On the other hand, we have $Z(\mathfrak{g}) \subset U(\mathfrak{g})_0 = \mathbb{C}[h, c]$ (Proposition 2.30(iii)). Let
$$g(h, c) = \sum_{i=0}^{k} g_i(h)c^i \in U(\mathfrak{g})_0.$$

If $g(h, c) \in Z(\mathfrak{g})$, then we must have $[e, g(h, c)] = 0$. On the other hand,

$$
\begin{aligned}
[e, g(h, c)] &= \left[e, \sum_{i=0}^{k} g_i(h) c^i \right] \\
&= \sum_{i=0}^{k} [e, g_i(h) c^i] \\
(\text{as } c^i \in Z(\mathfrak{g})) &= \sum_{i=0}^{k} c^i [e, g_i(h)] \\
(\text{by } (2.1)) &= e \sum_{i=0}^{k} c^i (g_i(h) - g_i(h+2)).
\end{aligned}
$$

Since $U(\mathfrak{g})$ is a domain (Corollary 2.26), we thus must have

$$
\sum_{i=0}^{k} c^i (g_i(h) - g_i(h+2)) = 0.
$$

From Proposition 2.30(iii) it then follows that $g_i(h) - g_i(h+2) = 0$ for all i, which means that for every i the polynomial g_i is a constant polynomial. This yields $g(h, c) \in \mathbb{C}[c]$ and completes the proof. $\qquad \square$

Theorem 2.33.

(i) *The algebra $U(\mathfrak{g})$ is free over $U(\mathfrak{g})_0$ both as a left and as a right module with basis*

$$
\mathbf{B}_1 = \{1, e, f, e^2, f^2, e^3, f^3, \dots\}.
$$

(ii) *The algebra $U(\mathfrak{g})$ is free over $Z(\mathfrak{g})$ both as a left and as a right module with basis*

$$
\mathbf{B}_2 = \{1, h, h^2, h^3, \dots\} \cdot \mathbf{B}_1
$$

and also with basis

$$
\mathbf{B}_3 = \mathbf{B}_1 \cdot \{1, h, h^2, h^3, \dots\}.
$$

Proof. Consider the decomposition (2.11). If $u \in U(\mathfrak{g})_0$ is a standard monomial, then for any $s \in \mathbb{N}_0$ we have that $u e^s \in U(\mathfrak{g})_{2s}$ is a standard monomial as well. On the other hand, every standard monomial $v \in U(\mathfrak{g})_{2s}$, $s \in \mathbb{N}_0$, has the form $u e^s$ for some standard monomial $u \in U(\mathfrak{g})_0$. Taking the PBW Theorem into account, we derive that $U(\mathfrak{g})_{2s}$, $s \in \mathbb{N}_0$, is a free left $U(\mathfrak{g})_0$-module of rank 1 with basis e^s.

Let us now prove that for $s \in \mathbb{Z}$, $s < 0$, the space $U(\mathfrak{g})_{2s}$ is a free left $U(\mathfrak{g})_0$-module of rank 1 with basis $f^{|s|}$. This is similar to the above, but requires extra computation. As $U(\mathfrak{g})$ is a domain, the left $U(\mathfrak{g})_0$-module, generated by $f^{|s|}$ is free. Hence we only have to check that $U(\mathfrak{g})_0 f^{|s|} = U(\mathfrak{g})_{2s}$. If $u \in U(\mathfrak{g})_0$ is a standard monomial, we have

$$uf^{|s|} = f^{|s|}u + [u, f^{|s|}].$$

Here $f^{|s|}u$ is a standard monomial in $U(\mathfrak{g})_{2s}$. Moreover, every standard monomial $v \in U(\mathfrak{g})_{2s}$ has the form $f^{|s|}u$ for some standard monomial $u \in U(\mathfrak{g})_0$. By Lemma 2.22, the element $[u, f^{|s|}]$ is a linear combination of standard monomials of strictly smaller degree (than that of $f^{|s|}u$). Hence, by induction on the degree of a monomial, it follows that all standard monomials from $U(\mathfrak{g})_{2s}$ belong to $U(\mathfrak{g})_0 f^{|s|}$. This proves the statement (i) for the left module structure. The statement (i) for the right module structure follows applying σ (see Exercise 2.3).

As $U(\mathfrak{g})_0 = \mathbb{C}[h, c]$ (Proposition 2.30(iii)) and $Z(\mathfrak{g}) = \mathbb{C}[c]$ (Theorem 2.32), we have that $U(\mathfrak{g})_0$ is a free $Z(\mathfrak{g})$-module with basis $\{1, h, h^2, \dots\}$. Now the statement (ii) follows from the statement (i) and the observation that every element of $Z(\mathfrak{g})$ commutes with every element of $U(\mathfrak{g})$. This completes the proof. $\square$

2.5 Harish-Chandra homomorphism

This section is motivated by the following observation: As we already know, the center of the algebra $U(\mathfrak{g})$ is generated by the Casimir element $c = (h+1)^2 + 4fe$. In Section 1.3 (namely Exercise 1.33) we saw that on the simple finite-dimensional module $\mathbf{V}^{(n)}$ the element c acts as the scalar n^2. The best way to compute this scalar is to apply c to the element v_0, for which we know that $E(v_0) = 0$ and $H(v_0) = (n-1)v_0$. The action of c on such v_0 reduces to the action of $(h+1)^2$. In other words, the action of an element from $Z(\mathfrak{g})$ reduces to the action of an element from $\mathbb{C}[h]$. This phenomenon can be given a neat theoretical description.

Lemma 2.34.

(i) For the set $I = U(\mathfrak{g})e \cap U(\mathfrak{g})_0$ we have $I = fU(\mathfrak{g}) \cap U(\mathfrak{g})_0$. In particular, I is an ideal of $U(\mathfrak{g})_0$.

(ii) We have $U(\mathfrak{g})_0 = \mathbb{C}[h] \oplus I$.

Proof. Every element in I is a linear combination of some elements of the form ue, where u is some standard monomial (note that in this case ue is a standard monomial as well). Since $U(\mathfrak{g})_0$ has a basis consisting of standard monomials, the whole linear combination belongs to $U(\mathfrak{g})_0$ if, and only if, each summand belongs to $U(\mathfrak{g})_0$. Furthermore, $ue \in U(\mathfrak{g})_0$ if, and only if, $u = f^{i+1}h^j e^i$ for some $i, j \in \mathbb{N}_0$. But then $ue = f(f^i h^j e^{i+1}) \in fU(\mathfrak{g})$, which implies the inclusion $I \subset fU(\mathfrak{g}) \cap U(\mathfrak{g})_0$. The opposite inclusion is proved similarly. The statement (i) follows. The statement (ii) follows from the statement (i) and the PBW Theorem. $\qquad\square$

Denote by κ the projection of $U(\mathfrak{g})_0$ onto $\mathbb{C}[h]$ with kernel I. Because of Lemma 2.34, the map κ is a homomorphism of associative algebras. It is called the *Harish-Chandra homomorphism*. From the definition we immediately have the following main property of κ:

Proposition 2.35. *Let V be a $\mathfrak{g}$-module and $v \in V$ be an element such that $E(v) = 0$. Then for any $g \in Z(\mathfrak{g})$ we have*

$$g(v) = \kappa(g) \cdot v.$$

Proof. By Theorem 2.32 we have $Z(\mathfrak{g}) = \mathbb{C}[c]$ and hence $g = g(c) \in \mathbb{C}[c]$. As $c = (h+1)^2 + 4fe$, from $E(v) = 0$ and the definition of κ we get

$$c(v) = (H+1)^2(v) = \kappa(c)(v).$$

Since κ is a homomorphism, it follows that $g(c)(v) = \kappa(g(c)) \cdot v$ as well. $\square$

Proposition 2.35 says that the restriction of κ to $Z(\mathfrak{g})$ is important in the study of $\mathfrak{g}$-modules. This restriction can be explicitly described:

Theorem 2.36. *The restriction of κ to $Z(\mathfrak{g})$ gives an isomorphism from $Z(\mathfrak{g})$ to $\mathbb{C}[(h+1)^2]$.*

Proof. Follows from the definitions and Theorem 2.32. $\qquad\square$

Although Theorem 2.36 looks extremely easy, it is very important (we will see some of its applications later on). One of the most interesting features of this theorem is the object $\mathbb{C}[(h+1)^2]$, which appears in the formulation and turns out to have a very neat interpretation in terms of invariant polynomials.

Consider the adjoint action of $\mathfrak{g}$ on itself. As we have already seen, the adjoint action of $\mathbf{h}$ (and more generally of the commutative Lie algebra $\mathfrak{h}$) is diagonalizable and $\mathbf{f}$, $\mathbf{h}$ and $\mathbf{e}$ form a basis of eigenvectors with eigenvalues -2, 0 and 2 respectively. We can consider these eigenvalues as elements of the vector space $\mathfrak{h}^* \cong \mathbb{C}$. The non-zero eigenvalues are called *roots* and the set of all non-zero eigenvalues is called the *root system* of $\mathfrak{g}$. Let W denote the subgroup of the general linear group on $\mathfrak{h}^*$ which preserves the set of all these eigenvalues. This group is called the *Weyl group* of $\mathfrak{g}$. Obviously, W consists of two elements: the identity transformation and the multiplication by -1. So, W is isomorphic to the symmetric group $\mathbf{S}_2$. The action of W on $\mathfrak{h}^*$ naturally induces an action of W on $\mathfrak{h}$. The algebra $\mathbb{C}[h]$ is then identified with the algebra of polynomial functions on $\mathfrak{h}^*$. Let $\delta = 1 = \frac{1}{2}2 \in \mathbb{C} = \mathfrak{h}^*$ be the half of the sum of all positive roots (we have only one positive root, namely 2). Denote by γ the automorphism of $\mathbb{C}[h]$, which maps the polynomial function g on $\mathfrak{h}^*$ to the function $\lambda \mapsto g(\lambda - \delta)$.

Corollary 2.37. *The restriction of* $\gamma \circ \kappa$ *to* $Z(\mathfrak{g})$ *is an isomorphism from* $Z(\mathfrak{g})$ *to the algebra* $\mathbb{C}[h]^W$ *of polynomials, invariant with respect to the action of* W.

Proof. By Theorem 2.36 we have that the restriction of κ to $Z(\mathfrak{g})$ gives an isomorphism from $Z(\mathfrak{g})$ to $\mathbb{C}[(h+1)^2]$. For any $g \in \mathbb{C}[(h+1)^2]$ we have
$$\gamma(g)((h+1)^2) = g((h-1+1)^2) = g(h^2).$$
Since $\mathbb{C}[h^2] = \mathbb{C}[h]^W$, the claim follows.   □

Alternatively we can say that the restriction of κ to $Z(\mathfrak{g})$ is an isomorphism from $Z(\mathfrak{g})$ to the algebra of polynomials, invariant with respect to the action of W, shifted by δ.

Exercise 2.38. Define the "other side versions" of κ associated with the ideal $I' = U(\mathfrak{g})f \cap U(\mathfrak{g})_0$ and then formulate and prove the corresponding analogs of Proposition 2.35, Theorem 2.36 and Corollary 2.37.

2.6 Noetherian property

Recall that an associative algebra A is called *left (or right) Noetherian* if it satisfies the ascending chain condition on left (or right) ideals, that is for any ascending chain
$$I_1 \subset I_2 \subset I_3 \subset \dots$$
of left (right) ideals in A there exists $n \in \mathbb{N}$ such that $I_n = I_{n+1} = \dots$.

Exercise 2.39. Show that A is left (right) Noetherian if and only if every left (right) ideal of A is finitely generated.

Theorem 2.40. *The algebra $U(\mathfrak{g})$ is both left and right Noetherian.*

Proof. Due to the existence of σ it is enough to show that $U(\mathfrak{g})$ is left Noetherian. By Exercise 2.39 it is enough to show that the every left ideal I of $U(\mathfrak{g})$ is finitely generated.

Because of the decomposition (2.11) every $u \in U(\mathfrak{g})$ can be written as $u = \sum_{i \in \mathbb{Z}} u_i$, where $u_i \in U(\mathfrak{g})_{2i}$ and only finitely many of u_i's are non-zero. By Theorem 2.33(i) we have

$$u_i = \begin{cases} e^i v_i, & i \geq 0; \\ f^{|i|} v_i, & i < 0; \end{cases}$$

where $v_i \in U(\mathfrak{g})_0$. For $u \neq 0$ let $d_+(u)$ and $d_-(u)$ denote the maximal and the minimal possible i such that $v_i \neq 0$, respectively. Set $k_+(u) = v_{d_+(u)}$ and $k_-(u) = v_{d_-(u)}$. For $i \in \mathbb{N}_0$ define

$$J_i = \{k_+(u) : u \in I, d_+(u) = i\}, \quad J_i' = \{k_-(u) : u \in I, d_-(u) = -i\}.$$
$$(2.12)$$

Lemma 2.41. *For every $i \in \mathbb{N}$ the sets J_i and J_i' are ideals of the algebra $U(\mathfrak{g})_0 = \mathbb{C}[h, c]$.*

Proof. If $u \in I$ is such that $d_+(u) = i$ and $v_i \in J_i$, then for any $g(h, c) \in \mathbb{C}[h, c]$ we have $g(h - 2i, c)u \in I$ (as I is a left ideal), and we also have $d_+(u) = d_+(g(h - 2i, c)u)$ (as $g(h - 2i, c) \in U(\mathfrak{g})_0$). At the same time

$$g(h - 2i, c)u_i = g(h - 2i, c)e^i v_i$$
$$\text{(by (2.1))} = e^i g(h, c) v_i.$$

This implies that $g(h, c)v_i \in J_i$. As it is obvious that J_i is a linear space, it follows that J_i is an ideal. Similarly one shows that J_i' is an ideal. $\square$

Furthermore, we observe that if $d_+(u) > 0$ then $d_+(eu) = d_+(u) + 1$ and $k_+(eu) = k_+(u)$. This, and a similar observation for k_-, imply that we have the following ascending chains of ideals in $U(\mathfrak{g})_0$:

$$J_1 \subset J_2 \subset J_3 \subset \dots, \qquad J_1' \subset J_2' \subset J_3' \subset \dots.$$

The algebra $U(\mathfrak{g})_0 = \mathbb{C}[h, c]$ is Noetherian by Hilbert's basis theorem. Hence there exists $m \in \mathbb{N}$ such that $J_m = J_{m+1} = \dots$ and also $J_m' = J_{m+1}' = \dots$, moreover, both J_m and J_m' are finitely generated. Let $a_1, \dots, a_s$ be elements of I such that $d_+(a_1) = \dots = d_+(a_s) = m$ and $k_+(a_1), \dots, k_+(a_s)$ generate J_m. Similarly, let $b_1, \dots, b_t$ be elements of I such that $d_-(b_1) = \dots = d_-(b_t) = -m$ and $k_-(b_1), \dots, k_-(b_t)$ generate J_m'.

Set

$$n = \max(m, |d_\pm(a_1)|, \ldots, |d_\pm(a_s)|, |d_\pm(b_1)|, \ldots, |d_\pm(b_t)|). \qquad (2.13)$$

By Theorem 2.33, the $\mathbb{C}[h, c]$-module

$$M = \bigoplus_{i=-n}^{n} U(\mathfrak{g})_{2i}$$

is a free $\mathbb{C}[h, c]$-module of finite rank, hence Noetherian (as $\mathbb{C}[h, c]$ is). Thus its submodule $I \cap M$ is finitely generated (over $\mathbb{C}[h, c]$). Let $p_1, \ldots, p_l$ be some set of generators for $I \cap M$.

Lemma 2.42. *The set* $X = \{a_1, \ldots, a_s, b_1, \ldots, b_s, p_1, \ldots, p_l\}$ *generates* I.

Proof. Let I' denote the ideal of $U(\mathfrak{g})$, generated by X. Let $u \in I$. If $d_-(u) \geq -n$ and $d_+(u) \leq n$, then $u \in I \cap M$ and hence $u \in I'$.

Otherwise, we show that $u \in I$ implies $u \in I'$ by induction on the number $N = \max(n, d_+(u)) + \max(n, |d_-(u)|)$. The basis of the induction, the case when $N = 2n$, is proved in the previous paragraph. Let now $u \in I$ be such that $N > 2n$. Assume $d_+(u) = i > n$ (the case $d_-(u) < n$ is treated similarly). Then the element $k_+(u)$ can be written in the form $k_+(u) = \sum_{j=1}^{s} g_j(h, c) k_+(a_j)$ for some $g_j(h, c) \in \mathbb{C}[h, c]$. Hence for the element

$$v = \sum_{j=1}^{s} g_j(h - 2i, c) e^{i-m} a_j \in I'$$

we have $d_+(v) = d_+(u)$ and $k_+(v) = k_+(u)$. This yields $d_+(u - v) < d_+(u)$. Thanks to (2.13) we also have $d_-(u - v) \geq \min(d_-(u), -n)$. Hence we conclude that $u - v \in I'$ by induction and thus $u \in I'$. This completes the proof. $\qquad \square$

The claim of Theorem 2.40 follows now from Lemma 2.42. $\qquad \square$

2.7 Addenda and comments

2.7.1

The material presented in this chapter can be found in several textbooks; see for example [23, 37, 46, 57, 106]. Some of the results (for example, the construction of the universal enveloping algebra and its basic properties, filtrations, PBW Theorem) are true for any Lie algebra with rather similar proofs. Commutativity of $U(\mathfrak{g})_0$ is a special feature of $\mathfrak{sl}_2$. For a simple finite-dimensional complex Lie algebra $\mathfrak{a}$ the center of $U(\mathfrak{a})$ is always a polynomial ring (in several variables) and can be given a description via invariant polynomials over the Cartan subalgebra. We refer the reader to [37] for details.

2.7.2

The universal enveloping algebra $U(\mathfrak{a})$ of a Lie algebra $\mathfrak{a}$ is usually described in the following way (see for example Subsection 2.1.1 of [37]): let $T^0 = \mathbb{C}$ and for $i \in \mathbb{N}$ set

$$T^i = \underbrace{\mathfrak{a} \otimes \mathfrak{a} \otimes \cdots \otimes \mathfrak{a}}_{i \text{ factors}}.$$

Consider the tensor algebra $T = T^0 \oplus T^1 \oplus T^2 \oplus \ldots$, in which the product is given by tensor multiplication. Let J denote the two-sided ideal of T, generated by all elements of the form

$$x \otimes y - y \otimes x - [x, y],$$

where $x, y \in \mathfrak{a}$. Then the algebra $U(\mathfrak{a})$ is defined as follows: $U(\mathfrak{a}) = T/J$. This can be easily reformulated in terms of generators and relations.

2.7.3

The most general formulation of the PBW Theorem for $\mathfrak{g}$ is the following: given any basis x_1, x_2 and x_3 of $\mathfrak{g}$, all monomials of the form $x_1^i x_2^j x_3^k$, where $i, j, k \in \mathbb{N}_0$, form a basis of $U(\mathfrak{g})$. This formulation (and the proof) generalizes to any Lie algebra in a straightforward way.

2.7.4

There is an alternative proof of the PBW Theorem, using the linear version of the *Diamond lemma*, proposed in [15]. The original combinatorial version of the Diamond lemma appears in [98] as a tool to prove the existence and uniqueness of certain normal forms in the following situation: Let X be a set and $\rightarrow$ a binary relation on X. Denote by $\twoheadrightarrow$ the transitive closure of the relation $\rightarrow$. Assume that X does not have infinite chains of the form $x_1 \rightarrow x_2 \rightarrow x_3 \rightarrow \ldots$. An element $y \in X$ is called a *normal form* for an element $x \in X$, if there exists a sequence $x = x_1 \rightarrow x_2 \rightarrow \cdots \rightarrow x_k = y$ and $y \nrightarrow z$ for any $z \in X$.

Theorem 2.43 (Diamond lemma). *The following conditions are equivalent:*

(i) *For all $x, y, z \in X$ such that $x \to y$ and $x \to z$, there exists $u \in X$ such that $y \twoheadrightarrow u$ and $z \twoheadrightarrow u$.*

(ii) *Every $x \in X$ has a unique normal form.*

The "Diamond lemma" is so-called due to the fact that the condition (i) of Theorem 2.43 can be drawn as the following "diamond":

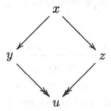

The main idea is that the binary relation $\to$ describes some "reduction process" and the condition from Theorem 2.43(i) basically says that if we start reducing x in two different ways, we can always reduce the results to the same element.

In [15] the reader will find a ring-theoretic reformulation of this lemma (which requires a number of new notions and notation) and an application to the proof of the PBW Theorem. The main idea is to consider the set of all monomials in $U(\mathfrak{a})$. Then the relation $xy - yx = [x, y]$, used in the definition of $U(\mathfrak{a})$, describes the reduction process, when we try to reduce our monomial to (a linear combination of) standard monomials. One has to verify that the "diamond condition" from Theorem 2.43(i) is satisfied for $U(\mathfrak{a})$, and this follows from the Jacobi identity. Then a direct application of the ring-theoretic Diamond lemma guarantees that the set of irreducible monomials (standard monomials) forms a basis. The advantage of this approach is that it can be generalized to many other algebras (for example to the algebras described in 2.7.5 below).

2.7.5

As already mentioned above, most of the results of this chapter generalize (in most cases with almost the same arguments) to universal enveloping algebras of arbitrary Lie algebras. However, they can be also generalized to several other classes of algebras.

The first such class of algebras is called algebras *similar to* $U(\mathfrak{sl}_2)$ and was introduced by S. Smith in [107]. For a fixed polynomial $g(x) \in \mathbb{C}[x]$ the algebra R_g, similar to $U(\mathfrak{sl}_2)$, is defined as the quotient of the free associative algebra R with generators e, f and h modulo the ideal generated by the relations

$$ef - fe = f(h), \quad he - eh = 2e, \quad hf - fh = -2f. \tag{2.14}$$

The universal enveloping algebra $U(\mathfrak{g})$ is obtained in the case $g(x) = x$. Most of the results from this chapter generalize to R_g. In particular, R_g is always a Noetherian domain and has a basis consisting of standard monomials. It contains an analog of the Casimir element, and its center is the polynomial ring in this element. Furthermore, R_g has a filtration by finite-dimensional subspaces such that the associated graded algebra is finitely generated and commutative. We refer the reader to [107] for details.

Recently, the definition of algebras similar to $U(\mathfrak{sl}_2)$ was further extended by S. Rueda in [103]. For a fixed polynomial $g(x) \in \mathbb{C}[x]$ and $\zeta \in \mathbb{C}$, $\zeta \neq 0$, we define the algebra $R_{g,\zeta}$ as the quotient of the free associative algebra with generators e, f and h modulo the ideal generated by the relations

$$ef - \zeta fe = f(h), \quad he - eh = 2e, \quad hf - fh = -2f. \tag{2.15}$$

Again, these are Noetherian domains with a basis consisting of standard monomials. However, the existence of an analog of the Casimir element actually depends on ζ: if ζ is a root of unity, then an analog of the Casimir element exists and the center of $R_{g,\zeta}$ is a polynomial ring in one variable; if ζ is not a root of unity, then the center of $R_{g,\zeta}$ is trivial. For details and further properties we refer the reader to [103, 104].

2.7.6

Another class of algebras with properties similar to those described in this chapter is the class of the so-called *generalized Weyl algebras*, introduced by V. Bavula in [9, 10]. The general definition of generalized Weyl algebras (of rank one) is the following: let D be a ring, τ a fixed automorphism of D and $a \in D$ a fixed central element. The generalized Weyl algebra $A = D[X, Y, \tau, a]$ is a D-algebra, generated over D by symbols X and Y subject to the following relations:

$$Xb = \tau(b)X, \quad Yb = \tau^{-1}(b)Y, \quad YX = a, \quad XY = \tau(a), \tag{2.16}$$

for all $b \in D$. Properties of A heavily depend on the original ring D and on the choice of τ and a. For example, the algebra A is a Noetherian domain provided that the ring D is a Noetherian domain.

Many known algebras can be realized as generalized Weyl algebras and hence studied from the unified point of view. For example, the algebra $\mathbb{C}[H][X, Y, \tau, H]$, where $\tau(H) = H - 1$, is isomorphic to the *first Weyl algebra*. The algebra $\mathbb{C}[H, C][X, Y, \tau, C - H(H+1)]$, where $\tau(H) = H - 1$ and $\tau(C) = C$ is isomorphic to the algebra $U(\mathfrak{sl}_2)$. Other classes of algebras that can be realized as generalized Weyl algebras include quantum $\mathfrak{sl}_2$ ([65]), algebras similar to $U(\mathfrak{sl}_2)$ as defined in 2.7.5 and down-up algebras from [14]. We refer the reader to [10] for details.

2.8 Additional exercises

Exercise 2.44. Define $U(\mathfrak{a})$ for any Lie algebra $\mathfrak{a}$ and show that $U(\mathfrak{a})$ is commutative if, and only if, $\mathfrak{a}$ is.

Exercise 2.45. Show that the left and the right multiplication with elements from $\mathfrak{g}$ define on $U(\mathfrak{g})$ the structure of a left and a right $\mathfrak{g}$-module, respectively.

Exercise 2.46. Show that the module $U(\mathfrak{g})$, considered as a left $\mathfrak{g}$-module as in 2.45, is not semi-simple.

Exercise 2.47. Let φ denote the unique linear map from $U(\mathfrak{g})$ to $\mathbb{C}[a, b, c]$ such that $\varphi(f^i h^j e^k) = a^i b^j c^k$. Show that φ is an isomorphism of $\mathfrak{g}$-modules, where the $\mathfrak{g}$-module structure on $U(\mathfrak{g})$ is given by the left multiplication and the $\mathfrak{g}$-module structure on $\mathbb{C}[a, b, c]$ is given by (2.5)–(2.7).

Exercise 2.48. Compute $\dim U(\mathfrak{g})^{(i)}/U(\mathfrak{g})^{(i-1)}$ for $i \in \mathbb{N}_0$.

Exercise 2.49. Consider on $U(\mathfrak{g})$ the structure of the adjoint $\mathfrak{g}$-module, given by Exercise 2.28. Prove that for every $i \in \mathbb{N}_0$ the space $U(\mathfrak{g})^{(i)}$ is a submodule of $U(\mathfrak{g})$. Deduce that, as the adjoint $\mathfrak{g}$-module, the module $U(\mathfrak{g})$ decomposes into an (infinite) direct sum of modules $\mathbf{V}^{(n)}$, $n \in \mathbb{N}$.

Exercise 2.50. Show that there exists a unique antiautomorphism ω of $U(\mathfrak{g})$ such that $\omega(x) = -x$ for all $x \in \mathfrak{g}$. The antiautomorphism ω is called the *principal antiautomorphism*.

Exercise 2.51. Show that for any $u, v \in U(\mathfrak{g})$ we have $uv - vu \neq 1$.

Exercise 2.52 ([24]). For every $i, j, k \in \mathbb{N}_0$ fix some monomial $m_{i,j,k}$ of the form $x_1 x_2 \cdots x_{i+j+k}$ such that $|\{s : x_s = f\}| = i$, $|\{s : x_s = h\}| = j$ and $|\{s : x_s = e\}| = k$. Show that $\{m_{i,j,k} : i, j, k \in \mathbb{N}_0\}$ is a basis of $U(\mathfrak{g})$.

Exercise 2.53. Show that the decomposition (2.11) makes $U(\mathfrak{g})$ into a graded algebra in the sense that

$$U(\mathfrak{g})_{2i} \, U(\mathfrak{g})_{2j} \subset U(\mathfrak{g})_{2(i+j)}$$

for all $i, j \in \mathbb{Z}$.

Exercise 2.54. Prove that the polynomial algebra $\mathbb{C}[e]$ is a canonical subalgebra of $U(\mathfrak{g})$. Prove further that the algebra $U(\mathfrak{g})$ is free both as a left and as a right $\mathbb{C}[e]$-module and determine some basis, which works for both the left and the right structure.

Exercise 2.55. Prove that every element from $Z(\mathfrak{g})$ acts as a scalar on every simple finite-dimensional $\mathfrak{g}$-module. Will the statement remain true if one drops the condition on the module to be simple?

Exercise 2.56. An associative algebra A is called *left (right) Artinian* if it satisfies the descending chain condition on left (right) ideals; that is for any descending chain

$$I_1 \supset I_2 \supset I_3 \supset \ldots$$

of left (right) ideals in A there exists $n \in \mathbb{N}$ such that $I_n = I_{n+1} = \ldots$. Show that the algebra $U(\mathfrak{g})$ is not Artinian.

Exercise 2.57. Show that for every $n \in \mathbb{N}$ the algebra of all $n \times n$ complex matrices is a quotient of the algebra $U(\mathfrak{g})$.

Exercise 2.58. Find some basis V of $U(\mathfrak{g})$ over $Z(\mathfrak{g})$ such that V is a submodule of the adjoint module $U(\mathfrak{g})$ and show that V is isomorphic to $\bigoplus_{i \in \mathbb{N}} \mathbf{V}^{(2i-1)}$.

Exercise 2.59. Let $u \in U(\mathfrak{g})$ be non-constant and $f(x) \in \mathbb{C}[x]$ be non-zero. Show that $f(u) \neq 0$ in $U(\mathfrak{g})$.

Chapter 3

Weight modules

3.1 Weights and weight modules

From the classification of simple finite-dimensional $\mathfrak{g}$-modules presented in Theorem 1.22, and the construction of the module $\mathbf{V}^{(n)}$ given by (1.9), we find that the action of the element $\mathbf{h}$ is diagonalizable on every simple finite-dimensional $\mathfrak{g}$-module. As any finite-dimensional $\mathfrak{g}$-module is a direct sum of simple modules by Weyl's theorem, we deduce that the action of $\mathbf{h}$ is diagonalizable on every finite-dimensional $\mathfrak{g}$-module. It seems natural, therefore, to study the class of those $\mathfrak{g}$-modules, the action of $\mathbf{h}$ on which is diagonalizable. Such modules are called *weight* modules and will be the main object of the study in this chapter.

Let V be a $\mathfrak{g}$-module (not necessarily finite-dimensional). For $\lambda \in \mathbb{C}$ denote

$$V_\lambda = \{v \in V \ : \ H(v) = \lambda \cdot v\}.$$

The number λ is called a *weight* and the space V_λ is called the corresponding *weight space*. The module V is called a *weight module* provided that

$$V = \bigoplus_{\lambda \in \mathbb{C}} V_\lambda. \tag{3.1}$$

For a weight $\mathfrak{g}$-module V we define the *support* $\operatorname{supp} V$ of V as follows:

$$\operatorname{supp}(V) = \{\lambda \in \mathbb{C} \ : \ V_\lambda \neq 0\}.$$

Example 3.1. From (1.9) we have that

$$\mathbf{V}^{(n)} = \bigoplus_{i=0}^{n-1} \mathbf{V}^{(n)}_{n-1-2i}.$$

Hence $\mathbf{V}^{(n)}$ is a weight module and

$$\operatorname{supp} \mathbf{V}^{(n)} = \{1-n, 3-n, \ldots, n-3, n-1\}.$$

Note that supp $\mathbf{V}^{(n)}$ is invariant with respect to the action of the Weyl group $\mathbf{S}_2$ of $\mathfrak{g}$ (see Section 2.5).

Exercise 3.2. Let V be a finite-dimensional $\mathfrak{g}$-module. Prove that V is a weight module and that supp V is invariant with respect to the action of $\mathbf{S}_2$ (as described in Section 2.5).

Exercise 3.3. Let V be a $\mathfrak{g}$-module. Identify $\mathfrak{h}^* = \mathbb{C}$ as in Section 2.5. Show that for every $x \in \mathfrak{h}$, $\lambda \in \mathfrak{h}^*$ and $v \in V_\lambda$ we have $x(v) = \lambda(x) v$.

Exercise 3.4. Let V and W be two weight $\mathfrak{g}$-modules and $\varphi : V \to W$ be a homomorphism. Show that $\varphi(V_\lambda) \subset W_\lambda$ for any $\lambda \in \mathbb{C}$.

Denote by $\mathfrak{W}$ the full subcategory of the category $\mathfrak{g}$-mod, consisting of all finitely generated weight modules. To understand the structure of $\mathfrak{W}$ we start with some elementary observations.

Lemma 3.5. *Let V be a weight $\mathfrak{g}$-module. Then for every $\lambda \in \mathbb{C}$ we have*

$$EV_\lambda \subset V_{\lambda+2}, \qquad FV_\lambda \subset V_{\lambda-2}.$$

Proof. This is analogous to Lemma 1.15: For $v \in V_\lambda$ we have

$$H(E(v)) \overset{(1.2)}{=} E(H(v)) + 2E(v) \overset{v \in V_\lambda}{=} \lambda E(v) + 2E(v) = (\lambda + 2)E(v).$$

The second inclusion is proved similarly. $\qquad\qquad\square$

Exercise 3.6. Let V be a weight module, $\lambda \in \mathbb{C}$ and $i \in \mathbb{Z}$. Show that $U(\mathfrak{g})_{2i}V_\lambda \subset V_{\lambda+2i}$.

Consider the additive subgroup $2\mathbb{Z}$ of $\mathbb{C}$ and the corresponding set $\mathbb{C}/2\mathbb{Z}$ of cosets. For a weight $\mathfrak{g}$-module V and $\xi \in \mathbb{C}/2\mathbb{Z}$ set

$$V^\xi = \bigoplus_{\lambda \in \xi} V_\lambda.$$

Denote by $\mathfrak{W}^\xi$ the full subcategory of $\mathfrak{W}$, consisting of all modules V such that supp $V \subset \xi$.

Corollary 3.7.

(i) *Let V be a weight $\mathfrak{g}$-module. Then for every $\xi \in \mathbb{C}/2\mathbb{Z}$ the subspace V^ξ is a submodule of V and we have*

$$V \cong \bigoplus_{\xi \in \mathbb{C}/2\mathbb{Z}} V^\xi. \tag{3.2}$$

(ii) We have

$$\mathfrak{W} \cong \bigoplus_{\xi \in \mathbb{C}/2\mathbb{Z}} \mathfrak{W}^{\xi}.$$

Proof. Let $\lambda \in \xi$. Then $\lambda \pm 2 \in \xi$ by definition. Thus Lemma 3.5 implies that V^{ξ} is invariant with respect to the action of both E and F. As $H = EF - FE$, V^{ξ} is also invariant with respect to the action of H. This means that V^{ξ} is a submodule of V. The decomposition (3.2) follows now from the decomposition (3.1). This proves the statement (i).

The statement (ii) follows from the statement (i) and Exercise 3.4. $\square$

Proposition 3.8.

(i) *Every submodule of a weight module is a weight module.*
(ii) *Every quotient of a weight module is a weight module.*
(iii) *Any direct sum of weight modules is a weight module.*
(iv) *Any finite tensor product of weight modules is a weight module.*

Proof. Let V be a weight module and $W \subset V$ be a submodule. Since V is a weight module, for any $w \in W \subset V$ we can write $w = w_1 + w_2 + \cdots + w_k$, where $w_i \in V$ are weight vectors for all $i = 1, \ldots, k$. Without loss of generality we may assume that all $w_i \neq 0$, that w_i has weight λ_i and that the λ_i's are pairwise different. For every $i \in \{1, \ldots, k\}$ consider the element

$$h_i = (h - \lambda_1)(h - \lambda_2) \ldots (h - \lambda_{i-1})(h - \lambda_{i+1}) \ldots (h - \lambda_k) \in U(\mathfrak{g}).$$

Then we have

$$h_i(w_j) = \begin{cases} 0, & i \neq j; \\ \prod_{s \neq i}(\lambda_i - \lambda_s)w_i & i = j. \end{cases}$$

Hence

$$W \ni h_i(w) = h_i(w_i) = \prod_{s \neq i}(\lambda_i - \lambda_s)w_i \neq 0.$$

This means that $w_i \in W$ and thus that every vector from W is a sum of weight vectors from W. This proves the statement (i).

Let V be a weight module and $W \subset V$ be a submodule. Then the image of any $\mathfrak{h}$-eigenbasis of V in V/W is, obviously, a generating system consisting of $\mathfrak{h}$-eigenvectors. This proves the statement (ii).

The statement (iii) follows from the observation that $H(v) = \lambda v$ and $H(w) = \lambda w$ implies $H(v \oplus w) = \lambda(v \oplus w)$.

Finally, the statement (iv) follows from the observation that $H(v) = \lambda v$ and $H(w) = \mu w$ implies

$$H(v \otimes w) \overset{(1.17)}{=} H(v) \otimes w + v \otimes H(w)$$
$$\text{(as } H(v) = \lambda v \text{ and } H(w) = \mu w) = \lambda v \otimes w + v \otimes \mu w$$
$$= (\lambda + \mu)v \otimes w.$$

This completes the proof. $\qquad\qquad\square$

Exercise 3.9. Let V and W be weight modules. Show that

$$(V \oplus W)_\lambda = V_\lambda \oplus W_\lambda, \qquad (V \otimes W)_\lambda = \bigoplus_{\mu+\nu=\lambda} V_\mu \otimes W_\nu.$$

Exercise 3.10. Let V and W be weight modules. Show that

$$\operatorname{supp} V \oplus W = \operatorname{supp} V \cup \operatorname{supp} W; \quad \operatorname{supp} V \otimes W = \operatorname{supp} V + \operatorname{supp} W.$$

Exercise 3.11. Let $\alpha : M \hookrightarrow N \twoheadrightarrow K$ be a short exact sequence of weight modules. Prove that for every $\lambda \in \mathbb{C}$ the sequence α induces a short exact sequence $M_\lambda \hookrightarrow N_\lambda \twoheadrightarrow K_\lambda$.

Proposition 3.12. *A module generated by weight vectors is a weight module.*

Proof. It is enough to prove the statement for a module V, which is generated by one weight vector v, say of weight λ. Then, by the PBW Theorem, the module V is generated, as a vector space, by the elements of the form $f^i h^j e^k(v)$, $i, j, k \in \mathbb{N}_0$. We have

$$h(f^i h^j e^k(v)) = f^i h^j e^k(h(v)) + [h, f^i h^j e^k](v)$$
$$\text{(by Lemma 2.29)} = \lambda f^i h^j e^k(v) + 2(k - i)f^i h^j e^k(v)$$
$$= (\lambda + 2(k - i))f^i h^j e^k(v).$$

Hence all vectors $f^i h^j e^k(v)$ are weight vectors. The claim follows. $\quad\square$

3.2 Verma modules

We already know many abstract properties of weight modules. However, we do not know any other example of weight modules apart from the modules $\mathbf{V}^{(n)}$, $n \in \mathbb{N}$, and their direct sums (possibly infinite). In particular, we have no idea whether the categories $\mathfrak{W}^\xi$, considered in the previous section, are nontrivial. So, it is a good time now to construct new examples of weight modules.

So far the only simple modules we know are the modules $\mathbf{V}^{(n)}$, $n \in \mathbb{N}$, given by (1.9). Let us try to think how we can extend the construction given by (1.9). A good suggestion is made by the proof of Lemma 1.18. Indeed, the induction described in this proof works for all $i \in \mathbb{N}$, not only for $i \in \{1, 2, \ldots, n-1\}$. This motivates to extend the picture (1.9) to simply consider vectors v_i such that $i \in \mathbb{N}_0$ (that is not only $i \in \{1, 2, \ldots, n-1\}$) and define on them the action of $\mathfrak{g}$ similar to (1.9). As we will see below, this works. It even admits a straightforward generalization, that the original vector v_0 may have any weight.

With this in mind, fix $\lambda \in \mathbb{C}$. Consider the vector space $M(\lambda)$ with the formal basis $\{v_i : i \in \mathbb{N}_0\}$. For $i \in \mathbb{N}$ set $a_i = i(\lambda - i + 1)$. Consider the linear operators E, F and H on $M(\lambda)$ defined as follows:

$$
\begin{aligned}
F(v_i) &= v_{i+1}; \\
H(v_i) &= (\lambda - 2i)v_i; \\
E(v_i) &= \begin{cases} a_i v_{i-1}, & i \neq 0; \\ 0, & i = 0. \end{cases}
\end{aligned}
\tag{3.3}
$$

This can be depicted as follows:

$$\tag{3.4}$$

Proposition 3.13.

(i) *The formulae* (3.3) *define on* $M(\lambda)$ *the structure of a weight* $\mathfrak{g}$-*module*.

(ii) $\operatorname{supp} M(\lambda) = \{\lambda - 2i : i \in \mathbb{N}_0\}$.

(iii) *The Casimir element* c *acts on* $M(\lambda)$ *as the scalar* $(\lambda + 1)^2$.

Proof. First we observe that, by definition, the vector v_i, $i \in \mathbb{N}_0$, is an eigenvector for H with eigenvalue $\lambda - 2i$. Since E increases the eigenvalue by 2 and F decreases the eigenvalue by 2, the relations $[H, E] = 2E$ and $[H, F] = -2F$ are obviously satisfied. Hence we need to check only the relation $[E, F] = H$. It is enough to check this relation on the elements v_i.

If we set $a_0 = 0$, then for every $i \in \mathbb{N}_0$ we have

$$
\begin{aligned}
(EF - FE)(v_i) &\overset{(3.3)}{=} (a_{i+1} - a_i)v_i \\
\text{(by definition of } a_k) &= ((i+1)(\lambda - (i+1)+1) - i(\lambda - i + 1))v_i \\
&= (\lambda - 2i)v_i \\
\text{(by (3.3))} &= H(v_i).
\end{aligned}
$$

Hence the formulae (3.3) define on $M(\lambda)$ the structure of a $\mathfrak{g}$-module. This module is a weight module as $\{v_i\}$ is an eigenbasis for H. The claim (i) follows. The claim (ii) then follows immediately from the definitions.

For the claim (iii) we note that $E(v_0) = 0$ and hence $c(v_0) = (\lambda+1)^2 v_0$ by Proposition 2.35. For $i \in \mathbb{N}$ we then have

$$
\begin{aligned}
c(v_i) &\overset{(3.3)}{=} c(F^i(v_0)) \\
\text{(by Exercise 2.1)} &= F^i(c(v_0)) \\
\text{(by above)} &= (\lambda + 1)^2 F^i(v_0) \\
\text{(by (3.3))} &= (\lambda + 1)^2 v_i.
\end{aligned}
$$

This completes the proof. $\qquad\qquad\qquad\qquad\qquad\qquad\qquad\qquad\square$

If V is a weight $\mathfrak{g}$-module and $\mu \in \operatorname{supp} V$ is such that $\mu + 2 \notin \operatorname{supp} V$, then the weight μ is called a *highest* weight and any non-zero $v \in V_\mu$ is called a *highest weight vector*. A module, generated by a highest weight vector, is called a *highest weight module*. For example, the weight λ is the unique highest weight of the module $M(\lambda)$ and the vector v_0 is a highest weight vector of weight λ. In particular, $M(\lambda)$ is a highest weight module. From the construction we have that the module $M(\lambda)$ is uniquely determined by its highest weight. The module $M(\lambda)$ is called the *Verma module* with highest weight λ.

By Proposition 3.13(iii) the Casimir element c acts on $M(\lambda)$ as the scalar $\tau = (\lambda + 1)^2$. Define the homomorphism

$$
\chi_\lambda : Z(\mathfrak{g}) = \mathbb{C}[c] \to \mathbb{C}
$$

via $\chi_\lambda(g(c)) = g(\tau)$, $g(c) \in \mathbb{C}[c]$. Then the central element $g(c)$ acts on $M(\lambda)$ as the scalar $g(\tau)$. The homomorphism χ_λ is called the *central character* of $M(\lambda)$. More generally, if every element $u \in Z(\mathfrak{g})$ acts on some module M as a scalar $\chi_M(u)$, then $\chi_M : Z(\mathfrak{g}) \to \mathbb{C}$ is a homomorphism, called the *central character* of M.

From the construction we have that the module $M(\lambda)$ is always generated by the highest weight vector v_0. In particular, if we consider $M(\lambda)$

as a $U(\mathfrak{g})$-module, the module $M(\lambda)$ must then be a quotient of the free $U(\mathfrak{g})$-module $U(\mathfrak{g})$. This gives us the following alternative description of $M(\lambda)$:

Proposition 3.14. *Let I denote the left ideal of $U(\mathfrak{g})$, generated by e and $h - \lambda$. Then $M(\lambda) \cong U(\mathfrak{g})/I$.*

Proof. As the $U(\mathfrak{g})$-module $U(\mathfrak{g})$ is free of rank one, the assignment $1 \mapsto v_0$ extends to a homomorphism $\varphi : U(\mathfrak{g}) \to M(\lambda)$. As $M(\lambda)$ is generated by v_0, this homomorphism is surjective. Let K denote the kernel of φ.

From (3.3) we have $E(v_0) = 0$ and $(H - \lambda)(v_0) = 0$. Hence $e, h - \lambda \in K$. This means $I \subset K$ and the homomorphism φ factors through $U(\mathfrak{g})/I$. Let $\overline{\varphi} : U(\mathfrak{g})/I \twoheadrightarrow M(\lambda)$ denote the induced epimorphism.

We claim that $U(\mathfrak{g})/I$ is spanned by the images of $\{f^i : i \in \mathbb{N}_0\}$. By the PBW Theorem it is enough to show that the image in $U(\mathfrak{g})/I$ of any standard monomial from $U(\mathfrak{g})$ can be written as a linear combination of (the images of) some f^i's. For $i, j \in \mathbb{N}_0$ and $k \in \mathbb{N}$, every standard monomial $f^i h^j e^k$ belongs to I. Hence such monomials are zero in $U(\mathfrak{g})/I$. For $i \in \mathbb{N}_0$ and $j \in \mathbb{N}$ we have

$$f^i h^j = f^i(h - \lambda + \lambda)^j = \sum_{s=0}^{j} \binom{j}{s} \lambda^{j-s} f^i (h - \lambda)^s = \lambda^j f^i + u,$$

where $u \in I$. Therefore $U(\mathfrak{g})/I$ is generated by the images of $\{f^i : i \in \mathbb{N}_0\}$. At the same time, the images of these generators under $\overline{\varphi}$ are exactly the v_i's, which are linearly independent. Hence $\{f^i : i \in \mathbb{N}_0\}$ is a basis of $U(\mathfrak{g})/I$ and $\overline{\varphi}$ is an isomorphism. This completes the proof. $\square$

Corollary 3.15 (Universal property of Verma modules).

(i) *Let V be a $\mathfrak{g}$-module and $v \in V$ be such that $E(v) = 0$ and $H(v) = \lambda v$. There then exists a unique homomorphism $\varphi \in \mathrm{Hom}_\mathfrak{g}(M(\lambda), V)$ such that $\varphi(v_0) = v$.*

(ii) *Let V be a $\mathfrak{g}$-module, generated by a highest weight vector of weight λ. Then V is a quotient of $M(\lambda)$.*

Proof. Consider $U(\mathfrak{g})$ as a free left $U(\mathfrak{g})$-module of rank one. Then we have a unique homomorphism $\psi \in \mathrm{Hom}_\mathfrak{g}(U(\mathfrak{g}), V)$ such that $\psi(1) = v$. The equalities $E(v) = 0$ and $H(v) = \lambda v$ imply that e and $h - \lambda$ belong to the kernel of ψ and hence ψ factors through the module $U(\mathfrak{g})/I$ from

Proposition 3.14. The statement (i) now follows from Proposition 3.14. The statement (ii) follows directly from the statement (i). $\quad\square$

Now we can describe the structure of Verma modules.

Theorem 3.16 (Structure of Verma modules).

(i) *The module $M(\lambda)$ is simple if, and only if, $\lambda \notin \mathbb{N}_0$.*

(ii) *For $n \in \mathbb{N}_0$ the module $M(n)$ is indecomposable. Furthermore, the module $M(-n-2)$ is a (unique) simple submodule of $M(n)$ and we have $M(n)/M(-n-2) \cong \mathbf{V}^{(n+1)}$.*

Proof. Let $\lambda \notin \mathbb{N}_0$, $V \subset M(\lambda)$ be a non-zero submodule and $v \in V$, $v \neq 0$. Then for some $k \in \mathbb{N}_0$ we have $v = \sum_{i=0}^{k} \alpha_i v_i$ and $\alpha_k \neq 0$. As $\lambda \notin \mathbb{N}_0$, in (3.3) we have $a_i \neq 0$ for all $i \in \mathbb{N}$. Hence from (3.3) we obtain that $E^k(v) = \alpha_k E^k(v_k)$ is a non-zero multiple of v_0. Therefore $v_0 \in V$ and thus $V = M(\lambda)$. This means that the module $M(\lambda)$ is simple for $\lambda \notin \mathbb{N}_0$.

Let now $n \in \mathbb{N}_0$. In (3.3) we have $a_{n+1} = 0$, which means that the vector v_{n+1} of $M(n)$ satisfies $E(v_{n+1}) = 0$ (by (3.3)). We also have $H(v_{n+1}) = (-n-2)v_{n+1}$ by (3.3). Hence, by the universal property of $M(-n-2)$ (Corollary 3.15), we have a non-zero homomorphism from $M(-n-2)$ to $M(n)$. In particular, this implies that $M(n)$ is not simple, proving (i). From the previous paragraph we have that the module $M(-n-2)$ is simple. The submodule $M(-n-2)$ of $M(n)$ has the basis $\{v_{n+1}, v_{n+2}, \dots\}$ and hence the quotient $M(n)/M(-n-2)$ has the basis $\{v_0, v_1, \dots, v_n\}$. The action of $\mathfrak{g}$ in this basis is given by (1.9). This means that $M(n)/M(-n-2) \cong \mathbf{V}^{(n+1)}$.

Now let V be any non-zero submodule of $M(n)$ and $v \in V$, $v \neq 0$. As the action of F on $M(n)$ is injective by (3.3), we have that $F^i(v) \neq 0$ for all i. On the other hand, the vector $F^{n+1}(v)$ is obviously a linear combination of $\{v_{n+1}, v_{n+2}, \dots\}$ and hence belongs to the submodule $M(-n-2)$. This implies that every non-zero submodule of $M(n)$ intersects $M(-n-2)$. In particular, the submodule $M(-n-2)$ is a unique simple submodule and the module $M(n)$ is indecomposable. This completes the proof. $\quad\square$

By Theorem 3.16 for $n \in \mathbb{N}_0$ the non-simple module $M(n)$ is uniserial and has the following tower of subquotients (say in the radical filtration):

$$M(n): \qquad \begin{array}{c} \mathbf{V}^{(n+1)} \\ | \\ M(-n-2). \end{array}$$

Exercise 3.17. For $\lambda, \mu \in \mathbb{C}$ prove that

$$
\mathrm{Hom}_{\mathfrak{g}}(M(\lambda), M(\mu)) = \begin{cases} \mathbb{C}\,\mathrm{id}_{M(\lambda)}, & \lambda = \mu; \\ \mathbb{C}\,\varphi_n, & \lambda = -n-2, \ \mu = n, \ n \in \mathbb{N}_0; \\ 0, & \text{otherwise}; \end{cases}
$$

where $\varphi_n : M(-n-2) \to M(n)$ is some fixed non-zero homomorphism. Deduce that every non-zero homomorphism between Verma modules is injective.

Corollary 3.18 (Classification of simple highest weight modules). *For every $\lambda \in \mathbb{C}$ there exists a unique simple weight module $L(\lambda)$ with highest weight λ. Moreover, we have*

$$
L(\lambda) = \begin{cases} M(\lambda), & \lambda \notin \mathbb{N}_0; \\ \mathbf{V}^{(n+1)}, & \lambda = n \in \mathbb{N}_0. \end{cases} \tag{3.5}
$$

Proof. Let V be a simple module with highest weight λ. The universal property of Verma modules gives us an epimorphism $M(\lambda) \twoheadrightarrow V$. From Theorem 3.16 we have that $M(\lambda)$ has a unique simple quotient, which is given by (3.5). $\qquad\square$

Corollary 3.19. *For $n \in \mathbb{N}_0$ we have a non-split short exact sequence*

$$
0 \to M(-n-2) \to M(n) \to \mathbf{V}^{(n+1)} \to 0.
$$

Proof. This follows immediately from Theorem 3.16. $\qquad\square$

Exercise 3.20. For $\lambda \in \mathbb{C}$ let $\overline{M}(\lambda)$ be the formal vector space with the basis $\{w_i : i \in \mathbb{N}_0\}$. For $i \in \mathbb{N}$ set $b_i = -i(\lambda + i - 1)$ and define operators E, F and H on $\overline{M}(\lambda)$ via:

$$
\begin{aligned}
E(w_i) &= w_{i+1}; \\
H(w_i) &= (\lambda + 2i)w_i; \\
F(w_i) &= \begin{cases} b_i w_{i-1}, & i \neq 0, \\ 0, & i = 0. \end{cases}
\end{aligned} \tag{3.6}
$$

Check that this defines on $\overline{M}(\lambda)$ the structure of a weight $\mathfrak{g}$-module with support $\{\lambda + 2i : i \in \mathbb{N}_0\}$.

Exercise 3.21. Show that the Casimir element c acts on $\overline{M}(\lambda)$ as the scalar $(\lambda - 1)^2$.

If V is a weight $\mathfrak{g}$-module and $\mu \in \operatorname{supp} V$ is such that $\mu - 2 \notin \operatorname{supp} V$, then the weight μ is called a *lowest* weight and any non-zero $v \in V_\mu$ is called a *lowest weight vector*. A module generated by a lowest weight vector is called a *lowest weight module*.

Exercise 3.22. Give an alternative description of the module $\overline{M}(\lambda)$, analogous to Proposition 3.14. Formulate and prove for $\overline{M}(\lambda)$ a lowest weight analog of the universal property.

Exercise 3.23. Show that $\overline{M}(\lambda)$ has a unique simple quotient $\overline{L}(\lambda)$ and that

$$\overline{L}(\lambda) = \begin{cases} \overline{M}(\lambda), & -\lambda \notin \mathbb{N}_0; \\ \mathbf{V}^{(n+1)}, & -\lambda = n \in \mathbb{N}_0. \end{cases}$$

Exercise 3.24. Show that modules $\{\overline{L}(\lambda) : \lambda \in \mathbb{C}\}$ classify all simple lowest weight modules.

3.3 Dense modules

In the previous section we have constructed many examples of simple weight $\mathfrak{g}$-modules. However, they all had either a highest or a lowest weight. A natural question is: Is it possible to construct a weight module without both highest and lowest weights? Later on in this section we will show that this is possible. A motivating example for our constructions is the following observation:

For some fixed $\lambda \in \mathbb{C}$ consider the vector space $V = M(\lambda) \oplus \overline{M}(\lambda + 2)$. Let the action of E, F and H on V be given as for the usual direct sum of modules $M(\lambda)$ and $\overline{M}(\lambda + 2)$ with one exception: instead of $F((0, w_0)) = 0$ we set $F((0, w_0)) = (v_0, 0)$.

Exercise 3.25. Check that the above defines on V the structure of an indecomposable $\mathfrak{g}$-module. Show further that $\operatorname{Supp} V = \lambda + 2\mathbb{Z}$.

There is, of course, a dual version of the above construction. Consider the vector space $V' = V$. Let the action of E, F and H on V' be given as for the usual direct sum of modules $M(\lambda)$ and $\overline{M}(\lambda + 2)$ with one exception: instead of $E((v_0, 0)) = 0$ we set $E((v_0, 0)) = (0, w_0)$.

Exercise 3.26. Check that the above defines on V' the structure of an indecomposable $\mathfrak{g}$-module, that $V' \not\cong V$ and that $\operatorname{Supp} V' = \lambda + 2\mathbb{Z}$.

From Corollary 3.7(i) we have that both modules V and V' constructed above have maximal possible supports for indecomposable modules. Modules with this property are called *dense* modules. The aim of this section is to construct many more examples of dense, especially of simple dense modules. To make things explicit, we call a weight $\mathfrak{g}$-module V *dense* provided that $\operatorname{Supp} V = \lambda + 2\mathbb{Z}$ for some $\lambda \in \mathbb{C}$.

Fix now $\xi \in \mathbb{C}/2\mathbb{Z}$ and $\tau \in \mathbb{C}$. Consider the vector space $\mathbf{V}(\xi, \tau)$ with the basis $\{v_\mu : \mu \in \xi\}$. Consider the linear operators E, F and H on $\mathbf{V}(\xi, \tau)$ defined as follows:

$$\begin{aligned}
F(v_\mu) &= v_{\mu-2}; \\
H(v_\mu) &= \mu v_\mu; \\
E(v_\mu) &= \tfrac{1}{4}(\tau - (\mu + 1)^2)v_{\mu+2}.
\end{aligned} \tag{3.7}$$

Setting $a_\mu = \tfrac{1}{4}(\tau - (\mu + 1)^2)$ for $\mu \in \xi$ and fixing some $\lambda \in \xi$, this can be depicted as follows:

$$\tag{3.8}$$

Lemma 3.27.

(i) *The formulae* (3.7) *define on the vector space* $\mathbf{V}(\xi, \tau)$ *the structure of a dense* $\mathfrak{g}$-*module with support* ξ.

(ii) *The Casimir element* c *acts on the module* $\mathbf{V}(\xi, \tau)$ *as the scalar* τ.

Proof. Just as in the proof of Proposition 3.13 we only have to check the relation $[E, F] = H$, when applied to the element v_λ. Formulae (3.7) reduce this to the following obvious identity: $a_{\lambda-2} - a_\lambda = \lambda$. That $\operatorname{Supp} \mathbf{V}(\xi, \tau) = \xi$ now follows from the definition. This and all definitions imply the statement (i). The statement (ii) follows then by a direct calculation. $\square$

Exercise 3.28. Let $\xi \in \mathbb{C}/2\mathbb{Z}$ and $\tau \in \mathbb{C}$. Show that every $\mathfrak{g}$-module V satisfying the following conditions is isomorphic to the module $\mathbf{V}(\xi, \tau)$:

(a) $\operatorname{Supp} V = \xi$;
(b) c acts on V as the scalar τ;
(c) $\dim V_\lambda = 1$ for some $\lambda \in \xi$;
(d) F acts bijectively on V.

For $\tau \in \mathbb{C}$ set $g_\tau(\lambda) = \tau - (\lambda + 1)^2 \in \mathbb{C}[\lambda]$. Note that $g_\tau(\lambda)$ is a quadratic polynomial, so it has at most two different complex roots. Now we are ready to describe the structure of the modules $\mathbf{V}(\xi, \tau)$.

Theorem 3.29 (Structure of $\mathbf{V}(\xi,\tau)$). *Let $\xi \in \mathbb{C}/2\mathbb{Z}$ and $\tau \in \mathbb{C}$.*

(i) *Every endomorphism of $\mathbf{V}(\xi,\tau)$ is a scalar multiple of the identity, in particular, the module $\mathbf{V}(\xi,\tau)$ is always indecomposable.*

(ii) *The module $\mathbf{V}(\xi,\tau)$ is simple if and only if ξ does not contain any root of the polynomial $g_\tau(\lambda)$.*

(iii) *If ξ contains exactly one root of the polynomial $g_\tau(\lambda)$, say μ, then the module $\mathbf{V}(\xi,\tau)$ contains a unique simple submodule $M(\mu)$ and the quotient $\mathbf{V}(\xi,\tau)/M(\mu) \cong \overline{M}(\mu+2)$ is also simple.*

(iv) *If ξ contains two different roots of the polynomial $g_\tau(\lambda)$, say μ_1 and μ_2, then $\tau = n^2$, $\mu_1 = n-1$ and $\mu_2 = -n-1$ for some $n \in \mathbb{N}$. Moreover, $\mathbf{V}(\xi,\tau)$ is a uniserial module of length three; it contains a unique simple submodule $M(-n-1)$, a unique non-simple proper submodule $M(n-1)$ (and hence the subquotient $M(n-1)/M(-n-1) \cong \mathbf{V}^{(n)}$) and the quotient $\mathbf{V}(\xi,\tau)/M(n-1) \cong \overline{M}(n+1)$ is simple.*

Proof. Let $\mu \in \xi$ and $\varphi \in \operatorname{End}_{\mathfrak{g}}(\mathbf{V}(\xi,\tau))$. Since φ commutes with the action of $\mathbf{h}$, and $\mathbf{V}(\xi,\tau)_\mu$ has basis v_μ by definition, we have $\varphi(v_\mu) = \alpha v_\mu$ for some $\alpha \in \mathbb{C}$. Using this, and the fact that φ commutes with the action of $\mathbf{f}$, for every $i > 0$ we have

$$\varphi(v_{\mu-2i}) = \varphi(F^i(v_\mu)) = F^i(\varphi(v_\mu)) = F^i(\alpha v_\mu) = \alpha F^i(v_\mu) = \alpha v_{\mu-2i}.$$

Similarly we have

$$F^i(\varphi(v_{\mu+2i})) = \varphi(F^i(v_{\mu+2i})) = \varphi(v_\mu) = \alpha v_\mu = \alpha F^i(v_{\mu+2i}).$$

As $F^i : \mathbf{V}(\xi,\tau)_{\mu+2i} \to \mathbf{V}(\xi,\tau)_\mu$ is bijective by our construction of $\mathbf{V}(\xi,\tau)$, we derive $\varphi(v_\lambda) = \alpha v_\lambda$ for all $\lambda \in \xi$. The statement (i) follows.

Assume that ξ does not contain any root of the polynomial $g_\tau(\lambda)$. Let $v \in \mathbf{V}(\xi,\tau)$, $v \neq 0$, and M denote the minimal submodule of $\mathbf{V}(\xi,\tau)$, containing v. Then we have $v = \sum_{\mu \in \xi} \alpha_\mu v_\mu$, where only finitely many α_μ's are non-zero. Let

$$\{\mu \in \xi : \alpha_\mu \neq 0\} = \{\mu_1, \ldots, \mu_k\},$$

where the μ_i's are different pairwise. Then $v \neq 0$ implies $k \geq 1$. If $k = 1$, then we have $v_{\mu_1} \in M$. Otherwise set

$$u = (h - \mu_2)(h - \mu_2)\cdots(h - \mu_k) \in U(\mathfrak{g}).$$

Then $u(v) \in M$ and

$$u(v) = \alpha_{\mu_1}(\mu_1 - \mu_2)(\mu_1 - \mu_2)\cdots(\mu_1 - \mu_k)v_{\mu_1} \neq 0.$$

This again implies that $v_{\mu_1} \in M$. Applying F inductively and using (3.7) we get $v_{\mu_1-2i} \in M$ for all $i \in \mathbb{N}_0$. As ξ does not contain any root of the polynomial $g_\tau(\lambda)$, applying E inductively and using (3.7) we get $v_{\mu_1+2i} \in M$ for all $i \in \mathbb{N}_0$. This yields $M = \mathbf{V}(\xi, \tau)$, which proves that $\mathbf{V}(\xi, \tau)$ is simple.

Assume that ξ contains exactly one root of the polynomial $g_\tau(\lambda)$, say μ. Then we have $E(v_\mu) = 0$ and $H(v_\mu) = \mu v_\mu$ by (3.7). Hence, by the universal property of Verma modules (Corollary 3.15), we have a non-zero homomorphism from $M(\mu)$ to $\mathbf{V}(\xi, \tau)$ (which sends the generator v_0 of $M(\mu)$ to the element v_μ of $\mathbf{V}(\xi, \tau)$). Note that $\mu \notin \mathbb{N}_0$ for otherwise $-\mu - 2$ would be a second root of $g_\tau(\lambda)$ in ξ. Hence $M(\mu)$ is simple by Theorem 3.16(i) and thus $M(\mu)$ is a simple submodule of $\mathbf{V}(\xi, \tau)$. The quotient $\mathbf{V}(\xi, \tau)/M(\mu)$ has a basis consisting of the images $\overline{v}_{\mu+2i}$ of $v_{\mu+2i}$, $i \in \mathbb{N}$. We have $F(\overline{v}_{\mu+2}) = 0$ and $H(\overline{v}_{\mu+2}) = (\mu + 2)\overline{v}_{\mu+2}$. Since μ was the only root of $g_\tau(\lambda)$ in ξ, we have $E^i(\overline{v}_{\mu+2}) \neq 0$ for all $i \in \mathbb{N}$ and hence the elements $E^i(\overline{v}_{\mu+2})$, $i \in \mathbb{N}_0$, form a basis of $\mathbf{V}(\xi, \tau)/M(\mu)$. Using Exercise 3.22 we get $\mathbf{V}(\xi, \tau)/M(\mu) \cong \overline{M}(\mu + 2)$, which is simple by Exercise 3.23. This proves the statement (iii).

Finally, assume that ξ contains two different roots of the polynomial $g_\tau(\lambda)$, say μ_1 and μ_2. Then we may assume $\mu_2 = \mu_1 - 2n$ for some $n \in \mathbb{N}$. This gives us

$$\tau - (\mu_1 + 1)^2 = \tau - (\mu_1 - 2n + 1)^2,$$

which yields $\mu_1 = n - 1 \in \mathbb{N}_0$, $\tau = n^2$ and $\mu_2 = -n - 1$. As in the previous paragraph, we get the inclusion $M(n - 1) \hookrightarrow \mathbf{V}(\xi, \tau)$ and the quotient $\mathbf{V}(\xi, \tau)/M(n - 1) \cong \overline{M}(n + 1)$, the latter being simple by Exercise 3.23. That $M(-n - 1) \hookrightarrow M(n - 1)$ follows from Theorem 3.16. This proves the statement (iv), and also completes the proof of the statement (ii) and thus of the whole theorem. $\square$

By Theorem 3.29, the module $\mathbf{V}(\xi, \tau)$ is always uniserial. If ξ contains exactly one root of the polynomial $g_\tau(\lambda)$, say μ, then $\mathbf{V}(\xi, \tau)$ has the following tower of simple subquotients in the radical filtration:

$$\mathbf{V}(\xi, \tau): \qquad \overline{M}(\mu + 2)$$
$$|$$
$$M(\mu).$$

For every $n \in \mathbb{N}_0$ the module $\mathbf{V}(n - 1 + 2\mathbb{Z}, n^2)$ has the following tower of simple subquotients in the radical filtration:

$$\mathbf{V}(n-1+2\mathbb{Z}, n^2): \qquad\qquad \overline{M}(n+1)$$

$$\mathbf{V}^{(n)}$$

$$M(-n-1).$$

Corollary 3.30.

(i) *If the coset* ξ *contains exactly one root of the polynomial* $g_\tau(\lambda)$*, say* μ*, then we have a non-split short exact sequence*

$$0 \to M(\mu) \to \mathbf{V}(\xi, \tau) \to \overline{M}(\mu+2) \to 0.$$

(ii) *For every* $n \in \mathbb{N}$ *we have a non-split short exact sequence*

$$0 \to M(n-1) \to \mathbf{V}(n-1+2\mathbb{Z}, n^2) \to \overline{M}(n+1) \to 0.$$

Proof.   This follows directly from Theorem 3.29.   □

Exercise 3.31. Let $n \in \mathbb{N}$. Then we have $M(-n-1) \subset \mathbf{V}(n-1+2\mathbb{Z}, n^2)$ by Theorem 3.29. Show that the quotient $\mathbf{V}(n-1+2\mathbb{Z}, n^2)/M(-n-1)$ has a lowest weight vector of weight $-n+1$ but is not isomorphic to $\overline{M}(-n+1)$.

3.4   Classification of simple weight modules

In the previous sections we constructed many examples of weight modules. Most of these examples were, in fact, simple weight modules. Now we are ready to give a complete classification of such modules.

Theorem 3.32 (Classification of simple weight $\mathfrak{sl}_2$-modules).
Each simple weight $\mathfrak{g}$*-module is isomorphic to one of the following (pairwise non-isomorphic) modules:*

(i) $\mathbf{V}^{(n)}$ *for some* $n \in \mathbb{N}$.
(ii) $M(\lambda)$ *for some* $\lambda \in \mathbb{C} \setminus \mathbb{N}_0$.
(iii) $\overline{M}(-\lambda)$ *for some* $\lambda \in \mathbb{C} \setminus \mathbb{N}_0$.
(iv) $\mathbf{V}(\xi, \tau)$ *for some* $\xi \in \mathbb{C}/2\mathbb{Z}$ *and* $\tau \in \mathbb{C}$ *such that* $\tau \neq (\mu+1)^2$ *for all* $\mu \in \xi$.

To prove this theorem we will need some preparation.

Lemma 3.33. *Let V be a simple weight $\mathfrak{g}$-module and $\lambda \in \operatorname{supp} V$. Then V_λ is a simple $\mathbb{C}[c]$-module.*

Proof. Assume that this is not the case and let $W' \subset V_\lambda$ be a proper $\mathbb{C}[c]$-submodule. Set $W = U(\mathfrak{g})W'$. We claim that W is a proper submodule of V. Obviously, W is a non-zero submodule as $W \supset W' \neq 0$. The module W is a weight module by Proposition 3.8(i). Let us show that $W_\lambda = W'$. The inclusion $W' \subset W_\lambda$ is obvious. From Exercise 3.6 we have $W_\lambda = U(\mathfrak{g})_0 W'$. By Proposition 2.30(iii), the algebra $U(\mathfrak{g})_0$ is generated by c and h. The space W' is invariant with respect to h, as V is a weight module and $W' \subset V_\lambda$. The space W' is invariant with respect to c by our assumption. Hence $W_\lambda = W' \neq V_\lambda$, which means that $W \neq V$. Therefore W is a proper submodule of V and hence V is not simple; a contradiction. This completes the proof. $\qquad\qquad\square$

Lemma 3.34. *Every simple $\mathbb{C}[c]$-module is one-dimensional.*

Proof. Let V be a simple $\mathbb{C}[c]$-module. If V is finite-dimensional, then the linear operator c on V has an eigenvector, which generates a one-dimensional $\mathbb{C}[c]$-submodule of V. As V is simple, it must therefore coincide with this submodule.

Let us now show that every infinite-dimensional $\mathbb{C}[c]$-module is not simple. Assume that V is a simple infinite-dimensional $\mathbb{C}[c]$-module. As the kernel of c is always a submodule, it must then either be V or 0. In the first case any subspace of V is a submodule and hence V is not simple. This means that c is injective. As the image of c is always a submodule, it must then either be V or 0. In the second case, any subspace of V is a submodule and hence V is not simple. This means that c is surjective and, in particular, bijective.

Let $v \in V$, $v \neq 0$. Consider $B = \{c^i(v) : c \in \mathbb{Z}\}$ (this is well-defined as c acts bijectively on V by the previous paragraph). Assume that the elements in B are linearly dependent. Then, applying some power of c, if necessary, we have

$$\alpha_0 v + \alpha_1 c(v) + \alpha_2 c^2(v) + \cdots + \alpha_k c^k(v) = 0$$

for some $k \in \mathbb{N}$ and $\alpha_0, \ldots, \alpha_k \in \mathbb{C}$, $\alpha_0, \alpha_k \neq 0$. It follows that the linear span of $\{v, c(v), \ldots, c^{k-1}(v)\}$ is a finite-dimensional submodule of V; a contradiction.

As a result, the elements in B are linearly independent. Their linear span is obviously invariant with respect to c and hence must coincide with V. Hence B is in fact a basis of V. But the linear span W of $\{v, c(v), c^2(v), \dots\}$ is then different from V and obviously invariant with respect to c, that is forms a proper submodule of V. This contradicts our assumption that V is simple and completes the proof. $\qquad\square$

Now we are ready to prove Theorem 3.32.

Proof. Let V be a simple $\mathfrak{g}$-module. In particular, V is indecomposable and hence $\operatorname{supp} V \subset \xi$ for some $\xi \in \mathbb{C}/2\mathbb{Z}$ by Corollary 3.7(i). Moreover, for any $\lambda \in \operatorname{supp} V$ we have $\dim V_\lambda = 1$ by Lemmas 3.33 and 3.34.

Consider the actions of E and F on V. If the action of E is not injective, then there must exist $v \in V$, $v \neq 0$, such that $E(v) = 0$. As non-zero elements in different weight spaces are linearly independent, we may assume that v is a weight vector. But then the universal property of Verma modules (Corollary 3.15) gives a nontrivial homomorphism from some Verma module to V. Since V is simple, this homomorphism must be an epimorphism, and V is thus a simple highest weight module. Thus V is either of the form (i) or of the form (ii) by Corollary 3.18. If the action of F on V is not injective, then we similarly get that V is a simple lowest weight module. In this case V is either of the form (i) or of the form (iii) by Exercise 3.23.

Assume now that the action of both E and F on V is injective. Let $\lambda \in \operatorname{supp} V$ and $v \in V_\lambda$, $v \neq 0$. Then $E^i(v) \neq 0$ and $F^i(v) \neq 0$ for all $i \in \mathbb{N}$ and we have $\operatorname{supp} V = \{\lambda + 2i : i \in \mathbb{Z}\} = \xi$ by Lemma 3.5. In particular, it follows that both E and F act bijectively on V. For $i \in \mathbb{Z}$ set $w_i = F^i(v)$, which is well-defined as F acts bijectively on V. Since V_λ is one-dimensional and the action of c commutes with the action of h, we have $c(v) = \tau v$ for some $\tau \in \mathbb{C}$. Since the action of c commutes with the action of F, we have

$$c(w_i) = c(F^i(v)) = F^i(c(v)) = F^i(\tau v) = \tau F^i(v) = \tau w_i$$

for any $i \in \mathbb{Z}$. Hence c acts on V as the scalar τ. From Exercise 3.28 we then derive $V \cong \mathbf{V}(\xi, \tau)$; that is V is given by (iv). The necessary restrictions on ξ and τ follow from Theorem 3.29(ii). This shows that every simple weight $\mathfrak{g}$-module is isomorphic to some module from the list (i)–(iv).

Now let us prove that the modules in the list (i)–(iv) are pairwise non-isomorphic. Let V and W be two different modules from the list and assume that they are isomorphic. Then, in particular, $\operatorname{supp} V = \operatorname{supp} W$.

From Example 3.1, Proposition 3.13(ii), Exercise 3.20 and Lemma 3.27(i) we have that $\operatorname{supp} V = \operatorname{supp} W$ is possible only in the case $V = \mathbf{V}(\xi, \tau)$ and $W = \mathbf{V}(\xi, \tau')$ for some $\tau \neq \tau'$. But then $V \not\cong W$ by Lemma 3.27(ii); a contradiction. Hence the modules in the list (i)–(iv) are pairwise non-isomorphic, which completes the proof. $\qquad\square$

Exercise 3.35. Show that every simple highest (or lowest) weight module is uniquely determined (up to isomorphism) by its support.

Exercise 3.36. Show that every simple weight module is uniquely determined (up to isomorphism) by its support and the eigenvalue of the Casimir element.

Exercise 3.37. Let $\xi \in \mathbb{C}/2\mathbb{Z}$ and $\tau \in \mathbb{C}$. Denote by $\mathbf{V}^{ss}(\xi, \tau)$ the same vector space as $\mathbf{V}(\xi, \tau)$. Define the linear operators E, F and H on $\mathbf{V}^{ss}(\xi, \tau)$ by (3.7) with the following exception: we set $F(v_\mu) = 0$ provided that $E(v_{\mu-2}) = 0$. Show that this defines on $\mathbf{V}^{ss}(\xi, \tau)$ the structure of a dense $\mathfrak{g}$-module with support ξ. Show further that the module $\mathbf{V}^{ss}(\xi, \tau)$ is semi-simple and has the same simple subquotients (with the same multiplicities) as the module $\mathbf{V}(\xi, \tau)$. The module $\mathbf{V}^{ss}(\xi, \tau)$ is called the *semi-simplification* of the module $\mathbf{V}(\xi, \tau)$.

Exercise 3.38. Show that every simple weight $\mathfrak{g}$-module is isomorphic to a simple subquotient of some (uniquely determined) module $\mathbf{V}(\xi, \tau)$ (or $\mathbf{V}^{ss}(\xi, \tau)$). Derive from this that the module $\oplus_{\xi, \tau} \mathbf{V}^{ss}(\xi, \tau)$ is a multiplicity-free direct sum of all simple weight $\mathfrak{g}$-modules (the so-called *Gelfand model* for weight modules).

Exercise 3.39. Show that every simple weight $\mathfrak{g}$-module has only scalar endomorphisms.

3.5 Coherent families

Fix $\tau \in \mathbb{C}$ and consider the module

$$\mathbf{V}(\tau) = \bigoplus_{\xi \in \mathbb{C}/2\mathbb{Z}} \mathbf{V}(\xi, \tau).$$

The module $\mathbf{V}(\tau)$ is called the *coherent family* corresponding to τ. The weight module $\mathbf{V}(\tau)$ has the following properties and is uniquely determined by them due to Exercise 3.28:

(I) $\dim \mathbf{V}(\tau)_\lambda = 1$ for all $\lambda \in \mathbb{C}$;
(II) the Casimir element c acts on $\mathbf{V}(\tau)$ as the scalar τ;
(III) F acts injectively (and hence bijectively) on $\mathbf{V}(\tau)$.

By definition, coherent families are indexed by the eigenvalues of the action of the Casimir element. By Theorem 3.29, all simple subquotients of every coherent family occur with multiplicity one, and, by Theorem 3.32, each simple weight $\mathfrak{g}$-module is a subquotient of some coherent family (see also Exercise 3.38). In this section we will show that coherent families can be defined in a natural way using only Verma modules.

Consider the associative algebra $U^{(f)}$, defined as the quotient of the free associative algebra R with generators f^{-1}, f, h and e modulo the ideal, generated by the relations

$$f^{-1}f = ff^{-1} = 1, \quad ef - fe = h, \quad he - eh = 2e, \quad hf - fh = -2f. \quad (3.9)$$

The algebra $U^{(f)}$ is called the *localization* of the algebra $U(\mathfrak{g})$ with respect to the multiplicative set $\{f^i : i \in \mathbb{N}\}$. As usual, abusing notation we will use the same notation for the elements of the original algebra R and of the quotient $U^{(f)}$.

Exercise 3.40. Show that the following relations hold in $U^{(f)}$:

$$hf^{-1} = f^{-1}(h+2), \quad f^{-1}e - ef^{-1} = f^{-2}(h+2) = (h-2)f^{-2}. \quad (3.10)$$

Consider the vector space $V = \mathbb{C}[a^{-1}, a, b, c]$ of polynomials, which are ordinary polynomials in b and c and Laurent polynomials in a. Define on V the linear operators F^{-1}, F, H and E by the following formulae (here $i \in \mathbb{Z}$ and $j, k \in \mathbb{N}_0$):

$$F(a^i b^j c^k) = a^{i+1} b^j c^k;$$
$$F^{-1}(a^i b^j c^k) = a^{i-1} b^j c^k;$$

$$H(a^i b^j c^k) = \begin{cases} b^{j+1} c^k, & i = 0, \\ F(H(a^{i-1} b^j c^k)) - 2F(a^{i-1} b^j c^k), & i > 0, \\ F^{-1}(H(a^{i+1} b^j c^k)) + 2F^{-1}(a^{i+1} b^j c^k), & i < 0; \end{cases}$$

$$E(a^i b^j c^k) = \begin{cases} c^{k+1}, & i, j = 0, \\ H(E(b^{j-1} c^k)) - 2E(b^{j-1} c^k), & i = 0, j \neq 0, \\ F(E(a^{i-1} b^j c^k)) + H(a^{i-1} b^j c^k), & i > 0, \\ F^{-1}(E(a^{i+1} b^j c^k)) - F^{-2}(H+2)(a^{i+1} b^j c^k), & i < 0. \end{cases}$$

$$(3.11)$$

Exercise 3.41. Check that the formulae (3.11) give well-defined linear operators on V. Check further that the linear operators F^{-1}, F, H and E satisfy the relations (3.9), in particular, that they define on V the structure of a $U^{(f)}$-module.

Theorem 3.42 (PBW Theorem for $U^{(f)}$). *The standard monomials*

$$\{f^i h^j e^k : i \in \mathbb{Z}; j, k \in \mathbb{N}_0\}$$

form a basis of the algebra $U^{(f)}$.

Proof. Using the relations (3.9) and (3.10) and induction similar to the one used in Lemma 2.14, one can show that every monomial in the generators f^{-1}, f, h and e can be written as a linear combination of standard monomials. In particular, standard monomials generate $U^{(f)}$. We leave the details to the reader.

On the other hand, as in the proof of Theorem 2.13, we can consider the $U^{(f)}$-module V, given by Exercise 3.41. Standard monomials of $U^{(f)}$ map the element $1 \in V$ to linearly independent monomials in the polynomial ring V. Hence standard monomials of $U^{(f)}$ must be linearly independent. This means that they form a basis in $U^{(f)}$, as asserted. □

Corollary 3.43. *There exists a unique homomorphism $\iota : U(\mathfrak{g}) \to U^{(f)}$ of associative algebras such that $\iota(f) = f$, $\iota(h) = h$ and $\iota(e) = e$. The homomorphism ι is injective.*

Proof. Mapping $f \in U(\mathfrak{g})$ to $f \in U^{(f)}$, $h \in U(\mathfrak{g})$ to $h \in U^{(f)}$ and $e \in U(\mathfrak{g})$ to $e \in U^{(f)}$ extends to a well-defined homomorphism ι from $U(\mathfrak{g})$ to $U^{(f)}$ as relations (3.9) contain relations (2.1) as a subset. The uniqueness of ι follows from the fact that f, h and e generate $U(\mathfrak{g})$. The homomorphism ι is injective as standard monomials from $U(\mathfrak{g})$, which form a basis of $U(\mathfrak{g})$ by the PBW Theorem (Theorem 2.13), are mapped to standard monomials in $U^{(f)}$, which are linearly independent by Theorem 3.42. This completes the proof. □

The homomorphism ι is called the *canonical embedding* of $U(\mathfrak{g})$ into $U^{(f)}$. For the moment we define a "bigger" algebra $U^{(f)}$, in which the element f is now invertible. How can we use this? Having an invertible element x of some algebra A we always have a family of automorphisms of A given by $a \mapsto x^i a x^{-i}$, $i \in \mathbb{Z}$. We will consider such automorphisms for $U^{(f)}$ and show that this discrete family can be embedded into a polynomial family of automorphisms indexed by complex numbers (that is pretending that $i \in \mathbb{C}$, despite the fact that x^i does not make sense for $i \notin \mathbb{Z}$). Having such a family of automorphisms of $U^{(f)}$ we can do the following trick with $U(\mathfrak{g})$-modules: Take some $U(\mathfrak{g})$-module, induce it up to $U^{(f)}$, twist by some automorphism from our family, and then restrict back to $U(\mathfrak{g})$. We will show that the whole coherent family can be obtained from one single Verma module using this procedure. Note that the appearance of the algebra $U^{(f)}$ is crucial, as the algebra $U(\mathfrak{g})$ does not have any nontrivial invertible elements. For $k \in \mathbb{Z}$ denote by Θ_k the automorphism of $U^{(f)}$, given by the assignment $\Theta_k(u) = f^k u f^{-k}$, $u \in U^{(f)}$.

Lemma 3.44. *For every $k \in \mathbb{Z}$ we have:*
$$\Theta_k(f) = f,$$
$$\Theta_k(f^{-1}) = f^{-1},$$
$$\Theta_k(h) = h + 2k,$$
$$\Theta_k(e) = e - kf^{-1}h - k(k+1)f^{-1}.$$

Proof. The first two equalities are obvious and the third equality follows from the first formula of (3.10). So, we have only to prove the last equality. We prove it for $k \in \mathbb{N}_0$ by induction on k. For $-k \in \mathbb{N}_0$ the arguments are similar. The basis of the induction is the case $k = 0$, which is obvious.
$$\Theta_{k+1}(e) = f^{k+1} e f^{-k-1}$$
$$= f f^k e f^{-k} f^{-1}$$
$$\text{(by induction)} = f(e - kf^{-1}h - k(k+1)f^{-1})f^{-1}$$
$$= fef^{-1} - kff^{-1}hf^{-1} - k(k+1)ff^{-1}f^{-1}$$
$$\text{(using (3.9))} = e - hf^{-1} - khf^{-1} - k(k+1)f^{-1}$$
$$= e - (k+1)hf^{-1} - k(k+1)f^{-1}$$
$$\text{(by (3.10))} = e - (k+1)f^{-1}(h+2) - k(k+1)f^{-1}$$
$$= e - (k+1)f^{-1}h - (k+1)(k+2)f^{-1}.$$
The claim follows. $\qquad\square$

Lemma 3.44 motivates the following definition:

Proposition 3.45. *For every $z \in \mathbb{C}$ there is a unique automorphism Θ_z of $U^{(f)}$ such that*
$$\Theta_z(f^{\pm 1}) = f^{\pm 1},$$
$$\Theta_z(h) = h + 2z, \tag{3.12}$$
$$\Theta_z(e) = e - zf^{-1}h - z(z+1)f^{-1}.$$
Moreover, we have $\Theta_z^{-1} = \Theta_{-z}$.

Proof. Let us first check that Θ_z extends uniquely to an endomorphism of $U^{(f)}$. For this we have to check the relations (3.9) for the elements f^{-1}, f, $h' = h + 2z$ and $e' = e - zf^{-1}h - z(z+1)f^{-1}$. The relation $f^{-1}f = ff^{-1} = 1$ is contained in (3.9). As $z \in \mathbb{C}$, we also have

$$[h', f] = [h + 2z, f] = [h, f] = -2f.$$

Similarly $[h', e'] = 2e'$ follows from $[h, f^{-1}] = 2f^{-1}$ (see (3.10)). Finally, we have

$$
\begin{aligned}
[e', f] &= [e - zf^{-1}h - z(z+1)f^{-1}, f] \\
\text{(as } f^{-1}f = ff^{-1}) &= [e - zf^{-1}h, f] \\
&= (e - zf^{-1}h)f - f(e - zf^{-1}h) \\
&= ef - fe - zf^{-1}hf + zh \\
\text{(using (3.10))} &= h + 2z \\
&= h'.
\end{aligned}
$$

This yields that Θ_z extends uniquely to an endomorphism of $U^{(f)}$.

To prove that Θ_z is an automorphism, it is enough to check the equality

$$\Theta_{-z}\Theta_z = \mathrm{id}_{U^{(f)}}. \tag{3.13}$$

Since both Θ_{-z} and Θ_z are homomorphisms (by the previous paragraph), it is sufficient to check when the equality (3.13) is applied to the generators. For the generators f and f^{-1} it is obvious and for the generator h it is straightforward. Using the fact that Θ_{-z} is a homomorphism, for the generator e we have:

$$
\begin{aligned}
\Theta_{-z}\Theta_z(e) &= \Theta_{-z}(e - zf^{-1}h - z(z+1)f^{-1}) \\
&= e + zf^{-1}h + z(-z+1)f^{-1} - zf^{-1}(h - 2z) - z(z+1)f^{-1} \\
&= e.
\end{aligned}
$$

This completes the proof. $\qquad\qquad\qquad\qquad\qquad\qquad\qquad\qquad\square$

Remark 3.46. One could give an alternative argument to prove Proposition 3.45 avoiding the direct computation. Observe that, written in some PBW basis, relations of Proposition 3.45 are given by some polynomials in z. For $z \in \mathbb{Z}$ all relations hold by Lemma 3.44. Hence they should hold for arbitrary $z \in \mathbb{C}$.

Exercise 3.47. Check that $\Theta_z\Theta_{z'} = \Theta_{z+z'}$ for all $z, z' \in \mathbb{C}$.

Fix $z \in \mathbb{C}$. Consider the vector space $B(z) = U^{(f)}$. For $x, y \in U(\mathfrak{g})$ and $u \in B$ set

$$x \cdot u \cdot y = \Theta_z(x) u y. \tag{3.14}$$

Exercise 3.48. Check that the assignment (3.14) defines on $B(z)$ the structure of a $U(\mathfrak{g})$-bimodule.

Consider the functor $\mathrm{B}_z : U(\mathfrak{g})\text{-mod} \to U(\mathfrak{g})\text{-mod}$ of tensoring with the bimodule $B(z)$, that is

$$\mathrm{B}_z M = B(z) \bigotimes_{U(\mathfrak{g})} M, \quad M \in U(\mathfrak{g})\text{-mod}.$$

This is well-defined by Exercise 3.48. The functors B_z are called *Mathieu's twisting* functors. Here are some basic properties of the functors B_z, $z \in \mathbb{C}$.

Proposition 3.49.

(i) *The functor* B_z *is exact for every* $z \in \mathbb{C}$.

(ii) *For* $z, z' \in \mathbb{C}$ *we have* $\mathrm{B}_z \circ \mathrm{B}_{z'} \cong \mathrm{B}_{z+z'}$.

(iii) *The inclusion* ι *induces a natural transformation* $\bar{\iota}$ *from the identity functor* $\mathrm{ID}_{U(\mathfrak{g})\text{-mod}}$ *to* B_0.

(iv) *The functor* B_0 *is isomorphic to the identity functor, when restricted to the full subcategory of* $U(\mathfrak{g})$-*mod, consisting of all modules* M, *on which the operator* F *acts bijectively.*

(v) *Assume that* M *admits a central character* χ_M *and that* $\mathrm{B}_z M \neq 0$. *Then* $\mathrm{B}_z M$ *admits a central character as well and* $\chi_M = \chi_{\mathrm{B}_z M}$.

Proof. By Theorem 3.42, the right $U(\mathfrak{g})$-module $B(z)$ is free. Hence the functor B_z is exact, proving the statement (i). The statement (ii) follows from the definitions and Exercise 3.47.

By Corollary 3.43, the map ι is a bimodule homomorphism from $U(\mathfrak{g})$ to $U^{(f)}$. Hence the statement (iii) follows from the obvious observation that the functor $U(\mathfrak{g}) \otimes_{U(\mathfrak{g})} \_$ is isomorphic to the identity functor on $U(\mathfrak{g})$-mod.

If F acts bijectively on the module M, then M caries the natural structure of a $U^{(f)}$-module, where f^{-1} acts via F^{-1}. Hence for such M the natural transformation $\bar{\iota}_M$, constructed in (iii), is an isomorphism. This implies the statement (iv).

Since the element c is central, we have $\Theta_z(c) = c$ by the definition of Θ_z. This implies (v) and completes the proof. $\qquad \square$

Exercise 3.50. Show that the functor B_0 is *not* isomorphic to the identity functor on $U(\mathfrak{g})$-mod.

Exercise 3.51. Let M be a $U(\mathfrak{g})$-module. Prove that the vector space
$$N = \{v \in M : F^i(v) = 0 \text{ for some } i \in \mathbb{N}\}$$
is in fact a submodule of M and that N coincides with the kernel of the natural transformation $\bar{\iota}_M$.

Now we are ready to describe our coherent families $\mathbf{V}(\tau)$ using Mathieu's twisting functors.

Theorem 3.52. *Let $\tau \in \mathbb{C}$.*

(i) *For every $\xi \in \mathbb{C}/2\mathbb{Z}$ and every $z \in \mathbb{C}$ we have an isomorphism $B_z \mathbf{V}(\xi, \tau) \cong \mathbf{V}(\xi + 2z, \tau)$.*

(ii) *For every $z \in \mathbb{C}$ we have $B_z \mathbf{V}(\tau) \cong \mathbf{V}(\tau)$.*

(iii) *Let $\lambda \in \mathbb{C}$ be such that $(\lambda + 1)^2 = \tau$. Then $B_0 M(\lambda) \cong \mathbf{V}(\lambda + 2\mathbb{Z}, \tau)$.*

Proof. First we recall that the action of F on $\mathbf{V}(\xi, \tau)$ is bijective. Therefore $\mathbf{V}(\xi, \tau)$ carries the natural structure of a $U^{(f)}$-module. By definition, the module $B_z \mathbf{V}(\xi, \tau)$ coincides with $\mathbf{V}(\xi, \tau)$ as the vector space, but the action of $U(\mathfrak{g})$ is twisted by Θ_z. Let $\lambda \in \xi$ and $v \in \mathbf{V}(\xi, \tau)$ be some weight vector. Then $h(v) = \lambda v$. Since $\Theta_z(h) = h + 2z$ (see (3.12)), we have
$$\Theta_z(h)(v) = (h + 2z)(v) = (\lambda + 2z)v.$$
This implies that the support of the module $B_z \mathbf{V}(\xi, \tau)$ equals $\xi + 2z$ and that all non-zero weight spaces of $B_z \mathbf{V}(\xi, \tau)$ are one-dimensional. The action of F on $B_z \mathbf{V}(\xi, \tau)$ is unchanged and hence obviously bijective. The element c acts on $B_z \mathbf{V}(\xi, \tau)$ as the scalar τ by Proposition 3.49(v). Hence the statement (i) follows from Exercise 3.28. The statement (ii) follows directly from the statement (i).

Assume that $M(\lambda)$ is given by (3.4). Then the module $M(\lambda)$ has a basis given by $\{F^i(v_0) : i \in \mathbb{N}_0\}$. From the definition of B_0 we get that the elements $\{F^i(v_0) : i \in \mathbb{Z}\}$ form a basis of $B_0 M(\lambda)$. Note that for $i \in \mathbb{Z}$ the element $F^i(v_0)$ has weight $\lambda - 2i$. This implies that the support of $B_0 M(\lambda)$ equals $\lambda + 2\mathbb{Z}$ and that all weight spaces of $B_0 M(\lambda)$ are one-dimensional. By definition, the element F acts bijectively on $B_0 M(\lambda)$. By Exercise 3.51, the natural transformation $\bar{\iota}_{M(\lambda)}$ is injective. The Casimir element c acts on $M(\lambda)$ as the scalar $(\lambda + 1)^2 = \tau$ by Proposition 3.13(iii). By Proposition 3.49(v), the element c acts on $B_0 M(\lambda)$ as the scalar τ. Thus $B_0 M(\lambda) \cong \mathbf{V}(\lambda + 2\mathbb{Z}, \tau)$ follows from Exercise 3.28. This proves the statement (iii) and completes the proof. $\square$

Theorem 3.52 allows us to produce coherent families from Verma modules in the following way:

Corollary 3.53. *Let* $\tau, \lambda \in \mathbb{C}$ *be such that* $(\lambda + 1)^2 = \tau$. *Then*

$$\mathbf{V}(\tau) = \bigoplus_{\substack{z \in \mathbb{C} \\ 0 \leq \Re(z) < 2}} \mathbf{B}_z \, M(\lambda).$$

Proof. This follows immediately from Theorem 3.52. $\square$

3.6 Category of all weight modules with finite-dimensional weight spaces

After the classification of all simple weight modules, obtained in Section 3.4 (Theorem 3.32), it is time to move on to the description of the whole category of weight modules. However, to make this description reasonable, we have to impose an extra restriction. We will limit our consideration to weight modules with finite-dimensional weight spaces. Note that all simple weight modules, classified by Theorem 3.32, do have finite-dimensional weight spaces. So, our restriction does not affect simple objects of our category.

Denote by $\overline{\mathfrak{W}}$ the full subcategory of $\mathfrak{W}$, which consists of all weight modules with finite-dimensional weight spaces. For every $\xi \in \mathbb{C}/2\mathbb{Z}$ we set $\overline{\mathfrak{W}}^\xi = \mathfrak{W}^\xi \cap \overline{\mathfrak{W}}$. The category $\overline{\mathfrak{W}}$ inherits from $\mathfrak{W}$ the following decomposition (given for $\mathfrak{W}$ by Corollary 3.7(ii)):

$$\overline{\mathfrak{W}} = \bigoplus_{\xi \in \mathbb{C}/2\mathbb{Z}} \overline{\mathfrak{W}}^\xi.$$

Firstly, we decompose the categories $\overline{\mathfrak{W}}^\xi$ further, using the action of the Casimir element c. Let $\xi \in \mathbb{C}/2\mathbb{Z}$ and $M \in \overline{\mathfrak{W}}^\xi$. Then for every $\lambda \in \mathbb{C}$ the action of the Casimir element c preserves the finite-dimensional vector space M_λ. Hence we have the following Jordan decomposition for c:

$$M_\lambda = \bigoplus_{\tau \in \mathbb{C}} M_\lambda(\tau),$$

where

$$M_\lambda(\tau) = \{ v \in M_\lambda \ : \ (c - \tau)^k(v) = 0 \text{ for some } k \in \mathbb{N} \}.$$

For $\tau \in \mathbb{C}$ set

$$M(\tau) = \bigoplus_{\lambda \in \mathbb{C}} M_\lambda(\tau).$$

Denote by $\overline{\mathfrak{W}}^{\xi,\tau}$ the full subcategory of $\overline{\mathfrak{W}}^{\xi}$ consisting of all $M \in \overline{\mathfrak{W}}^{\xi}$ such that $M = M(\tau)$.

Lemma 3.54. *Let $\xi \in \mathbb{C}/2\mathbb{Z}$.*

(i) *For every $M \in \overline{\mathfrak{W}}^{\xi}$ and $\tau \in \mathbb{C}$ the space $M(\tau)$ is a submodule of M and we have*

$$M = \bigoplus_{\tau \in \mathbb{C}} M(\tau). \tag{3.15}$$

(ii) *We have*

$$\overline{\mathfrak{W}}^{\xi} = \bigoplus_{\tau \in \mathbb{C}} \overline{\mathfrak{W}}^{\xi,\tau}.$$

Proof. That $M(\tau)$ is a submodule of M follows from Exercise 1.31 since c commutes with the operators E, F and H. This implies the decomposition (3.15) and the claim (i) follows.

If M and N are two $\mathfrak{g}$-modules and $\varphi \in \mathrm{Hom}_{\mathfrak{g}}(M, N)$, then the operator φ intertwines, by definition, the action of the Casimir element c on M and N. This yields $\varphi(M(\tau)) \subset N(\tau)$ for any $\tau \in \mathbb{C}$. Hence the claim (ii) follows from the claim (i). This completes the proof. □

The aim of the next few sections is to describe the categories $\overline{\mathfrak{W}}^{\xi,\tau}$ for all $\xi \in \mathbb{C}/2\mathbb{Z}$ and $\tau \in \mathbb{C}$. Some elementary properties of these categories are given by the following proposition:

Proposition 3.55. *Let $\xi \in \mathbb{C}/2\mathbb{Z}$ and $\tau \in \mathbb{C}$.*

(i) *The category $\overline{\mathfrak{W}}^{\xi,\tau}$ is an abelian category.*

(ii) *If $\tau \neq (\mu + 1)^2$ for all $\mu \in \xi$, then $\overline{\mathfrak{W}}^{\xi,\tau}$ has only one simple object, namely $\mathbf{V}(\xi, \tau)$.*

(iii) *If $\tau = (\mu+1)^2$ for exactly one $\mu \in \xi$, then $\overline{\mathfrak{W}}^{\xi,\tau}$ has two simple objects, namely $M(\mu)$ and $\overline{M}(\mu + 2)$.*

(iv) *If $\tau = (\mu + 1)^2 = (\mu + 2n + 1)^2$ for some $n \in \mathbb{N}$, then $\mu = -n - 1$, $\tau = n^2$ and $\overline{\mathfrak{W}}^{\xi,\tau}$ has three simple objects, namely $M(-n - 1)$, $\mathbf{V}^{(n)}$ and $\overline{M}(n + 1)$.*

(v) *Every object in $\overline{\mathfrak{W}}^{\xi,\tau}$ has finite length.*

(vi) *$\dim \mathrm{Hom}_{\mathfrak{g}}(M, N) < \infty$ for all $M, N \in \overline{\mathfrak{W}}^{\xi,\tau}$.*

(vii) *The category $\overline{\mathfrak{W}}^{\xi,\tau}$ is a Krull–Schmidt category, that is, every object in $\overline{\mathfrak{W}}^{\xi,\tau}$ decomposes into a finite direct sum of indecomposable objects, moreover, such decomposition is unique up to isomorphism and permutation of summands.*

Proof. The statement (i) follows from Proposition 3.8 and definitions. The statements (ii)–(iv) follow from the classification of simple weight modules (Theorem 3.32).

Let $M \in \overline{\mathfrak{W}}^{\xi,\tau}$ be arbitrary and $L \in \overline{\mathfrak{W}}^{\xi,\tau}$ be a simple module. Then all non-zero weight spaces of L are one-dimensional (see Theorem 3.32) and hence the multiplicity $[M : L]$ of L in M cannot exceed $\dim M_\lambda$ for any $\lambda \in \mathbb{C}$ such that $\dim L_\lambda \neq 0$. The statement (v) now follows from the fact that M has finite-dimensional weight spaces and the fact that we have only finitely many simple objects in $\overline{\mathfrak{W}}^{\xi,\tau}$ by statements (ii)–(iv).

Let $A, B, C, D \in \overline{\mathfrak{W}}^{\xi,\tau}$ are such that there is a short exact sequence $A \hookrightarrow B \twoheadrightarrow C$. Then the left exactness of the $\mathrm{Hom}_\mathfrak{g}(\_, \_)$ bifunctor gives us the following exact sequences:

$$0 \to \mathrm{Hom}_\mathfrak{g}(D, A) \to \mathrm{Hom}_\mathfrak{g}(D, B) \to \mathrm{Hom}_\mathfrak{g}(D, C),$$
$$0 \to \mathrm{Hom}_\mathfrak{g}(C, D) \to \mathrm{Hom}_\mathfrak{g}(B, D) \to \mathrm{Hom}_\mathfrak{g}(A, D).$$

It follows that

$$\begin{aligned} \dim \mathrm{Hom}_\mathfrak{g}(D, B) &\leq \dim \mathrm{Hom}_\mathfrak{g}(D, A) + \dim \mathrm{Hom}_\mathfrak{g}(D, C), \\ \dim \mathrm{Hom}_\mathfrak{g}(B, D) &\leq \dim \mathrm{Hom}_\mathfrak{g}(A, D) + \dim \mathrm{Hom}_\mathfrak{g}(C, D). \end{aligned} \tag{3.16}$$

By Exercise 3.39, the endomorphism algebra of any simple weight $\mathfrak{g}$-module is isomorphic to $\mathbb{C}$ and hence is finite-dimensional. Since every object in $\overline{\mathfrak{W}}^{\xi,\tau}$ has finite length by the statement (v), the statement (vi) now follows from (3.16) by induction on the sum of the lengths of M and N.

If $M \in \overline{\mathfrak{W}}^{\xi,\tau}$ and φ is an idempotent endomorphism of M, then the morphism $\mathrm{id}_M - \varphi$ is also an idempotent endomorphism of M and we have the obvious decomposition $M \cong \varphi(M) \oplus (\mathrm{id}_M - \varphi)(M)$ into a direct sum of $\mathfrak{g}$-modules. Now the statement (vi) follows from the abstract Krull–Schmidt Theorem (see for example Theorem 3.6 in [7]). $\qquad\square$

Although the categories $\overline{\mathfrak{W}}^{\xi,\tau}$ are abelian and have finitely many simple objects, later on we will see that they do not have projective objects and hence cannot be described as categories of modules over some finite-dimensional complex associative algebras. However, a satisfactory description can be provided using the (infinite-dimensional) algebra $\mathbb{C}[[x]]$ of formal power series in x with complex coefficients. The description of $\overline{\mathfrak{W}}^{\xi,\tau}$ obviously depends on the number of simple objects in $\overline{\mathfrak{W}}^{\xi,\tau}$.

3.7 Structure of $\overline{\mathfrak{W}}^{\xi,\tau}$ in the case of one simple object

First we consider the case when the category $\overline{\mathfrak{W}}^{\xi,\tau}$ has only one simple object; that is $\tau \neq (\mu+1)^2$ for all $\mu \in \xi$. In this case, the description of $\overline{\mathfrak{W}}^{\xi,\tau}$ will be especially nice. Fix some $\lambda \in \xi$. Consider the category $\mathbb{C}[[x]]$-mod of all *finite-dimensional* $\mathbb{C}[[x]]$-modules. For the action of any power series on such module to be well-defined, the action of the element x must be given by a nilpotent matrix. From the Jordan decomposition theorem it follows that indecomposable objects in $\mathbb{C}[[x]]$-mod naturally correspond to nilpotent Jordan cells (which represent the action of x).

Let $V \in \mathbb{C}[[x]]$-mod be such that the action of the power series $x \in \mathbb{C}[[x]]$ on V is given by the linear operator X. We define on V the structure of a $\mathbb{C}[h,c]$-module by saying that h acts on V as the scalar λ and c acts on V as the linear operator $\tau \mathrm{id}_V + X$. This allows us to consider the following functor:

$$\mathrm{F} = U(\mathfrak{g}) \bigotimes_{\mathbb{C}[h,c]} \_ : \mathbb{C}[[x]]\text{-mod} \to \overline{\mathfrak{W}}^{\xi,\tau},$$

which is defined on morphisms in the natural way.

Exercise 3.56. Check that for any $V \in \mathbb{C}[[x]]$-mod we indeed have that $\mathrm{F}V \in \overline{\mathfrak{W}}^{\xi,\tau}$.

From the usual adjunction between induction and restriction, we have the right adjoint functor $\mathrm{G} : \overline{\mathfrak{W}}^{\xi,\tau} \to \mathbb{C}[[x]]$-mod. For $M \in \overline{\mathfrak{W}}^{\xi,\tau}$ the module $\mathrm{G}M$ is given by M_λ, where the action of x on M_λ is given by the nilpotent linear operator $c - \tau$.

Exercise 3.57. Prove that the functor G is exact.

Theorem 3.58. *The functors F and G are mutually inverse equivalences of categories. In particular, the categories $\overline{\mathfrak{W}}^{\xi,\tau}$ and $\mathbb{C}[[x]]$-mod are equivalent.*

Proof. Consider the adjunction morphisms

$$\alpha : \mathrm{ID}_{\mathbb{C}[[x]]\text{-mod}} \to \mathrm{GF}, \qquad \beta : \mathrm{FG} \to \mathrm{ID}_{\overline{\mathfrak{W}}^{\xi,\tau}}. \qquad (3.17)$$

First we observe from the definition of G and the construction of $\mathbf{V}(\xi,\tau)$ that the module $\mathrm{GV}(\xi,\tau)$ is one-dimensional hence simple.

Let $V = \mathbb{C}$ be the (unique) simple $\mathbb{C}[[x]]$-module with generator v. Then from Theorem 2.33(i) we have that the module $\mathrm{F}V$ has the basis $\{v, e^i(v), f^i(v) : i \in \mathbb{N}\}$. This means that all non-zero weight spaces of

FV are one-dimensional. We have $FV \in \overline{\mathfrak{W}}^{\xi,\tau}$ by Exercise 3.56. Since $\mathbf{V}(\xi,\tau)$ is the only simple object in $\overline{\mathfrak{W}}^{\xi,\tau}$, the module $\mathbf{V}(\xi,\tau)$ must be a submodule of FV. Comparing the dimensions of the weight spaces we get $FV \cong \mathbf{V}(\xi,\tau)$.

The above means that both F and G send simple modules to simple modules. As both adjunction morphisms α and β from (3.17) must be non-zero, when evaluated at modules which are not annihilated by our functors, we get that both α and β are isomorphisms, when evaluated at simple modules.

Now we prove that α is an isomorphism when evaluated at all modules by induction on the length of a module. Note that, by Proposition 3.55(v), every object in $\overline{\mathfrak{W}}^{\xi,\tau}$ has finite length. Let $V \in \overline{\mathfrak{W}}^{\xi,\tau}$ be an arbitrary non-simple module. Then there is a short exact sequence $\mathbf{V}(\xi,\tau) \hookrightarrow V \twoheadrightarrow V'$, where the length of V' is strictly smaller than that of V. Observe that the functor F is exact by Theorem 2.33, and that the functor G is exact by Exercise 3.57. Hence we have the following commutative diagram with exact rows:

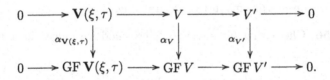

From the first part of the proof we have that the map $\alpha_{\mathbf{V}(\xi,\tau)}$ is an isomorphism. By induction, the map $\alpha_{V'}$ is an isomorphism. Hence α_V is an isomorphism as well by the Five Lemma. This proves that α is an isomorphism of functors. Similarly one proves that β is an isomorphism of functors. The claim follows.   □

3.8 Structure of $\overline{\mathfrak{W}}^{\xi,\tau}$ in the case of two simple objects

Now let us consider the case where there exists a unique element $\lambda \in \xi$ such that $(\lambda + 1)^2 = \tau$. Recall that a category $\mathcal{A}$ is called an R-*category*, where R is a commutative ring, provided that for any two objects $a, b \in \mathcal{A}$ the set $\mathcal{A}(a,b)$ has a fixed R-module structure and the multiplication of morphisms is R-bilinear. Consider the $\mathbb{C}[[x]]$-category $\mathfrak{A}$ with two objects $\mathbf{p}$ and $\mathbf{q}$, generated by morphisms $\mathbf{a} \in \mathfrak{A}(\mathbf{p},\mathbf{q})$ and $\mathbf{b} \in \mathfrak{A}(\mathbf{q},\mathbf{p})$ subject to the relations $\mathbf{ab} = x1_{\mathbf{q}}$ and $\mathbf{ba} = x1_{\mathbf{p}}$:

$$\mathfrak{A} : \qquad p \underset{b}{\overset{a}{\rightleftarrows}} q , \qquad ab = x1_q, \ ba = x1_p.$$

Exercise 3.59. Show that the path algebra of $\mathfrak{A}$ is isomorphic to the algebra of all those 2×2-matrices $\begin{pmatrix} a & b \\ c & d \end{pmatrix}$ with coefficients from $\mathbb{C}[[x]]$, which satisfy the condition $b \in x\mathbb{C}[[x]]$.

As usual, a finite-dimensional $\mathfrak{A}$-*module* V is given by a functor from $\mathfrak{A}$ to the category of all finite-dimensional vector spaces. Hence V consists of two finite-dimensional $\mathbb{C}[[x]]$-modules V_p and V_q (on which the action of x is given by the linear operators X_p and X_q respectively) and two $\mathbb{C}[[x]]$-homomorphisms $V(a) : V_p \to V_q$ and $V(b) : V_q \to V_p$, satisfying the conditions $V(a)V(b) = X_q$ and $V(b)V(a) = X_p$.

$$V : \qquad X_p \overset{\curvearrowright}{\underset{}{}} V_p \underset{V(b)}{\overset{V(a)}{\rightleftarrows}} V_q \overset{\curvearrowleft}{\underset{}{}} X_q , \qquad \begin{matrix} V(a)V(b) = X_q \\ V(b)V(a) = X_p \end{matrix}. \qquad (3.18)$$

A homomorphism of $\mathfrak{A}$-modules is, then, just a natural transformation of functors. In other words, if V and W are two $\mathfrak{A}$-modules, then a homomorphism $\varphi : V \to W$ consists of two linear maps $\varphi_p : V_p \to W_p$ and $\varphi_q : V_q \to W_q$, which intertwine the action of $\mathfrak{A}$ on V and W. The latter means that the following diagram commutes:

$$
\begin{array}{ccc}
X_p \ V_p & \underset{V(b)}{\overset{V(a)}{\rightleftarrows}} & V_q \ X_q \\
\varphi_p \downarrow & & \varphi_q \downarrow \\
X_p \ W_p & \underset{W(b)}{\overset{W(a)}{\rightleftarrows}} & W_q \ X_q
\end{array}
\qquad (3.19)
$$

The category of all finite-dimensional $\mathfrak{A}$-modules is denoted by $\mathfrak{A}$-mod.

For $V \in \overline{\mathfrak{W}}^{\xi,\tau}$ consider the following diagram:

$$\mathsf{G}\, V : \qquad (c-\tau) \overset{\curvearrowright}{\underset{}{}} V_\lambda \underset{2F}{\overset{2E}{\rightleftarrows}} V_{\lambda+2} \overset{\curvearrowleft}{\underset{}{}} (c-\tau) .$$

Exercise 3.60. Check that the diagram $\mathsf{G}\, V$ defines an $\mathfrak{A}$-module by setting $V_p = V_\lambda$, $V_q = V_{\lambda+2}$, $X_p = (c-\tau)|_{V_\lambda}$, $X_q = (c-\tau)|_{V_{\lambda+2}}$, $V(a) = 2E|_{V_\lambda}$ and $V(b) = 2F|_{V_{\lambda+2}}$.

Exercise 3.61. Let $V, W \in \overline{\mathfrak{W}}^{\xi,\tau}$ and $\varphi \in \mathrm{Hom}_{\mathfrak{g}}(V, W)$. Show that the restriction of φ to V_λ and $V_{\lambda+2}$ defines an $\mathfrak{A}$-module homomorphism from GV to GW, and derive from this that $G : \overline{\mathfrak{W}}^{\xi,\tau} \to \mathfrak{A}$-mod is a functor.

For an $\mathfrak{A}$-module V, given by (3.18), consider the following diagram:

$$\cdots \underset{\mathrm{id}}{\overset{a_{\lambda-6}}{\rightleftarrows}} V_p \underset{\mathrm{id}}{\overset{a_{\lambda-4}}{\rightleftarrows}} V_p \underset{\mathrm{id}}{\overset{a_{\lambda-2}}{\rightleftarrows}} V_p \underset{\frac{1}{2}V(b)}{\overset{\frac{1}{2}V(a)}{\rightleftarrows}} V_q \underset{\mathrm{id}}{\overset{a_{\lambda+2}}{\rightleftarrows}} V_q \underset{\mathrm{id}}{\overset{a_{\lambda+4}}{\rightleftarrows}} V_q \underset{\mathrm{id}}{\overset{a_{\lambda+6}}{\rightleftarrows}} \cdots ,$$

$$\tag{3.20}$$

where

$$a_{\lambda+2i} = \frac{1}{4}X_q - i(\lambda + i + 1), \quad a_{\lambda-2i} = \frac{1}{4}X_p + i(\lambda - i + 1)$$

for all $i \in \mathbb{N}$. Denote by FV the vector space underlying the diagram (3.20) (that is the direct sum of all vector spaces on the diagram). Define the linear operator H on FV as indicated by the dotted arrows (each such arrow denotes a scalar operator on the corresponding vector space). Define the linear operator E on FV as indicated by the regular arrows (going from the left to the right). Finally, define the linear operator F on FV as indicated by the double arrow.

Lemma 3.62. *The above assignment defines on FV the structure of a $\mathfrak{g}$-module. Moreover, $FV \in \overline{\mathfrak{W}}^{\xi,\tau}$.*

Proof.   To prove the lemma we have to check the relations (1.2). By definition we have that the operator H is diagonalizable on FV, that E increases the H-eigenvalues by 2 and that F decreases the H-eigenvalues by 2. This means that the relations $[H, F] = -2F$ and $[H, E] = 2E$ are satisfied. So, it remains to check the relation $[E, F] = H$, which we can check separately on eigenspaces of H.

For the eigenspaces corresponding to the eigenvalues $\lambda + 2i$, $i \in \mathbb{N}$, $i > 1$, the relation $[E, F] = H$ follows from the relation $a_{\lambda+2i} - a_{\lambda+2(i+1)} = \lambda + 2(i + 1)$, which, in turn, follows from the definition of $a_{\lambda+2i}$, $i \in \mathbb{N}$. For the eigenvalue $\lambda + 2$, the relation $[E, F] = H$ reduces to the following computation:

$$\frac{1}{4}V(a)V(b) - a_{\lambda+2} = \frac{1}{4}X_q - \frac{1}{4}X_q + (\lambda + 2) = \lambda + 2.$$

Similarly, we can check the relation $[E, F] = H$ for all eigenvalues $\lambda - 2i$, $i \in \mathbb{N}_0$. Thus FV is indeed a $\mathfrak{g}$-module.

As $\operatorname{supp} F V \subset \xi$ by definition, to check that $F V \in \overline{\mathfrak{W}}^{\xi,\tau}$ we have to compute all eigenvalues of the operator c on $F V$. It suffices to restrict c to V_{p} and V_{q}. From the definition of $F V$ we have that the restriction of c to V_{p} equals $(\lambda+1)^2 + X_{\mathrm{p}}$ and that the restriction of c to V_{q} equals $(\lambda+1)^2 + X_{\mathrm{q}}$. As both X_{p} and X_{q} are nilpotent, we derive that the only eigenvalue of c is $(\lambda+1)^2 = \tau$. Hence $F V \in \overline{\mathfrak{W}}^{\xi,\tau}$ and the proof is complete. $\qquad\square$

Let V and W be two $\mathfrak{A}$-modules and $\varphi : V \to W$ a homomorphism. Repeating φ_{p} on all components V_{p} in (3.20) and repeating φ_{q} on all components V_{q} in (3.20) we extend φ to a linear map $F(\varphi) : F V \to F W$.

Exercise 3.63. Check that the linear map $F(\varphi)$ defined above is a $\mathfrak{g}$-homomorphism and derive from this that $F : \mathfrak{A}\text{-mod} \to \overline{\mathfrak{W}}^{\xi,\tau}$ is a functor.

Theorem 3.64. *The functors F and G are mutually inverse equivalences of categories, in particular, the categories $\overline{\mathfrak{W}}^{\xi,\tau}$ and $\mathfrak{A}$-mod are equivalent.*

Proof. That $GF \cong \mathrm{ID}_{\mathfrak{A}\text{-mod}}$ follows immediately from the construction.

By Proposition 3.55, every module in $\overline{\mathfrak{W}}^{\xi,\tau}$ has finite length with simple modules $M(\lambda)$ and $\overline{M}(\lambda+2)$ as subquotients. From the construction of these modules we see that in both $M(\lambda)$ and $\overline{M}(\lambda+2)$ the restriction of F to any weight space $\lambda+2i$, $i \in \mathbb{Z}$, $i \neq 1$, is an isomorphism to the weight space $\lambda + 2(i-1)$. Hence, for any $V \in \overline{\mathfrak{W}}^{\xi,\tau}$, we can identify all $V_{\lambda+2i}$, $i \in \mathbb{N}$, with $V_{\lambda+2}$ using the action of F. Similarly we can identify all $V_{\lambda-2i}$, $i \in \mathbb{N}_0$, with V_λ. Since the Casimir element c commutes with F, the above identification also identifies the actions of c on all $V_{\lambda+2i}$, $i \in \mathbb{N}$, with the action of c on $V_{\lambda+2}$; and the actions of c on all $V_{\lambda-2i}$, $i \in \mathbb{N}_0$, with the action of c on V_λ. Thus, using $c = (\mu+1)^2 + 4FE$, we can uniquely determine the action of E when restricted to all $V_{\lambda+2i}$, $i \in \mathbb{N}$, by the formula $E = \frac{1}{4}F^{-1}(c - (\lambda+2i+1)^2)$. Similarly, the action of E on all $V_{\lambda-2i}$, $i \in \mathbb{N}$ is also uniquely determined.

This shows that for any $\mathfrak{g}$-module V there is a unique way to reconstruct the $\mathfrak{g}$-module $FG V$ from the $\mathfrak{A}$-module $G V$, in particular, that $FG V \cong V$. This and the construction of G and F yield the existence of an isomorphism $FG \cong \mathrm{ID}_{\overline{\mathfrak{W}}^{\xi,\tau}}$, which completes the proof. $\qquad\square$

3.9 Structure of $\overline{\mathfrak{W}}^{\xi,\tau}$ in the case of three simple objects

Finally, in this section we are going to describe the category $\overline{\mathfrak{W}}^{n-1+2\mathbb{Z},n^2}$ for each (fixed) $n \in \mathbb{N}$. By Proposition 3.55(iv), it has three simple modules, namely $M(-n-1)$, $\mathbf{V}^{(n)}$ and $\overline{M}(n+1)$.

Consider the $\mathbb{C}[[x]]$-category $\mathfrak{B}$ with three objects $\mathbf{p}$, $\mathbf{q}$ and $\mathbf{r}$, generated by morphisms

$$\mathbf{a} \in \mathfrak{B}(\mathbf{p},\mathbf{q}), \quad \mathbf{b} \in \mathfrak{B}(\mathbf{q},\mathbf{p}), \quad \mathbf{c} \in \mathfrak{B}(\mathbf{q},\mathbf{r}), \quad \mathbf{d} \in \mathfrak{B}(\mathbf{r},\mathbf{q}),$$

subject to the relations

$$\mathbf{ba} = x\mathbf{1_p}, \quad \mathbf{ab} = \mathbf{dc} = x\mathbf{1_q}, \quad \mathbf{cd} = x\mathbf{1_r}.$$

The category $\mathfrak{B}$ can be depicted as follows:

$$\mathfrak{B}: \quad \mathbf{p} \underset{b}{\overset{a}{\rightleftarrows}} \mathbf{q} \underset{d}{\overset{c}{\rightleftarrows}} \mathbf{r}, \qquad \begin{aligned} \mathbf{ba} &= x\mathbf{1_p}, \\ \mathbf{ab} &= \mathbf{dc} = x\mathbf{1_q}, \\ \mathbf{cd} &= x\mathbf{1_r}. \end{aligned}$$

The notions of $\mathfrak{B}$-modules, their homomorphisms and the category $\mathfrak{B}$-mod are defined in the usual way (as in the case of the category $\mathfrak{A}$ from Section 3.8).

Exercise 3.65. Show that the path algebra of $\mathfrak{B}$ is isomorphic to the algebra of all those 3×3-matrices $\begin{pmatrix} a_{11} & a_{12} & a_{13} \\ a_{21} & a_{22} & a_{23} \\ a_{31} & a_{32} & a_{33} \end{pmatrix}$ with coefficients from $\mathbb{C}[[x]]$, which satisfy the condition $a_{12}, a_{23} \in x\mathbb{C}[[x]]$, $a_{13} \in x^2\mathbb{C}[[x]]$.

For $V \in \overline{\mathfrak{W}}^{n-1+2\mathbb{Z},n^2}$ consider the following diagram:

$$GV: \quad (c-n^2) \overset{>}{\underset{}{\dashrightarrow}} V_{-n-1} \underset{2F}{\overset{2E}{\rightleftarrows}} V_{-n+1} \overset{\scriptstyle (c-n^2)}{\underset{2F^n}{\overset{2EF^{1-n}}{\rightleftarrows}}} V_{n+1} \overset{<}{\underset{}{\dashleftarrow}} (c-n^2).$$

$$(3.21)$$

First we check that the diagram (3.21) makes sense.

Lemma 3.66. *The linear operator* $EF^{1-n} : V_{-n+1} \to V_{n+1}$ *on the diagram* (3.21) *is well defined.*

Proof. By Proposition 3.55, every object in $\overline{\mathfrak{W}}^{n-1+2\mathbb{Z},n^2}$ has finite length with subquotients of the form $M(-n-1)$, $\mathbf{V}^{(n)}$ and $\overline{M}(n+1)$. Note that

$$\dim M(-n-1)_{-n+1} = \dim \overline{M}(n+1)_{-n+1} = 0,$$
$$\dim M(-n-1)_{n-1} = \dim \overline{M}(n+1)_{n-1} = 0$$

by Proposition 3.13(ii) and Exercise 3.20. On the other hand, from the picture (1.9) we have that the restriction of the linear operator F^{n-1} to $\mathbf{V}^{(n)}_{n-1}$ is a bijection to $\mathbf{V}^{(n)}_{-n+1}$. Hence for $V \in \overline{\mathfrak{W}}^{n-1+2\mathbb{Z},n^2}$ we can use the induction on the length of V and conclude that the restriction of F^{n-1} to V_{n-1} is a bijection to V_{-n+1}. This means that the desired linear operator $EF^{1-n} : V_{-n+1} \to V_{n+1}$ is well-defined. $\qquad\square$

Exercise 3.67. Show that for every $V \in \overline{\mathfrak{W}}^{n-1+2\mathbb{Z},n^2}$ the diagram (3.21) defines a $\mathfrak{B}$-module GV. Show further that defining G on homomorphisms via restriction produces a functor $G : \overline{\mathfrak{W}}^{n-1+2\mathbb{Z},n^2} \to \mathfrak{B}$-mod.

For a $\mathfrak{B}$-module V, given by

$$
X_p \; \overset{\nearrow}{\big(}\; V_p \underset{V(b)}{\overset{V(a)}{\rightleftarrows}} V_q \underset{V(d)}{\overset{V(c)}{\rightleftarrows}} V_r \;\overset{\nearrow}{\big)}\; X_r
$$

(with X_q above V_q)

consider the following diagram:

$$(3.22)$$

where

$$a_{-n-1-2i} = \frac{1}{4}X_p - i(n+i), \quad a_{n-1+2i} = \frac{1}{4}X_r - i(n+i)$$

for all $i \in \mathbb{N}$, and

$$a_{-n-1+2i} = \frac{1}{4}X_q + i(n-i)$$

for all $i \in \{1, 2, \ldots, n-1\}$.

Exercise 3.68. Show that, as in Lemma 3.62, the diagram (3.22) defines the structure of a $\mathfrak{g}$-module on the underlying vector space FV. Show further that, similarly to Exercise 3.63, the map F extends to a functor $F : \mathfrak{B}$-mod $\to \overline{\mathfrak{W}}^{n-1+2\mathbb{Z},n^2}$.

Theorem 3.69. *The functors* F *and* G *are mutually inverse equivalences of categories. In particular, the categories* $\overline{\mathfrak{W}}^{n-1+2\mathbb{Z},n^2}$ *and* $\mathfrak{B}$-mod *are equivalent.*

Proof. That $GF \cong ID_{\mathfrak{A}\text{-mod}}$ follows immediately by construction. To prove that the isomorphism $FG \cong ID_{\overline{\mathfrak{W}}^{n-1+2\mathbb{Z},n^2}}$ holds, one must show that for any $V \in \overline{\mathfrak{W}}^{n-1+2\mathbb{Z},n^2}$ there is a unique extension of the $\mathfrak{g}$-module structure from the $\mathfrak{B}$-module GV. This is done analogously to the proof of Theorem 3.64. We leave the details to the reader. □

3.10 Tensoring with a finite-dimensional module

For a finite-dimensional $\mathfrak{g}$-module V and an arbitrary $\mathfrak{g}$-module M we can consider the tensor product $V \otimes M$, which becomes a $\mathfrak{g}$-module using (1.17). If M and N are two $\mathfrak{g}$-modules and $\varphi : M \to N$ is a homomorphism, then the mapping

$$\varphi_V : V \otimes M \to V \otimes N,$$
$$v \otimes m \mapsto v \otimes \varphi(m)$$

is a $\mathfrak{g}$-module homomorphism. In particular, we get a functor

$$V \otimes {}_- : \mathfrak{g}\text{-mod} \to \mathfrak{g}\text{-mod}.$$

Functors of the form $V \otimes {}_-$ (also on subcategories of $\mathfrak{g}$-mod) and their direct summands are called *projective functors*.

Let V be a finite-dimensional $\mathfrak{g}$-module. Using the principal antiautomorphism ω from Exercise 2.50 we define on the dual space V^* the structure of a $\mathfrak{g}$-module via $(x \cdot g)(v) = g(\omega(x)(v))$ for $x \in \mathfrak{g}$, $g \in V^*$ and $v \in V$. This dual module will be denoted V^ω.

Exercise 3.70. Check that the assignment above defines on V^ω the structure of a $\mathfrak{g}$-module and that $(V^\omega)^\omega \cong V$.

Lemma 3.71. *Let* V *be a finite-dimensional* $\mathfrak{g}$-module. *Then the endofunctor* $V^\omega \otimes {}_-$ *of* $\mathfrak{g}$-mod *is both left and right adjoint to* $V \otimes {}_-$.

Proof. Let $v_1, \ldots, v_k$ be a basis of V and $g_1, \ldots, g_k$ be the dual basis of V^ω. Assume that $\varphi \in \text{Hom}_\mathfrak{g}(M, V \otimes N)$ is given by

$$\varphi(m) = \sum_i v_i \otimes n_i^{(m)}, \quad m \in M, \tag{3.23}$$

for some $n_i^{(m)} \in N$. Define the linear map $\psi : V^\omega \otimes M \to N$ as follows:

$$\psi(g \otimes m) = \sum_i g(v_i) n_i^{(m)}, \quad g \in V^\omega. \tag{3.24}$$

First we check that the map ψ, defined by (3.24), is a homomorphism of $\mathfrak{g}$-modules. Since φ is a $\mathfrak{g}$-module homomorphism, for every $x \in \mathfrak{g}$ and $m \in M$ we have, using (3.23) and (1.17), the following:

$$\varphi(x(m)) = x(\varphi(m)) = \sum_i x(v_i) \otimes n_i^{(m)} + \sum_i v_i \otimes x(n_i^{(m)}). \tag{3.25}$$

Set $A = \sum_i g(x(v_i)) n_i^{(m)}$. Now we compute:

$$\psi(x(g \otimes m)) = \psi(x(g) \otimes m + g \otimes x(m))$$

$$\text{(linearity of } \psi) = \psi(x(g) \otimes m) + \psi(g \otimes x(m))$$

$$\text{(definition of } \psi) = \sum_i x(g)(v_i) n_i^{(m)} + \psi(g \otimes x(m))$$

$$\text{(definition of } V^\omega) = \sum_i g(\omega(x)(v_i)) n_i^{(m)} + \psi(g \otimes x(m))$$

$$\text{(definition of } \omega) = -A + \psi(g \otimes x(m))$$

$$\text{(definition of } \psi \text{ and (3.25))} = -A + A + \sum_i g(v_i) x(n_i^{(m)})$$

$$= \sum_i g(v_i) x(n_i^{(m)})$$

$$\text{(linearity of } x) = x(\sum_i g(v_i) n_i^{(m)})$$

$$\text{(definition of } \psi) = x(\psi(g \otimes m)).$$

Hence we obtain a map

$$\alpha : \text{Hom}_{\mathfrak{g}}(M, V \otimes N) \to \text{Hom}_{\mathfrak{g}}(V^\omega \otimes M, N),$$

$$\varphi \mapsto \psi.$$

If $\psi \in \text{Hom}_{\mathfrak{g}}(V^\omega \otimes M, N)$ and $m \in M$, set $n_i^{(m)} = \psi(g_i \otimes m)$. This allows us to define the map $\varphi : M \to V \otimes N$ via

$$\varphi(m) = \sum_i v_i \otimes n_i^{(m)}. \tag{3.26}$$

We claim that φ is a homomorphism of $\mathfrak{g}$-modules. Indeed, by (1.17), for $x \in \mathfrak{g}$ we have

$$x(\varphi(m)) = \sum_i x(v_i) \otimes n_i^{(m)} + \sum_i v_i \otimes x(n_i^{(m)}). \tag{3.27}$$

Applying the $\mathfrak{g}$-homomorphism ψ to $x(g_i \otimes m)$ we obtain

$$
\begin{aligned}
x(n_i^{(m)}) &= x(\psi(g_i \otimes m)) \\
&= \psi(x(g_i \otimes m)) \\
\text{(by 1.17)} \ &= \psi(x(g_i) \otimes m + g_i \otimes x(m)) \\
&= \psi(x(g_i) \otimes m) + n_i^{(x(m))}.
\end{aligned} \tag{3.28}
$$

Hence for $\varphi(x(m))$ we get

$$
\begin{aligned}
\varphi(x(m)) \ &\overset{\text{(def)}}{=} \ \sum_i v_i \otimes n_i^{(x(m))} \\
\text{(by (3.28))} \ &= \ \sum_i v_i \otimes (x(n_i^{(m)}) - \psi(x(g_i) \otimes m)) \\
&= \ \sum_i v_i \otimes x(n_i^{(m)}) - \sum_i v_i \otimes \psi(x(g_i) \otimes m).
\end{aligned} \tag{3.29}
$$

Comparing (3.27) and (3.29), the equality $\varphi(x(m)) = x(\varphi(m))$ reduces to

$$
-\sum_i v_i \otimes \psi(x(g_i) \otimes m) = \sum_i x(v_i) \otimes n_i^{(m)}. \tag{3.30}
$$

Let $x(g_i) = \sum_j x_{ij} g_j$. Then $x(v_i) = -\sum_j x_{ji} v_j$ by the definition of ω. From this observation and the definition of ψ we obtain that both sides of the equality (3.30) equal

$$
-\sum_{i,j} x_{ij} v_i \otimes n_j^{(m)}.
$$

This proves the equality (3.30) and shows that φ is a homomorphism of $\mathfrak{g}$-modules. Hence we obtain a map

$$
\beta : \mathrm{Hom}_{\mathfrak{g}}(V^\omega \otimes M, N) \to \mathrm{Hom}_{\mathfrak{g}}(M, V \otimes N),
$$
$$
\psi \mapsto \varphi.
$$

By construction, $\beta = \alpha^{-1}$ and both α and β are natural. Hence we have a natural isomorphism

$$
\alpha : \mathrm{Hom}_{\mathfrak{g}}(M, V \otimes N) \cong \mathrm{Hom}_{\mathfrak{g}}(V^\omega \otimes M, N),
$$

which means that $V^\omega \otimes \_$ is left adjoint to $V \otimes \_$. Substituting V by V^ω and using Exercise 3.70 we also get that $V \otimes \_$ is left adjoint to $V^\omega \otimes \_$, which completes the proof. $\qquad\square$

Exercise 3.72. Show that for every finite-dimensional $\mathfrak{g}$-module V we have $V^\omega \cong V$. Derive from this that the functor $V \otimes \_$ is self-adjoint.

From Lemma 3.71 it follows that the functor $V \otimes \_$ is exact.

Lemma 3.73. *For any finite-dimensional module V, the functor $V \otimes \_$ preserves both the category $\mathfrak{W}$ and the category $\overline{\mathfrak{W}}$.*

Proof. A tensor product of a finitely-generated and a finite-dimensional module is obviously finitely generated. Hence the fact that V preserves $\mathfrak{W}$ follows directly from Proposition 3.8(iv).

Let $v_1, \ldots, v_k$ be a weight basis of V and assume that the element v_i has weight λ_i, $i = 1, \ldots, k$. If $M \in \overline{\mathfrak{W}}$, then from Exercise 3.9 we obtain that for all $\mu \in \mathbb{C}$ we have

$$(V \otimes M)_\mu = \sum_{i=1}^{k} \mathbb{C} v_i \otimes M_{\mu - \lambda_i}.$$

Since all weight spaces of M are finite-dimensional, it follows that all weight spaces of $V \otimes M$ are finite-dimensional as well. $\square$

Unfortunately, the functor $V \otimes \_$ does not preserve the categories $\overline{\mathfrak{W}}^{\xi,\tau}$ in general, which follows, for example, from Theorem 1.39 and the definition of $\overline{\mathfrak{W}}^{\xi,\tau}$. By Weyl's Theorem (Theorem 1.29), every functor $V \otimes \_$ decomposes into a direct sum of functors of the form $\mathbf{V}^{(n)} \otimes \_$, $n \in \mathbb{N}$. Hence to understand projective functors we can restrict our consideration to the functors $\mathbf{V}^{(n)} \otimes \_$, $n \in \mathbb{N}$.

Exercise 3.74. Let V and W be finite-dimensional modules. Show that the composition of functors $V \otimes \_$ and $W \otimes \_$ is isomorphic to the functor $(V \otimes W) \otimes \_$.

Exercise 3.75. Show that the endofunctor $\mathbf{V}^{(1)} \otimes \_$ of $\mathfrak{g}$-mod is isomorphic to the identity functor.

From Theorem 1.39 we have that every $\mathbf{V}^{(n)}$, $n \in \mathbb{N}$, $n > 1$, occurs as a direct summand of $(\mathbf{V}^{(2)})^{\otimes n-1}$. From Exercise 3.74 it follows that to understand projective functors it is enough to understand the functor $\mathbf{V}^{(2)} \otimes \_$ and its powers. Let us try to describe what $\mathbf{V}^{(2)} \otimes \_$ does to modules $\mathbf{V}(\xi, \tau)$.

Fix $\xi \in \mathbb{C}/2\mathbb{Z}$, $\tau \in \mathbb{C}$, and for $\mu \in \xi$ set $a_\mu = \frac{1}{4}(\tau - (\mu + 1)^2)$. Consider the vector space $\mathbf{W}(\xi, \tau)$ with the basis $\{v_\mu, w_\mu \ : \ \mu \in \xi\}$ and the linear operators E, F and H on $\mathbf{W}(\xi, \tau)$ defined as follows:

$$\begin{aligned} F(v_\mu) &= v_{\mu-2}; & F(w_\mu) &= w_{\mu-2}; \\ H(v_\mu) &= \mu v_\mu; & H(w_\mu) &= \mu w_\mu; \\ E(v_\mu) &= a_\mu v_{\mu+2}; & E(w_\mu) &= a_\mu w_{\mu+2} + v_{\mu+2}. \end{aligned} \tag{3.31}$$

This can be depicted as follows:

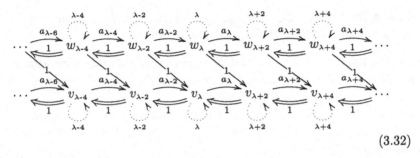

$$(3.32)$$

Exercise 3.76. Check that the formulae (3.31) define on $\mathbf{W}(\xi, \tau)$ the structure of a weight $\mathfrak{g}$-module with support ξ.

Proposition 3.77 (Structure of $\mathbf{W}(\xi,\tau)$). *Let $\xi \in \mathbb{C}/2\mathbb{Z}$, $\tau \in \mathbb{C}$.*

(i) *There is a short exact sequence*

$$0 \to \mathbf{V}(\xi,\tau) \to \mathbf{W}(\xi,\tau) \to \mathbf{V}(\xi,\tau) \to 0.$$

(ii) *The module $\mathbf{W}(\xi,\tau)$ is generated by any w_μ, $\mu \in \xi$, such that $a_{\mu+2i} \neq 0$ for all $i \in \mathbb{N}_0$.*

(iii) *The endomorphism algebra of $\mathbf{W}(\xi,\tau)$ is isomorphic to $\mathbb{C}[x]/(x^2)$; in particular, the module $\mathbf{W}(\xi,\tau)$ is indecomposable.*

(iv) *The action of the Casimir element c on the weight space $\mathbf{W}(\xi,\tau)_\mu$, $\mu \in \xi$, is given in the basis v_μ, w_μ of $\mathbf{W}(\xi,\tau)_\mu$ by the matrix*

$$\begin{pmatrix} \tau & 4 \\ 0 & \tau \end{pmatrix},$$

in particular $(c-\tau)V \neq 0$ and $(c-\tau)^2 V = 0$.

Proof. Comparing (3.8) and (3.32) we see that sending the basis element $v_\mu \in \mathbf{V}(\xi,\tau)$ to the basis element $v_\mu \in \mathbf{W}(\xi,\tau)$ defines an injective homomorphism φ from $\mathbf{V}(\xi,\tau)$ to $\mathbf{W}(\xi,\tau)$. Furthermore, sending the basis element $w_\mu \in \mathbf{W}(\xi,\tau)$ to the basis element $v_\mu \in \mathbf{W}(\xi,\tau)$ defines a surjective homomorphism ψ from $\mathbf{W}(\xi,\tau)$ to $\mathbf{V}(\xi,\tau)$, whose kernel coincides with the image of φ. This gives the short exact sequence as required in the claim (i).

Let $\mu \in \xi$ be such that $a_{\mu+2i} \neq 0$ for all $i \in \mathbb{N}_0$. Denote by M the submodule of $\mathbf{W}(\xi,\tau)$ generated by w_μ. From (3.32) we have that the vectors $FE(w_\mu)$ and w_μ are linearly independent, which yields $M_\mu = \mathbf{W}(\xi,\tau)_\mu$. As the action of F on $\mathbf{W}(\xi,\tau)$ is injective by (3.32), we get $M_{\mu-2i} = \mathbf{W}(\xi,\tau)_{\mu-2i}$ for all $i \in \mathbb{N}_0$. Finally, from $a_{\mu+2i} \neq 0$ for all $i \in \mathbb{N}_0$ it follows that the action of E on all $\mathbf{W}(\xi,\tau)_{\mu+2i}$, $i \in \mathbb{N}_0$, is injective as well. Hence we have that $M_{\mu+2i} = \mathbf{W}(\xi,\tau)_{\mu+2i}$ for all $i \in \mathbb{N}_0$. This means that $M = \mathbf{W}(\xi,\tau)$ and proves (ii).

Let $\mu \in \xi$ be such that $a_{\mu+2i} \neq 0$ for all $i \in \mathbb{N}_0$ (this can always be chosen since $\tau^2 - (\mu+1)^2$ is a quadratic polynomial in μ). Then the element w_μ generates $\mathbf{W}(\xi,\tau)$ by (ii). Let η be an endomorphism of $\mathbf{W}(\xi,\tau)$. Then η is uniquely determined by $\eta(w_\mu) \in \mathbf{W}(\xi,\tau)_\mu$, the latter being a vector space of dimension two. On the other hand, we do have two linearly independent endomorphisms of $\mathbf{W}(\xi,\tau)$; namely the identity and the endomorphism $\varphi\psi$ (from the proof of (i)). The endomorphism $\varphi\psi$ satisfies $(\varphi\psi)^2 = 0$. Hence the endomorphism algebra of $\mathbf{W}(\xi,\tau)$ is the local algebra $\mathbb{C}[\varphi\psi]/((\varphi\psi)^2)$. This implies that $\mathbf{W}(\xi,\tau)$ is indecomposable and proves (iii). The statement (iv) follows by a direct computation. $\qquad\square$

Exercise 3.78. Let $\xi \in \mathbb{C}/2\mathbb{Z}$ and $\tau \in \mathbb{C}$. Assume that V is a $\mathfrak{g}$-module such that

(a) $\operatorname{Supp} V = \xi$;
(b) $\dim V_\lambda = 2$ for some $\lambda \in \xi$;
(c) F acts bijectively on V;
(d) $(c - \tau)V \neq 0$, while $(c - \tau)^2 V = 0$.

Show that $V \cong \mathbf{W}(\xi,\tau)$.

Exercise 3.79. Let $\xi \in \mathbb{C}/2\mathbb{Z}$ and $\tau \in \mathbb{C}$. Show that

$$\operatorname{Hom}_\mathfrak{g}(\mathbf{W}(\xi,\tau), \mathbf{V}(\xi,\tau)) = \mathbb{C}.$$

Proposition 3.80. *Let $\xi \in \mathbb{C}/2\mathbb{Z}$.*

(i) For $\tau \in \mathbb{C}$ we have

$$\mathbf{V}^{(2)} \otimes \mathbf{V}(\xi,\tau) \cong \begin{cases} \mathbf{V}(\xi+1, (\sqrt{\tau}+1)^2) \oplus \mathbf{V}(\xi+1, (\sqrt{\tau}-1)^2), & \tau \neq 0; \\ \mathbf{W}(\xi+1, 1^2), & \tau = 0. \end{cases}$$

(ii) For $n \in \mathbb{N}$ we also have

$$\mathbf{V}^{(2)} \otimes \mathbf{W}(\xi,n^2) \cong \begin{cases} \mathbf{V}(\xi+1,0) \oplus \mathbf{V}(\xi+1,0) \oplus \mathbf{W}(\xi+1,2^2), & n = 1; \\ \mathbf{W}(\xi+1,(n-1)^2) \oplus \mathbf{W}(\xi+1,(n+1)^2), & n \neq 1. \end{cases}$$

Proof. Assume that the module $\mathbf{V}^{(2)}$ is given by (1.9) and the module $\mathbf{V}(\xi, \tau)$ is given by (3.8). Then from Exercise 3.10 we have

$$\text{supp}\, \mathbf{V}^{(2)} \otimes \mathbf{V}(\xi, \tau) = \xi + 1.$$

Moreover, for every $\mu \in \xi$ we get $\dim(\mathbf{V}^{(2)} \otimes \mathbf{V}(\xi, \tau))_{\mu+1} = 2$ by Exercise 3.9. Fix some $\mu \in \xi$. Then the vectors $v_0 \otimes v_\mu$ and $v_1 \otimes v_{\mu+2}$ form a basis of $(\mathbf{V}^{(2)} \otimes \mathbf{V}(\xi, \tau))_{\mu+1}$. Using (1.17), we have

$$
\begin{aligned}
F(v_0 \otimes v_\mu) &= F(v_0) \otimes v_\mu + v_0 \otimes F(v_\mu) &&= v_1 \otimes v_\mu + v_0 \otimes v_{\mu-2}, \\
F(v_1 \otimes v_{\mu+2}) &= F(v_1) \otimes v_{\mu+2} + v_1 \otimes F(v_{\mu+2}) &&= v_1 \otimes v_\mu.
\end{aligned}
$$

Hence the vectors $F(v_0 \otimes v_\mu)$ and $F(v_1 \otimes v_{\mu+2})$ are linearly independent and thus the action of F on $\mathbf{V}^{(2)} \otimes \mathbf{V}(\xi, \tau)$ is bijective.

A direct calculation shows that in the basis $v_0 \otimes v_\mu$, $v_1 \otimes v_{\mu+2}$ of the space $(\mathbf{V}^{(2)} \otimes \mathbf{V}(\xi, \tau))_{\mu+1}$ the action of the Casimir element c is given by the matrix

$$
\begin{pmatrix}
\tau - (\mu+1)^2 + (\mu+2)^2 & 4 \\
\tau - (\mu+1)^2 & 4 + \tau - (\mu+3)^2 + (\mu+2)^2
\end{pmatrix}. \tag{3.33}
$$

The characteristic polynomial of this matrix equals $t^2 - 2(\tau+1)t + (\tau-1)^2$ and hence the eigenvalues are $(\sqrt{\tau}+1)^2$ and $(\sqrt{\tau}-1)^2$ (this does not depend on the choice of μ).

If $\tau \neq 0$, then $(\sqrt{\tau}+1)^2 \neq (\sqrt{\tau}-1)^2$ and hence the action of c on $\mathbf{V}^{(2)} \otimes \mathbf{V}(\xi, \tau)$ is diagonalizable. Considering the corresponding eigenspaces for c and using Exercise 3.28 we get

$$\mathbf{V}^{(2)} \otimes \mathbf{V}(\xi, \tau) \cong \mathbf{V}(\xi+1, (\sqrt{\tau}+1)^2) \oplus \mathbf{V}(\xi+1, (\sqrt{\tau}-1)^2).$$

If $\tau = 0$, then the linear operator c has only one eigenvalue, namely $(0-1)^2 = (0+1)^2 = 1$. However, the action of c on $(\mathbf{V}^{(2)} \otimes \mathbf{V}(\xi, \tau))_{\mu+1}$ is not scalar (since the matrix (3.33) is not scalar). This implies that $c - 1$ does not annihilate $(\mathbf{V}^{(2)} \otimes \mathbf{V}(\xi, \tau))_{\mu+1}$, while $(c-1)^2$ does. Thus from Exercise 3.78 we obtain

$$\mathbf{V}^{(2)} \otimes \mathbf{V}(\xi, 0) \cong \mathbf{W}(\xi+1, 1).$$

This proves the statement (i).

To prove the statement (ii) we consider the short exact sequence

$$0 \to \mathbf{V}(\xi, n^2) \to \mathbf{W}(\xi, n^2) \to \mathbf{V}(\xi, n^2) \to 0, \tag{3.34}$$

given by Proposition 3.77(i). Since $n \neq 0$, by (i) we have

$$\mathbf{V}^{(2)} \otimes \mathbf{V}(\xi, n^2) \cong \mathbf{V}(\xi+1, (n-1)^2) \oplus \mathbf{V}(\xi+1, (n+1)^2),$$

where the decomposition is given by different eigenvalues (namely $(n-1)^2$ and $(n+1)^2$) of the Casimir element c. Since the functor $\mathbf{V}^{(2)} \otimes \_$ is exact, applying this to (3.34) says that $\mathbf{V}^{(2)} \otimes \mathbf{W}(\xi, n^2) \cong M \oplus N$, where for the module M there is a short exact sequence

$$0 \to \mathbf{V}(\xi+1, (n-1)^2) \to M \to \mathbf{V}(\xi+1, (n-1)^2) \to 0 \qquad (3.35)$$

and for the module N there is a short exact sequence

$$0 \to \mathbf{V}(\xi+1, (n+1)^2) \to N \to \mathbf{V}(\xi+1, (n+1)^2) \to 0. \qquad (3.36)$$

As with the proof of (i), we obtain that the action of F on both M and N is bijective.

Consider first the module N and the action of the Casimir element c on some weight space of N, which generates N (that is on which the action of every E^i, $i \in \mathbb{N}$, is bijective). Since F acts on N bijectively, the action of the element c on all other weight spaces is similar and can be obtained using some F^i, $i \in \mathbb{Z}$. If c would act as a scalar, then every weight element in N would be an eigenvector for both c and h. Hence, by Theorem 2.33 and Exercise 3.28, every such element would generate a submodule of N, isomorphic to the module $\mathbf{V}(\xi+1, (n+1)^2)$. In particular, we would have the isomorphism $N \cong \mathbf{V}(\xi+1, (n+1)^2) \oplus \mathbf{V}(\xi+1, (n+1)^2)$, which would imply the following:

$$\dim \mathrm{Hom}_{\mathfrak{g}}(\mathbf{V}^{(2)} \otimes \mathbf{W}(\xi, n^2), \mathbf{V}(\xi+1, (n+1)^2))$$
$$= \dim \mathrm{Hom}_{\mathfrak{g}}(N, \mathbf{V}(\xi+1, (n+1)^2))$$
$$= 2.$$

However, using the self-adjointness of $\mathbf{V}^{(2)} \otimes \_$ we have

$$\dim \mathrm{Hom}_{\mathfrak{g}}(\mathbf{V}^{(2)} \otimes \mathbf{W}(\xi, n^2), \mathbf{V}(\xi+1, (n+1)^2))$$
$$= \dim \mathrm{Hom}_{\mathfrak{g}}(\mathbf{W}(\xi, n^2), \mathbf{V}^{(2)} \otimes \mathbf{V}(\xi+1, (n+1)^2))$$
$$\overset{(i)}{=} \dim \mathrm{Hom}_{\mathfrak{g}}(\mathbf{W}(\xi, n^2), \mathbf{V}(\xi, n^2) \oplus \mathbf{V}(\xi, (n+2)^2))$$
$$= 1,$$

where the last equality follows from Exercise 3.79. This is a contradiction. Hence c does not act on N as a scalar. Therefore we can apply Exercise 3.78 to conclude $N \cong \mathbf{W}(\xi+1, (n+1)^2)$.

Exactly the same arguments show that $M \cong \mathbf{W}(\xi+1, (n-1)^2))$ in the case $n \neq 1$. On the other hand, for $n = 1$ a similar computation using the self-adjointness of $\mathbf{V}^{(2)} \otimes {}_-$ gives

$$\dim \mathrm{Hom}_{\mathfrak{g}}(\mathbf{V}^{(2)} \otimes \mathbf{W}(\xi, 1), \mathbf{V}(\xi + 1, 0))$$
$$= \dim \mathrm{Hom}_{\mathfrak{g}}(\mathbf{W}(\xi, 1), \mathbf{V}^{(2)} \otimes \mathbf{V}(\xi + 1, 0))$$
$$\overset{(\mathrm{i})}{=} \dim \mathrm{Hom}_{\mathfrak{g}}(\mathbf{W}(\xi, 1), \mathbf{W}(\xi, 1))$$
$$= 2,$$

where the last equality is given by Proposition 3.77(iii). Hence, due to Exercise 3.79, the situation where the Casimir element c acts on M by some nontrivial Jordan cell is not possible. This means that c acts on M as the scalar 0. The arguments of the previous paragraph imply

$$M \cong \mathbf{V}(\xi + 1, 0) \oplus \mathbf{V}(\xi + 1, 0).$$

This completes the proof. $\square$

Theorem 3.81. *Let* $n \in \mathbb{N}$, $\xi \in \mathbb{C}/2\mathbb{Z}$ *and* $\tau \in \mathbb{C}$.

(i) *If* $\tau \neq k^2$ *for all* $k = 0, 1, \ldots, n - 2$, *then*

$$\mathbf{V}^{(n)} \otimes \mathbf{V}(\xi, \tau) \cong \bigoplus_{i=0}^{n-1} \mathbf{V}(\xi + n - 1, (\sqrt{\tau} + n - 1 - 2i)^2).$$

(ii) *If* $\tau = k^2$ *for some* $k \in \{0, 1, \ldots, n - 2\}$ *and* $n - k$ *is even, then*

$$\mathbf{V}^{(n)} \otimes \mathbf{V}(\xi, \tau) \cong \bigoplus_{i=0}^{k-1} \mathbf{V}(\xi + n - 1, (n - k + 1 + 2i)^2)$$

$$\oplus \bigoplus_{i=0}^{\frac{n-k-2}{2}} \mathbf{W}(\xi + n - 1, (1 + 2i)^2).$$

(iii) *If* $\tau = k^2$ *for some* $k \in \{0, 1, \ldots, n - 2\}$ *and* $n - k$ *is odd, then*

$$\mathbf{V}^{(n)} \otimes \mathbf{V}(\xi, \tau) \cong \bigoplus_{i=0}^{k-1} \mathbf{V}(\xi + n - 1, (n - k + 1 + 2i)^2)$$

$$\oplus \mathbf{V}(\xi + n - 1, 0) \oplus \bigoplus_{i=0}^{\frac{n-k-3}{2}} \mathbf{W}(\xi + n - 1, (2 + 2i)^2).$$

Proof. We proceed by induction on n. For $n = 1$ we have the obvious isomorphism of $\mathfrak{g}$-modules as follows: $\mathbf{V}^{(1)} \otimes \mathbf{V}(\xi, \tau) \cong \mathbf{V}(\xi, \tau)$, since $\mathbf{V}^{(1)}$ is the trivial $\mathfrak{g}$-module (see Exercise 3.75). This agrees with (i) (note that for $n = 1$ we always have the case (i)). For $n = 2$ the statement of the theorem reduces to Proposition 3.80(i).

For $n > 2$, using Theorem 1.39, we write

$$\mathbf{V}^{(2)} \otimes \mathbf{V}^{(n-1)} \cong \mathbf{V}^{(n)} \oplus \mathbf{V}^{(n-2)}.$$

In the case (i), making use of the inductive assumption, Proposition 3.80(i) and Exercise 3.74 we find that the module $\mathbf{V}^{(2)} \otimes \mathbf{V}^{(n-1)} \otimes \mathbf{V}(\xi, \tau)$ is isomorphic to

$$\bigoplus_{i=0}^{n-1} \mathbf{V}(\xi + n - 1, (\sqrt{\tau} + n - 1 - 2i)^2) \oplus \bigoplus_{i=0}^{n-3} \mathbf{V}(\xi + n - 1, (\sqrt{\tau} + n - 3 - 2i)^2),$$
(3.37)

while the module $(\mathbf{V}^{(n)} \oplus \mathbf{V}^{(n-2)}) \otimes \mathbf{V}(\xi, \tau)$ is isomorphic to

$$X \oplus \bigoplus_{i=0}^{n-3} \mathbf{V}(\xi + n - 3, (\sqrt{\tau} + n - 3 - 2i)^2).$$
(3.38)

Note that $\xi + n - 3 = \xi + n - 1$. Since $\overline{\mathfrak{W}}$ is a Krull–Schmidt category (Proposition 3.55(vii)), the claim (i) follows by comparing (3.37) with (3.38).

The proof of the claims (ii) and (iii) are similar; we leave the details to the reader. □

3.11 Duality

Let $V \in \overline{\mathfrak{W}}$ and

$$V = \bigoplus_{\lambda \in \mathbb{C}} V_\lambda$$

be the corresponding decomposition into a direct sum of finite-dimensional weight spaces. Consider the vector space

$$V^\circledast = \bigoplus_{\lambda \in \mathbb{C}} V_\lambda^*.$$

For $g_\lambda \in V_\lambda^* \subset V^\circledast$, $v \in V$ and $x \in \mathfrak{g}$ set

$$(x \cdot g_\lambda)(v) = g_\lambda(x^\star(v)).$$
(3.39)

See Section 1.5 for the definition of the anti-involution $\star$.

Proposition 3.82.

(i) The formula (3.39) defines on $V^\circledast$ the structure of a $\mathfrak{g}$-module.
(ii) $V^\circledast \in \overline{\mathfrak{W}}$, moreover $\operatorname{supp} V^\circledast = \operatorname{supp} V$.
(iii) $(V^\circledast)^\circledast \cong V$ canonically.

Proof. For $v \in V_\mu$ and $x \in \mathfrak{g}$ we have $x(v) \in V_{\mu-2} \oplus V_\mu \oplus V_{\mu+2}$ by Lemma 3.5. Hence $(x \cdot g_\lambda)(w) = 0$ for all $w \in V_\mu$ such that we have $\mu \notin \{\lambda - 2, \lambda, \lambda + 2\}$, which implies that $x \cdot g_\lambda \in V^\circledast$. Thus the formula (3.39) does define on $V^\circledast$ the structure of a $\mathfrak{g}$-module. This proves (i).

Since $\mathbf{h}^\star = \mathbf{h}$ by the definition of $\star$, for $v \in V_\mu$ and $g_\lambda \in V_\lambda^*$ we have

$$((\mathbf{h} - \lambda)g_\lambda)(v) = g_\lambda((\mathbf{h} - \lambda)(v)).$$

If $\mu = \lambda$, then $(\mathbf{h} - \lambda)(v) = 0$. If $\mu \neq \lambda$, then $(\mathbf{h} - \lambda)(v) \in V_\mu$ and thus $g_\lambda((\mathbf{h} - \lambda)(v)) = 0$ again. Hence g_λ is a weight vector in $V^\circledast$ of weight λ. Since all weight spaces of V are finite-dimensional, all weight spaces of $V^\circledast$ are finite-dimensional as well. The claim (ii) follows.

The claim (iii) is now obtained using the canonical isomorphism

$$(V_\lambda^*)^* \cong V_\lambda$$

for the finite-dimensional vector space V_λ. This completes the proof. $\qquad\square$

Exercise 3.83. Let $V, W \in \overline{\mathfrak{W}}$ and $\varphi \in \mathrm{Hom}_\mathfrak{g}(V, W)$. Then $\varphi = \bigoplus_{\lambda \in \mathbb{C}} \varphi_\lambda$, where $\varphi_\lambda : V_\lambda \to W_\lambda$. For every $\lambda \in \mathbb{C}$ denote the dual map of φ_λ by $\varphi_\lambda^* : W_\lambda^* \to V_\lambda^*$. Show that the map

$$\varphi^\circledast = \bigoplus_{\lambda \in \mathbb{C}} \varphi_\lambda^* : W^\circledast \to V^\circledast$$

is a homomorphism of $\mathfrak{g}$-modules. Derive from this that $\circledast$ defines a contravariant endofunctor of $\overline{\mathfrak{W}}$.

The functor $\circledast$ is called the *duality* on $\overline{\mathfrak{W}}$. Important properties of $\circledast$ are collected in the following statement:

Theorem 3.84.

(i) *The functor $\circledast$ is an exact contravariant and involutive equivalence on the category $\overline{\mathfrak{W}}$.*

(ii) *The functor $\circledast$ preserves the isomorphism classes of simple objects in the category $\overline{\mathfrak{W}}$.*

(iii) *The functor $\circledast$ preserves the category $\overline{\mathfrak{W}}^{\xi,\tau}$ for any $\xi \in \mathbb{C}/2\mathbb{Z}$ and $\tau \in \mathbb{C}$.*

Proof. That $\circledast$ is exact and contravariant follows directly from the definition. The canonical isomorphism $(V^\circledast)^\circledast \cong V$, given by Proposition 3.82(iii), is natural. From this we obtain an isomorphism of functors as follows: $\circledast \circ \circledast \cong \mathrm{ID}_{\overline{\mathfrak{W}}}$. This implies the statement (i).

Let $V \in \overline{\mathfrak{W}}$ be simple. From Proposition 3.82(ii) we have $\operatorname{supp} V^{\circledast} = \operatorname{supp} V$. A direct computation also shows that for the Casimir element c we have $c^* = c$. This implies that the eigenvalues of c on V and $V^{\circledast}$ coincide. From Exercise 3.36 we have that every simple weight $\mathfrak{g}$-module is uniquely determined by its support and the eigenvalue of c. The claim (ii) follows. The claim (iii) follows immediately from the claims (i) and (ii). This completes the proof. $\qquad\square$

Exercise 3.85. Let $\xi \in \mathbb{C}/2\mathbb{Z}$ and $\tau \in \mathbb{C}$. Show that the module $\mathbf{V}(\xi, \tau)$ is *self-dual* (that is satisfies $\mathbf{V}(\xi, \tau)^{\circledast} \cong \mathbf{V}(\xi, \tau)$) if, and only if, $\mathbf{V}(\xi, \tau)$ is simple.

3.12 Addenda and comments

3.12.1

Literature on the material presented in this section is harder to find than for that in the previous sections. Basic generalities on weight modules (over all simple finite-dimensional Lie algebras) are included in most of the textbooks (see for example [37, 46, 49, 57]). For the general theory of Verma modules (again, over all simple finite-dimensional Lie algebras) the reader could consult [37, 63, 64, 96]. Classification of weight $\mathfrak{sl}_2$-modules can be found in Subsection 7.8.16 of [37] or in [82].

3.12.2

Classification of all simple weight modules with finite-dimensional weight spaces over all simple finite-dimensional complex Lie algebras was recently completed by O. Mathieu (see [80]). Such modules can be of two types: dense modules and simple quotients of modules, parabolically induced from dense modules. The most difficult aspect is the classification of simple dense modules, which is done in [80]. For some time it had been known that such modules exist only for Lie algebras of type A and C (see [47]). Using the twisting functor, which appears in [80], Mathieu reduces the classification of simple dense modules to that of coherent families. The latter classification is then reduced to the classification of simple highest weight modules, whose dimensions of weight spaces are uniformly bounded. All the details can be found in [80]. For the special case of the algebra $\mathfrak{sl}_2$, the classification of simple weight modules is usually attributed to the unpublished lecture notes by P. Gabriel [50], although Dixmier's book [37] also mentions [95]. On the other hand, the classification result for $\mathfrak{sl}_2$ can also be viewed as a special case of the classification of simple weight modules that have one-dimensional weight spaces, obtained in [25].

Classification of all simple weight modules over a simple finite-dimensional complex Lie algebra $\mathfrak{a}$ is still an open problem for every $\mathfrak{a} \neq \mathfrak{sl}_2$.

3.12.3

Description of the categories $\overline{\mathfrak{W}}^{\xi,\tau}$ and indecomposable objects of these categories is also usually attributed to [50]. Here we mostly follow Y. Drozd's paper [39], where several more general results are obtained. In particular, Drozd describes not only the category of all weight $\mathfrak{sl}_2$-modules but also some other categories, including the category of all modules on which the action of the algebra $\mathbb{C}[h,c]$ is locally finite (that is, the space $\mathbb{C}[h,c](v)$ is finite-dimensional for all elements v), and does all this not only over $\mathbb{C}$ but also over fields of positive characteristics.

For the case of complex numbers, Drozd's description of these other categories looks rather similar to the one presented in Sections 3.7–3.9 for the category $\overline{\mathfrak{W}}^{\xi,\tau}$. Let V be a $\mathfrak{g}$-module. For $\lambda \in \mathbb{C}$ denote

$$V(\lambda) = \{v \in V \,:\, (H - \lambda)^k(v) = 0 \text{ for some } k \in \mathbb{N}\}.$$

Note that $V_\lambda \subset V(\lambda)$ and $V(\lambda) \neq 0$ implies $V_\lambda \neq 0$. The module V is called a *generalized weight* module provided that

$$V = \bigoplus_{\lambda \in \mathbb{C}} V(\lambda).$$

The subspace $V(\lambda)$ is called a *generalized weight* space. Denote by $\mathfrak{G}\mathfrak{W}$ the full subcategory of $\mathfrak{g}$-mod, consisting of all finitely generated generalized weight modules and by $\overline{\mathfrak{G}\mathfrak{W}}$ the full subcategory of $\mathfrak{G}\mathfrak{W}$, consisting of all modules with finite-dimensional generalized weight spaces. For $\xi \in \mathbb{C}/2\mathbb{Z}$ and $\tau \in \mathbb{C}$, we also define the full subcategories $\mathfrak{G}\mathfrak{W}^{\xi,\tau}$ and $\overline{\mathfrak{G}\mathfrak{W}}^{\xi,\tau}$ using the restrictions that the support of the module should belong to ξ, and that τ should be the only eigenvalue of the Casimir element c. Then one immediately gets the decompositions

$$\mathfrak{G}\mathfrak{W} = \bigoplus_{\xi,\tau} \mathfrak{G}\mathfrak{W}^{\xi,\tau}, \qquad \overline{\mathfrak{G}\mathfrak{W}} = \bigoplus_{\xi,\tau} \overline{\mathfrak{G}\mathfrak{W}}^{\xi,\tau}. \qquad (3.40)$$

One can show (see Exercise 3.90) that simple objects in the categories $\overline{\mathfrak{G}\mathfrak{W}}^{\xi,\tau}$ and $\overline{\mathfrak{W}}^{\xi,\tau}$ coincide. When $\overline{\mathfrak{G}\mathfrak{W}}^{\xi,\tau}$ has only one simple object, it is equivalent to the category of modules of the $\mathbb{C}[[x,y]]$-category with one object (and no nontrivial morphisms). When the category $\overline{\mathfrak{G}\mathfrak{W}}^{\xi,\tau}$ has two simple objects, it is equivalent to the category of modules over the category $\mathfrak{A}$ from Section 3.8, which is considered as a $\mathbb{C}[[x,y]]$-category (and not as a $\mathbb{C}[[x]]$-category). Finally, when $\overline{\mathfrak{G}\mathfrak{W}}^{\xi,\tau}$ has three simple objects, it is equivalent to the category of modules over the $\mathbb{C}[[x,y]]$-category, defined on the underlying quiver of the category $\mathfrak{B}$ from Section 3.9, via relations

$$\mathsf{ba} = x\mathbf{1}_\mathsf{p}, \quad \mathsf{ab} = x\mathbf{1}_\mathsf{q}, \quad \mathsf{dc} = (x+y)\mathbf{1}_\mathsf{q}, \quad \mathsf{cd} = (x+y)\mathbf{1}_\mathsf{r}.$$

The basic idea of the proof is to consider the weight picture (3.8) of a module as a category, whose morphisms are supposed to satisfy the defining relations of $\mathfrak{g}$. This category will have a lot of isomorphic objects. Taking one representative from each isomorphism class, one is left either with one object, or with a category similar to the category $\mathfrak{A}$ or the category $\mathfrak{B}$. As an application, Drozd also describes the full subcategories of $\mathfrak{G}\mathfrak{W}$, which consist of modules on which the Casimir element c acts diagonalizably, and on which both h and c act diagonalizably (see Exercises 3.107 and 3.108).

3.12.4

The description of $\overline{\mathfrak{W}}^{\xi,\tau}$, presented in Sections 3.7–3.9, enables a complete classification of indecomposable objects in $\overline{\mathfrak{W}}^{\xi,\tau}$. This classification reduces to the classification problem, solved in [97]. The categories $\overline{\mathfrak{G}\mathfrak{W}}^{\xi,\tau}$ turn out to be much more complicated. Even in the case of one simple object the category $\overline{\mathfrak{G}\mathfrak{W}}^{\xi,\tau}$ is *wild*, which, roughly speaking, means that the description of indecomposable objects in this category is as complicated as the description of all indecomposable modules over all finitely generated algebras. For the moment no complete description (classification) of all indecomposable modules for wild categories is known, and the problem to give such a description is considered very difficult. Indecomposable (generalized) weight modules can be described in a more general setup of generalized Weyl algebras (see [11, 41]).

3.12.5

One observes that the statement of Theorem 3.81 can be formulated using the support of the module $\mathbf{V}^{(n)}$. For example, the formula from Theorem 3.81(i) could be written as follows:

$$\mathbf{V}^{(n)} \otimes \mathbf{V}(\xi, \tau) \cong \bigoplus_{\lambda \in \mathrm{supp}\, \mathbf{V}^{(n)}} \mathbf{V}(\xi + \lambda, (\sqrt{\tau} + \lambda)^2).$$

Similarly, the formulae from Theorem 3.81(ii) and (iii) could be uniformly written as follows:

$$\mathbf{V}^{(n)} \otimes \mathbf{V}(\xi, \tau) \cong \bigoplus_{\lambda \in X} \mathbf{V}(\xi + \lambda, (k + \lambda)^2) \oplus \bigoplus_{\lambda \in Y} \mathbf{W}(\xi + \lambda, (k + \lambda)^2),$$

where

$$X = \{\lambda \in \mathrm{supp}\, \mathbf{V}^{(n)} \,:\, k + \lambda = 0 \text{ or } 2k + \lambda \notin \mathrm{supp}\, \mathbf{V}^{(n)}\}$$

and

$$Y = \{\lambda \in \mathrm{supp}\, \mathbf{V}^{(n)} \,:\, 2k + \lambda \in \mathrm{supp}\, \mathbf{V}^{(n)} \text{ and } |\lambda| < |2k + \lambda|\}.$$

This is closely related to the following result from [77]:

Theorem 3.86 (Kostant's Theorem). *Let V be a* $\mathfrak{g}$*-module, on which the Casimir element c acts as the scalar τ. Then for any $n \in \mathbb{N}$ the element*

$$\prod_{\lambda \in \mathrm{supp}\, \mathbf{V}^{(n)}} (c - (\sqrt{\tau} + \lambda)^2)$$

annihilates the module $\mathbf{V}^{(n)} \otimes V$.

For modules $\mathbf{V}(\xi, \tau)$ the statement of Theorem 3.86 follows immediately from Theorem 3.81 and Proposition 3.77(iv). The really interesting thing is that the statement is true even in the general case. The idea of the proof is that, regardless of V, the action of c on $\mathbf{V}^{(n)} \otimes V$ is given by some polynomial formula. If one could find "sufficiently many" modules V, for which the statement of Theorem 3.86 would be true, a density argument would prove the statement in the general case. Then one can show that already the set of all modules of the form $\mathbf{V}(\xi, \tau)$ gives us "sufficiently many" modules to check the statement of Theorem 3.86. Alternatively one could use all Verma modules.

An appropriate generalization of Theorem 3.86 is true for any finite-dimensional semi-simple complex Lie algebra, giving very important information about the way the central character of a module behaves under tensor product with a finite-dimensional module. Since $\mathrm{supp}\, \mathbf{V}^{(n)}$ is invariant with respect to the map $\lambda \mapsto -\lambda$, the statement of the theorem does not depend on the choice of $\sqrt{\tau}$. The latter applies also to Theorem 3.81.

3.12.6

In the formula (3.39) one could use the principal anti-involution ω instead of $\star$. In this case, one obtains another duality on $\overline{\mathfrak{W}}$. Let us denote it by $\odot$. Similarly to the proof of Proposition 3.82 and Theorem 3.84 one shows that $\odot$ is an exact contravariant and involutive equivalence on $\overline{\mathfrak{W}}$. However, since $\omega(\mathbf{h}) = -\mathbf{h}$, the duality $\odot$ satisfies

$$\operatorname{supp} V^{\odot} = -\operatorname{supp} V$$

and hence $\odot$ does not preserve the isomorphism classes of simple objects in $\overline{\mathfrak{W}}$ in general.

3.12.7

Consider the set $\mathbb{N}_0^\infty = \mathbb{N}_0 \cup \{\infty\}$. For $i, j \in \mathbb{N}_0^\infty$ set

$$i + j = \begin{cases} i + j, & i, j \in \mathbb{N}_0; \\ \infty, & \text{otherwise;} \end{cases} \qquad i \cdot j = \begin{cases} ij, & i, j \in \mathbb{N}_0; \\ \infty, & i, j \neq 0 \text{ and } \{i, j\} \not\subset \mathbb{N}_0; \\ 0, & \text{otherwise.} \end{cases}$$

This equips $\mathbb{N}_0^\infty$ with the structure of a semiring. Consider now the set Char of all functions from $\mathbb{C}$ to $\mathbb{N}_0^\infty$. The elements of Char are called *characters*. The set Char has the natural structure of a commutative monoid with respect to the usual pointwise addition of functions.

For $\lambda \in \mathbb{C}$ we denote by $\exp(\lambda)$ the character defined as follows:

$$\exp(\lambda)(\mu) = \begin{cases} 1, & \lambda = \mu; \\ 0, & \text{otherwise.} \end{cases}$$

Using this notation every character $\eta \in$ Char can be uniquely written in the following form:

$$\eta = \sum_{\lambda \in \mathbb{C}} a_\lambda \exp(\lambda),$$

where $a_\lambda \in \mathbb{N}_0^\infty$. We define on Char an associative multiplication as follows:

$$\left(\sum_{\lambda \in \mathbb{C}} a_\lambda \exp(\lambda) \right) \cdot \left(\sum_{\mu \in \mathbb{C}} b_\mu \exp(\mu) \right) = \sum_{\nu \in \mathbb{C}} \left(\sum_{\lambda + \mu = \nu} a_\lambda b_\mu \right) \exp(\nu)$$

(here if the sum $\sum_{\lambda+\mu=\nu} a_\lambda b_\mu$ contains infinitely many non-zero summands, its value is postulated to be ∞). It is straightforward to verify that this equips Char with the structure of a semiring.

Now, let V be a weight $\mathfrak{g}$-module. The *character* of V is the function $\mathrm{ch}_V \in \mathrm{Char}$ defined as follows: $\mathrm{ch}_V(\lambda) = \dim V_\lambda$, $\lambda \in \mathbb{C}$. From Exercise 3.9 for $\mathfrak{g}$-modules V and W we have

$$\mathrm{ch}_{V \oplus W} = \mathrm{ch}_V + \mathrm{ch}_W, \quad \mathrm{ch}_{V \otimes W} = \mathrm{ch}_V \cdot \mathrm{ch}_W.$$

Imposing some restrictions on the support of a module (this works, for example, for highest weight modules) it is possible also to define a subtraction of the corresponding characters and thus embed the set of characters into a ring. This can be then used to produce nice combinatorial formulae for characters of highest weight modules; particularly simple finite-dimensional modules (the so-called *Weyl's character formula*). We will not go into the corresponding technicalities and instead refer the reader to Section 7.5 of [37] for details.

3.13   Additional exercises

Exercise 3.87. Prove the decompositions (3.40).

Exercise 3.88. Prove the decompositions from Theorem 3.81(ii) and (iii).

Exercise 3.89. Prove that the left regular $U(\mathfrak{g})$-module is not a weight $\mathfrak{g}$-module.

Exercise 3.90. Prove that every simple generalized weight module is in fact a weight module.

Exercise 3.91. Construct an example of a generalized weight module, which is not a weight module.

Exercise 3.92. Show that the category $\mathfrak{W}$ is not closed with respect to extensions.

Exercise 3.93. Show that there exist weight $\mathfrak{g}$-modules of infinite length all weight spaces of which are finite-dimensional.

Exercise 3.94. Let V be a $\mathfrak{g}$-module, generated by some elements $v_1, \ldots, v_k$. Assume that $\dim \mathbb{C}[h](v_i) < \infty$ for all $i = 1, \ldots, k$. Show that V is a generalized weight module.

Exercise 3.95. Let V be a highest weight $\mathfrak{g}$-module on which the operator F acts injectively. Show that V is a Verma module.

Exercise 3.96 ([17]). Show that for every finite-dimensional $\mathfrak{g}$-module V and any $\lambda \in \mathbb{C}$ the module $V \otimes M(\lambda)$ has a filtration, whose subquotients are Verma modules.

Exercise 3.97. Show that for any $\lambda \in \mathbb{C}$ there is a unique (up to scalar) non-zero homomorphism from $M(\lambda)$ to $M(\lambda)^{\circledast}$, and that the image of this homomorphism is exactly $L(\lambda)$.

Exercise 3.98. Prove that $\mathrm{Hom}_{\mathfrak{g}}(M(\lambda), \overline{M}(\mu)) = 0$ for all $\lambda, \mu \in \mathbb{C}$.

Exercise 3.99. Consider the one-dimensional $\mathbb{C}[h, c]$-module V given by $c(v) = \tau v$ and $h(v) = \lambda v$ for some $\tau, \lambda \in \mathbb{C}$. Set $\xi = \lambda + 2\mathbb{Z}$. Define the $\mathfrak{g}$-module

$$\tilde{\mathbf{V}}(\xi, \tau) = U(\mathfrak{g}) \bigotimes_{\mathbb{C}[h,c]} V.$$

(a) Prove that $\tilde{\mathbf{V}}(\xi, \tau)$ is a weight module with support ξ.
(b) Prove that all non-zero weight spaces of $\tilde{\mathbf{V}}(\xi, \tau)$ are one-dimensional.
(c) Prove that the Casimir element c acts on $\tilde{\mathbf{V}}(\xi, \tau)$ as the scalar τ.
(d) Prove that $\tilde{\mathbf{V}}(\xi, \tau) \cong \mathbf{V}(\xi, \tau)$ if and only if all operators E^i, $i \in \mathbb{N}$, act injectively on $\mathbf{V}(\xi, \tau)_{\lambda}$.
(e) Prove that the module $\tilde{\mathbf{V}}(\xi, \tau)$ has a unique simple quotient, namely a simple weight module W, which is uniquely defined by the conditions that $W_{\lambda} \neq 0$ and that c acts on W as the scalar τ.

Exercise 3.100 (Gelfand–Zetlin model, [40]). For $x, y, z \in \mathbb{C}$ consider the set $\mathbf{T}_{x,y,z}$ consisting of all tableaux

$$\mathfrak{t}(x, y, a) = \begin{array}{|c|c|}\hline x & y \\\hline \multicolumn{2}{c}{} \end{array} \begin{array}{|c|}\hline a \\\hline\end{array},$$

where $z - a \in \mathbb{Z}$. Let $V = V_{x,y,z}$ denote the linear span of all elements from $\mathbf{T}_{x,y,z}$. Define the linear operators E, F and H on V as follows:

$$F(\mathfrak{t}(x, y, a)) = \mathfrak{t}(x, y, a - 1);$$
$$E(\mathfrak{t}(x, y, a)) = -(x - a)(y - a) \cdot \mathfrak{t}(x, y, a + 1);$$
$$H(\mathfrak{t}(x, y, a)) = (2a - x - y - 1) \cdot \mathfrak{t}(x, y, a).$$

Show that this turns V into a $\mathfrak{g}$-module, which is isomorphic to the module $\mathbf{V}(2z - x - y - 1 + 2\mathbb{Z}, (x - y)^2)$.

Exercise 3.101. Show that for every $n \in \mathbb{N}$ there exists a short exact sequence

$$0 \to M(-n-1) \to \mathbf{V}(n+1+2\mathbb{Z}, n^2) \to \overline{M}(-n+1)^\circledast \to 0.$$

Exercise 3.102. Let $V \in \overline{\mathfrak{W}}^{\xi,\tau}$ for some $\xi \in \mathbb{C}/2\mathbb{Z}$ and $\tau \in \mathbb{C}$. Assume that for any $\lambda \in \mathbb{C}$ the $\mathbb{C}[c,h]$-module V_λ is either zero or simple. Show that this assumption is not enough to guarantee that V is simple.

Exercise 3.103. Let $\xi \in \mathbb{C}/2\mathbb{Z}$ and $\tau \in \mathbb{C}$. Show that $B_0(\mathbf{V}(\xi,\tau)) \cong B_0(\mathbf{V}(\xi,\tau)^\circledast)$. At the same time, show that the natural transformation $\bar{\iota}$ does not have to be injective on $\mathbf{V}(\xi,\tau)^\circledast$, while it it always injective on $\mathbf{V}(\xi,\tau)$.

Exercise 3.104. Show that in Corollary 3.53 the module $M(\lambda)$ can be replaced by $\mathbf{V}(\xi,(\lambda+1)^2)$ for an arbitrary $\xi \in \mathbb{C}/2\mathbb{Z}$.

Exercise 3.105. Let $\xi \in \mathbb{C}/2\mathbb{Z}$. Let V be a weight $\mathfrak{g}$-module with finite-dimensional weight spaces, such that $\operatorname{supp} V \subset \xi$. Show that V is of finite length if and only if there exists $n \in \mathbb{N}$ such that $\dim V_\lambda < n$ for every $\lambda \in \xi$.

Exercise 3.106. Let $\xi \in \mathbb{C}/2\mathbb{Z}$ and $\tau \in \mathbb{C}$. Denote by $\mathfrak{X}$ the full subcategory of $\overline{\mathfrak{W}}^{\xi,\tau}$, which consists of all module V on which the operator F acts bijectively.

(a) Show that $\mathfrak{X}$ is an abelian category.
(b) Show that $\mathbf{V}(\xi,\tau)$ is the unique simple object of $\mathfrak{X}$.
(c) Show that every $V \in \mathfrak{X}$ has a filtration, whose quotients are isomorphic to $\mathbf{V}(\xi,\tau)$.
(d) Show that $\mathfrak{X}$ is equivalent to $\mathbb{C}[[x]]$-mod.

Exercise 3.107 ([39]). For $\xi \in \mathbb{C}/2\mathbb{Z}$ and $\tau \in \mathbb{C}$ denote by $\overline{\mathfrak{C}}^{\xi,\tau}$ the category of all generalized weight modules V satisfying the following conditions:

- $\dim V(\lambda) < \infty$ for all $\lambda \in \mathbb{C}$;
- $\operatorname{supp} V \subset \xi$;
- c acts on V as the scalar τ.

Show that

(a) If $\overline{\mathfrak{C}}^{\xi,\tau}$ has one simple object, then $\overline{\mathfrak{C}}^{\xi,\tau} \cong \mathbb{C}[[x]]$-mod.

(b) If $\overline{\mathfrak{C}}^{\xi,\tau}$ has two simple objects and $\tau \neq 0$, then $\overline{\mathfrak{C}}^{\xi,\tau} \cong \mathfrak{A}$.

(c) If $\overline{\mathfrak{C}}^{\xi,\tau}$ has three simple objects, then $\overline{\mathfrak{C}}^{\xi,\tau} \cong \mathfrak{B}$.

(d) If $\overline{\mathfrak{C}}^{\xi,\tau}$ has two simple objects and $\tau = 0$, then $\overline{\mathfrak{C}}^{\xi,\tau}$ is equivalent to the $\mathbb{C}[[x]]$-category, defined on the underlying quiver of the category $\mathfrak{A}$ from Section 3.8 via relations

$$\mathsf{ba} = x^2 1_{\mathsf{p}}, \quad \mathsf{ab} = x^2 1_{\mathsf{q}}.$$

Exercise 3.108 ([39]). Let $\xi \in \mathbb{C}/2\mathbb{Z}$ and $\tau \in \mathbb{C}$. Show that

(a) If $\overline{\mathfrak{W}}^{\xi,\tau}$ has one simple object, then $\overline{\mathfrak{W}}^{\xi,\tau} \cap \overline{\mathfrak{C}}^{\xi,\tau} \cong \mathbb{C}$-mod.

(b) If $\overline{\mathfrak{W}}^{\xi,\tau}$ has two simple objects, then the category $\overline{\mathfrak{W}}^{\xi,\tau} \cap \overline{\mathfrak{C}}^{\xi,\tau}$ is equivalent to the $\mathbb{C}$-category, defined on the underlying quiver of the category $\mathfrak{A}$ from Section 3.8 via relations

$$\mathsf{ba} = \mathsf{ab} = 0.$$

(c) If $\overline{\mathfrak{W}}^{\xi,\tau}$ has three simple objects, then the category $\overline{\mathfrak{W}}^{\xi,\tau} \cap \overline{\mathfrak{C}}^{\xi,\tau}$ is equivalent to the $\mathbb{C}$-category, defined on the underlying quiver of the category $\mathfrak{B}$ from Section 3.9 via relations

$$\mathsf{ba} = \mathsf{ab} = \mathsf{cd} = \mathsf{dc} = 0.$$

Exercise 3.109. Show that all categories appearing in Exercise 3.108 have only finitely many indecomposable objects and determine all such objects.

Exercise 3.110. Let $n \in \mathbb{N}$ and $\lambda \in \mathbb{C} \setminus X_n$, where

$$X_n = \begin{cases} \varnothing, & n = 1; \\ \{-n+1, -n+2, \ldots, n-3\}, & n \neq 1. \end{cases}$$

Show that

$$\mathbf{V}^{(n)} \otimes M(\lambda) \cong \bigoplus_{\mu \in \operatorname{supp} \mathbf{V}^{(n)}} M(\lambda + \mu).$$

Exercise 3.111. Let $N(-2) = \mathbf{V}^{(2)} \otimes M(-1)$.

(a) Show that the module $N(-2)$ is indecomposable and uniserial.

(b) Show that there exists a short exact sequence

$$0 \to M(0) \to N(-2) \to M(-2) \to 0.$$

(c) Show that c does not act on $N(-2)$ as a scalar.

(d) Show that the multiplication with $c - 1$ defines a non-zero nilpotent endomorphism ψ of $N(-2)$ and that every endomorphism of $N(-2)$ is a linear combination of ψ and the identity map.

(e) Show that $N(-2)^{\circledast} \cong N(-2)$.

Exercise 3.112. Show that for every $n \in \mathbb{N}$ there exists a unique weight module $N(-n-1)$ such that there exists a non-split short exact sequence

$$0 \to M(n-1) \to N(-n-1) \to M(-n-1) \to 0.$$

Show that the module $N(-n-1)$ is indecomposable, uniserial and satisfies $N(-n-1)^{\circledast} \cong N(-n-1)$.

Exercise 3.113. Show that for every $n \in \mathbb{N}$ and $m \in \mathbb{Z}$ the $\mathfrak{g}$-module $\mathbf{V}^{(n)} \otimes M(m)$ is a direct sum of Verma modules and modules of the form $N(-k-1)$ for some $k \in \mathbb{N}$ and determine this decomposition.

Exercise 3.114. Show that for every $n \in \mathbb{N}$ there is a filtration

$$0 = M_0 \subset M_1 \subset M_2 \subset \cdots \subset M_n = \mathbf{V}^{(n)} \otimes M(-1)$$

such that $M_{i+1}/M_i \cong M(n-2-2i)$ for all $i \in \{0, 2, \ldots, n-1\}$.

Exercise 3.115. For any $V \in \overline{\mathfrak{W}}$ set

$$\mathcal{E}(V) = \{v \in V \,:\, E^i(v) = 0 \text{ for some } i \in \mathbb{N}\}.$$

Show that $\mathcal{E}(V)$ is a submodule of V and that for any finite-dimensional module W we have $\mathcal{E}(W \otimes V) \cong W \otimes \mathcal{E}(V)$.

Exercise 3.116. Let $n \in \mathbb{N}$ and V be a simple weight $\mathfrak{g}$-module. Show that the following conditions are equivalent:

(a) $\operatorname{Hom}_{\mathfrak{g}}(V, V \otimes \mathbf{V}^{(n)}) \neq 0$.
(b) $\operatorname{Hom}_{\mathfrak{g}}(V \otimes \mathbf{V}^{(n)}, V) \neq 0$.
(c) $\operatorname{supp} \mathbf{V}^{(n)} \cap \{0\} \neq 0$.
(d) n is odd.

Exercise 3.117. Describe tensor products of $\overline{M}(\lambda)$, $\lambda \in \mathbb{C}$, with finite-dimensional $\mathfrak{g}$-modules in a similar way to Exercises 3.110–3.113.

Exercise 3.118. Let $V \in \overline{\mathfrak{W}}$ be arbitrary and W be a finite-dimensional $\mathfrak{g}$-module. Show that $V^{\circledast} \cong V$ implies $(V \otimes W)^{\circledast} \cong V \otimes W$.

Exercise 3.119. Show that any finite-dimensional $\mathfrak{g}$-module V satisfies $V^{\circledast} \cong V$.

Exercise 3.120. Let $\xi \in \mathbb{C}/2\mathbb{Z}$ and $\tau \in \mathbb{C}$. Assume that V is a $\mathfrak{g}$-module such that

(a) $\operatorname{Supp} V = \xi$;
(b) c acts on V as the scalar τ;
(c) $\dim V_\lambda = 1$ for some $\lambda \in \xi$;
(d) E acts bijectively on V.

Show that $V \cong \mathbf{V}(\xi, \tau)^\circledast$.

Exercise 3.121. Let $\lambda \in \mathbb{C}$ and $\tau \in \mathbb{C}$. Show that there is a unique indecomposable $\mathfrak{g}$-module V such that

(a) $\dim V_\lambda = 1$;
(b) c acts on V as the scalar τ;
(c) F acts bijectively on V.

Show further that $V \cong \mathbf{V}(\xi, \tau)$.

Exercise 3.122. Compute characters of Verma modules and of all simple weight $\mathfrak{g}$-modules.

Exercise 3.123. Let V be a finite-dimensional $\mathfrak{g}$-module and $n \in \mathbb{N}$. Show that both $E^n : V_{-n} \to V_n$ and $F^n : V_n \to V_{-n}$ are isomorphisms.

Exercise 3.124. Let $\lambda, \tau \in \mathbb{C}$. Prove that the category $\mathbb{C}[[x]]$-mod is equivalent to the full subcategory of the category of all finite-dimensional $\mathbb{C}[h, c]$-modules on which h acts as the scalar λ and c acts with the unique eigenvalue τ.

Chapter 4

The primitive spectrum

4.1 Annihilators of Verma modules

In this chapter we return to the study of the universal enveloping algebra $U(\mathfrak{g})$, the understanding of which is very important for the study of $\mathfrak{g}$-modules. Our first step is to give some description of primitive ideals of $U(\mathfrak{g})$.

Let us recall the basic definitions. Let A be an associative algebra and M be an A-module. Then the *annihilator* $\text{Ann}_A(M)$ of M in A is defined as follows:

$$\text{Ann}_A(M) = \{a \in A \ : \ a(m) = 0 \text{ for all } m \in M\}.$$

Note that $\text{Ann}_A(M)$ always contains the zero element; it is always non-empty.

Exercise 4.1. Show that $\text{Ann}_A(M)$ is a two-sided ideal of A.

An ideal $I \subset A$ is called *primitive* provided that $I = \text{Ann}_A(L)$ for some simple A-module L. From the definitions, it is evident that primitive ideals play an important role in the study of A-modules. The set of all primitive ideals of A is called the *primitive spectrum* of A and is denoted by $\text{Prim}(A)$.

Our main goal in this chapter is to give a complete and explicit description of all primitive ideals of $U(\mathfrak{g})$. We will start by obtaining a description of annihilators of Verma modules. Note that Theorem 3.16 says that almost all Verma modules are simple, therefore, in describing annihilators of Verma modules we will get many primitive ideals. In fact, later on we will see that every annihilator of a Verma module is a primitive ideal.

Theorem 4.2. *Let* $\lambda \in \mathbb{C}$. *Then the annihilator* $\mathrm{Ann}_{U(\mathfrak{g})}(M(\lambda))$ *of the Verma module* $M(\lambda)$ *in* $U(\mathfrak{g})$ *is the two-sided ideal* $\mathcal{I}_\lambda$ *of* $U(\mathfrak{g})$, *generated by the element* $c - (\lambda+1)^2$.

To prove this theorem we will need the following lemma:

Lemma 4.3. *Let* $g(h) \in \mathbb{C}[h]$ *be a non-zero polynomial. Then for every* $k \in \mathbb{N}_0$ *the action of both* $e^k g(h)$ *and* $f^k g(h)$ *on* $M(\lambda)$ *is non-zero.*

Proof. Assume the the polynomial $g(h)$ has degree n and that $M(\lambda)$ is given by (3.4). Consider first the case where $\lambda \notin \mathbb{N}_0$. Let $i \in \mathbb{N}_0$ be such that $i > k$ and that $\lambda - 2i$ is not a root of $g(h)$. Using $h(v_i) = (\lambda - 2i)v_i$, we have

$$e^k g(h)(v_i) = g(\lambda - 2i)e^k(v_i), \quad f^k g(h)(v_i) = g(\lambda - 2i)f^k(v_i).$$

As $g(\lambda - 2i) \neq 0$ by our choice of i, we immediately have $g(\lambda - 2i)f^k(v_i) \neq 0$ as the action of f on $M(\lambda)$ is injective. Similarly, $g(\lambda - 2i)e^k(v_i) \neq 0$ follows from our restrictions $i > k$ and $\lambda \notin \mathbb{N}_0$ using (3.4).

For the case $\lambda \in \mathbb{N}_0$ the proof is completely similar. The only difference is that one has to choose $i > 2\lambda + k$. $\qquad\square$

Now we are ready to prove Theorem 4.2.

Proof. Denote by I the two-sided ideal of $U(\mathfrak{g})$, generated by the element $c - (\lambda+1)^2$. From Proposition 3.13(iii) we get $c - (\lambda+1)^2 \in \mathrm{Ann}_{U(\mathfrak{g})}(M(\lambda))$. Hence $I \subset \mathrm{Ann}_{U(\mathfrak{g})}(M(\lambda))$ by Exercise 4.1.

Now, to prove the equality $I = \mathrm{Ann}_{U(\mathfrak{g})}(M(\lambda))$ it is sufficient to show that the action of every element from the algebra $U(\mathfrak{g})/I$ on $M(\lambda)$ is non-zero. From Theorem 2.33(ii) it follows that the algebra $U(\mathfrak{g})/I$ has a basis of the form

$$\{h^i, e^k h^i, f^k h^i : i \in \mathbb{N}_0, k \in \mathbb{N}\}.$$

This means that every $u \in U(\mathfrak{g})/I$ can be written in the following form:

$$u = g(h) + \sum_{k=1}^{j} e^k g_k(h) + \sum_{k=1}^{j} f^k \tilde{g}_k(h) \qquad (4.1)$$

for some $j \in \mathbb{N}_0$ and polynomials $g, g_k, \tilde{g}_k$, $k = 1, \ldots, j$. We have $u \neq 0$ if, and only if, at least one summand in (4.1) is non-zero.

Let now $v \in M(\lambda)$ be a weight vector. From Exercise 3.6 it follows that different summands in the decomposition (4.1) map v to weight vectors of different weights. Since non-zero weight vectors of different weights are linearly independent, we only get $u(v) \neq 0$ if at least one summand from the decomposition (4.1) on v is non-zero. At the same time, by Lemma 4.3, for every non-zero summand from the decomposition (4.1) there is a weight vector in $M(\lambda)$ on which this summand acts as non-zero. This completes the proof. $\qquad\square$

Corollary 4.4.

(i) For all $\lambda, \mu \in \mathbb{C}$ we have $\mathcal{I}_\lambda = \mathcal{I}_\mu$ if, and only if, $\mu = \lambda$ or $\mu = -\lambda - 2$.

(ii) For all $\lambda \in \mathbb{C}$ the ideal $\mathcal{I}_\lambda$ is primitive.

Proof. The statement (i) follows immediately from the definition of $\mathcal{I}_\lambda$. If $\lambda \notin \mathbb{N}_0$, then, by Theorem 4.2, the ideal $\mathcal{I}_\lambda$ is the annihilator of the module $M(\lambda)$, the latter being simple by Theorem 3.16(i). Hence $\mathcal{I}_\lambda$ is primitive in this case.

If $\lambda \in \mathbb{N}_0$, then $\mathcal{I}_\lambda = \mathcal{I}_{-\lambda-2}$ by statement (i). However, $-\lambda - 2 \notin \mathbb{N}_0$ and hence the ideal $\mathcal{I}_{-\lambda-2}$ is primitive by above. This completes the proof of our statement. $\qquad\square$

Exercise 4.5. Let $\xi \in \mathbb{C}/2\mathbb{Z}$ and $\tau \in \mathbb{C}$. Show that

$$\mathrm{Ann}_{U(\mathfrak{g})} \mathbf{V}(\xi, \tau) = \mathcal{I}_\lambda,$$

where $\lambda \in \mathbb{C}$ is such that $(\lambda + 1)^2 = \tau$.

Exercise 4.6. For $\lambda \in \mathbb{C}$ show that

$$\mathrm{Ann}_{U(\mathfrak{g})} \overline{M}(\lambda) = \mathcal{I}_{\lambda-2}.$$

4.2 Simple modules and central characters

In the previous section we constructed primitive ideals $\mathcal{I}_\lambda$ in $U(\mathfrak{g})$, indexed by $\lambda \in \mathbb{C}$. The aim of this section is to show that the ideals $\mathcal{I}_\lambda$ are exactly the minimal primitive ideals of $U(\mathfrak{g})$. The main subject of this section is the following theorem:

Theorem 4.7. Let L be a simple $U(\mathfrak{g})$-module.

(i) The Casimir element c acts on L as some scalar.

(ii) There exists $\lambda \in \mathbb{C}$ such that $\mathcal{I}_\lambda \subset \mathrm{Ann}_{U(\mathfrak{g})}(L)$.

We emphasize that the module L from the formulation of Theorem 4.7 is an arbitrary simple module; it is not assumed to be a weight module. Another way to formulate Theorem 4.7(i) is to say that every simple $\mathfrak{g}$-module has a central character.

Remark 4.8. Observe that for every central character, that is for every homomorphism $\chi : \mathbb{C}[c] \to \mathbb{C}$, there exists $\lambda \in \mathbb{C}$ such that $\chi = \chi_\lambda$.

Now let us prove Theorem 4.7.

Proof.   Assume first that there exists a non-zero polynomial $g(c) \in \mathbb{C}[c]$ and a non-zero element $v \in L$ such that $g(c)(v) = 0$. The polynomial $g(c)$ is then obviously non-constant. Since $\mathbb{C}$ is algebraically closed, we can factor $g(c)$ into linear factors as follows:

$$g(c) = a(c - \tau_1)(c - \tau_2) \cdots (c - \tau_k),$$

where $a, \tau_i \in \mathbb{C}$, $a \neq 0$. For $j = 1, \ldots, k$ set $g_j(c) = \prod_{i=j}^{k}(c - \tau_i)$ and $v_j = g_j(c)(v)$. We also set $v_{k+1} = av$. Then $v_{k+1} \neq 0$ and $v_1 = 0$ imply that there exists some $l \in \{1, 2, \ldots, k\}$ such that $v_l = 0$ while $v_{l+1} \neq 0$. In particular, $v_l = (c - \tau_l)v_{l+1} = 0$, and hence τ_l is an eigenvalue of c. Since c is central, by Exercise 1.31 the vector space of all eigenvectors for c with the eigenvalue τ_l is a non-zero submodule N of L. As L is simple, we get $N = L$ and thus c acts on L as the scalar τ_l.

Now assume that $g(c)(v) \neq 0$ for all non-zero $g(c) \in \mathbb{C}[c]$ and $v \in L$. Consider the quotient field $Q = \mathrm{Quot}(\mathbb{C}[c])$ (the field of rational functions in c). As c is central, we deduce that for every non-zero polynomial $g(c) \in \mathbb{C}[c]$ the linear map $g(c)$ is a non-zero endomorphism of L. Since L is simple, from Lemma 1.14 it follows that the linear transformation $g(c)$ is invertible. Hence L is equipped with the natural structure of a Q-module; in particular, L is a vector space over Q. However, the $\mathbb{C}$-vector space Q is of uncountable dimension over $\mathbb{C}$ (since $\mathbb{C}$ is uncountable and thus we have uncountably many linear factors for denominators in Q). At the same time L has countable dimension as a quotient of the left regular module $U(\mathfrak{g})$, which has countable dimension by the PBW Theorem (Theorem 2.13). This contradiction shows that the case $g(c)(v) \neq 0$ for all non-zero $g(c) \in \mathbb{C}[c]$ and $v \in L$ is not possible and proves the statement (i).

To prove the statement (ii) one now has only to observe that if c acts on L as the scalar $(\lambda + 1)^2 \in \mathbb{C}$ (the latter is true for some $\lambda \in \mathbb{C}$ by (i)), then we have $\mathcal{I}_\lambda \subset \mathrm{Ann}_{U(\mathfrak{g})}(L)$. This completes the proof. $\square$

Corollary 4.9. *The ideals* $\mathcal{I}_\lambda$, $\lambda \in \mathbb{C}$, *are exactly the primitive ideals of* $U(\mathfrak{g})$ *which are minimal with respect to the inclusion order.*

Proof. This follows directly from Theorem 4.7(ii). □

Exercise 4.10. Let L be a simple $\mathfrak{g}$-module. Show that $\mathrm{End}_{\mathfrak{g}}(L) \cong \mathbb{C}$.

4.3 Classification of primitive ideals

The aim of this section is to give a complete classification of all primitive ideals of $U(\mathfrak{g})$. From the previous sections we already know about the primitive ideals $\mathcal{I}_\lambda$, $\lambda \in \mathbb{C}$, which are annihilators of Verma modules. Let us define a few other primitive ideals.

For $n \in \mathbb{N}$ denote by $\mathcal{J}_n$ the annihilator of the simple $\mathfrak{g}$-module $\mathbf{V}^{(n)}$. Then $\mathcal{J}_n$ is a primitive ideal of finite codimension in $U(\mathfrak{g})$. Now we are ready to classify all primitive ideals in $U(\mathfrak{g})$.

Theorem 4.11 (Classification of primitive ideals).

(i) Every primitive ideal of $U(\mathfrak{g})$ coincides with the ideal $\mathcal{I}_\lambda$ for some $\lambda \in \mathbb{C}$ or with the ideal $\mathcal{J}_n$ for some $n \in \mathbb{N}$. The only nontrivial equalities in this list are given by Corollary 4.4(i).

(ii) The only nontrivial inclusions between the primitive ideals of $U(\mathfrak{g})$ are the inclusions $\mathcal{I}_{-n-1} = \mathcal{I}_{n-1} \subset \mathcal{J}_n$, $n \in \mathbb{N}$.

Proof. Let L be a simple $U(\mathfrak{g})$-module and $I = \mathrm{Ann}_{U(\mathfrak{g})}(L)$. By Corollary 4.9, there exists $\lambda \in \mathbb{N}$ such that $\mathcal{I}_\lambda \subset I$. Assume that $\mathcal{I}_\lambda \neq I$. Consider the quotient algebra $U(\mathcal{I}_\lambda) = U(\mathfrak{g})/\mathcal{I}_\lambda$. Then the image $\overline{I}$ of I in $U(\mathcal{I}_\lambda)$ is a non-zero two-sided ideal of $U(\mathcal{I}_\lambda)$. Moreover, the module L is a simple $U(\mathcal{I}_\lambda)$-module and $\overline{I} = \mathrm{Ann}_{U(\mathcal{I}_\lambda)}(L)$. Abusing notation, we will denote the elements of $U(\mathfrak{g})$ and $U(\mathcal{I}_\lambda)$ in the same way.

With respect to the decomposition (2.11), the ideal $\mathcal{I}_\lambda$ is homogeneous (as it is generated by the homogeneous element $c - (\lambda + 1)^2$). Hence the quotient $U(\mathcal{I}_\lambda)$ inherits from (2.11) the graded decomposition

$$U(\mathcal{I}_\lambda) = \bigoplus_{s \in \mathbb{Z}} U(\mathcal{I}_\lambda)_{2s}, \qquad (4.2)$$

where

$$U(\mathcal{I}_\lambda)_{2s} = \{u \in U(\mathcal{I}_\lambda) : [h, u] = 2su\}.$$

Lemma 4.12. *The ideal $\overline{I}$ is homogeneous with respect to the grading (4.2).*

Proof. Consider the adjoint $\mathfrak{g}$-module $U(\mathfrak{g})$. This module is a weight module by Lemma 2.29. Since $\mathcal{I}_\lambda$ is a two-sided ideal, it is a submodule of this module, and hence we get the induced adjoint $\mathfrak{g}$-module structure on the quotient $U(\mathcal{I}_\lambda)$. The module $U(\mathcal{I}_\lambda)$ is weight by Proposition 3.8(ii). As $\overline{I}$ is a two-sided ideal of $U(\mathcal{I}_\lambda)$, it is a submodule of $U(\mathcal{I}_\lambda)$ and hence is a weight module by Proposition 3.8(i). The claim follows. $\square$

By Theorem 2.33(ii), the set $\mathbf{B}_2$ is a basis of $U(\mathcal{I}_\lambda)$. Hence, from Lemma 4.12 we obtain that the ideal $\overline{I}$ contains a non-zero element of the form $g(h)x$, where $g(h) \in \mathbb{C}[h]$ and $x \in \mathbf{B}_1$.

Assume first that $x \notin \overline{I}$. There then exists $v \in L$ such that $x(v) = w \neq 0$, while $g(h)(w) = 0$. Since $\mathbb{C}$ is algebraically closed, we can factor $g(h)$ into linear factors and, as done for Theorem 4.7, obtain that h has a non-zero eigenvector in L. This eigenvector is then a weight vector by definition. However, L is simple and thus is generated by any vector, hence L is a weight module by Proposition 3.12. All simple weight modules are classified in Theorem 3.32. For modules listed in Theorem 3.32(ii)–(iv) we know that their annihilators have the form $\mathcal{I}_\mu$ for some $\mu \in \mathbb{C}$ (see Theorem 4.2 and Exercises 4.5 and 4.6). This yields $L \cong \mathbf{V}^{(n)}$ for some $n \in \mathbb{N}$ and hence $I = \mathcal{J}_n$.

Assume now that $x = e^k \in \overline{I}$, where $k \in \mathbb{N}$. Then there exists $v \in L$, $v \neq 0$, such that $e(v) = 0$. From this and $c(v) = (\lambda + 1)^2 v$ we get $((h+1)^2 - (\lambda+1)^2)(v) = 0$ and we can use the same arguments as in the previous paragraph to conclude that $I = \mathcal{J}_n$ for some $n \in \mathbb{N}$. The case $x = f^k \in \overline{I}$, where $k \in \mathbb{N}$, is dealt with similarly.

To complete the proof of the statement (i) we now have to determine which ideals listed in (i) coincide. For ideals of the form $\mathcal{I}_\lambda$ the claim follows from Corollary 4.4(i). Every $\mathcal{I}_\lambda$ has infinite codimension in $U(\mathfrak{g})$ and hence is different from any of $\mathcal{J}_n$, the latter being of infinite codimension. That $\mathcal{J}_n \neq \mathcal{J}_m$ for $n > m$ follows from the fact that $e^m \in \mathcal{J}_m \setminus \mathcal{J}_n$, which, in turn, follows from (1.9). This completes the proof of the statement (i).

For $n \in \mathbb{N}$ the Casimir element c acts on $\mathbf{V}^{(n)}$ as the scalar n^2 by Exercise 1.33. Hence

$$\mathcal{I}_{-n-1} \overset{\text{Corollary 4.4}}{=} \mathcal{I}_{n-1} \subset \mathcal{J}_n. \tag{4.3}$$

If we have two different primitive ideals which do not form a pair of the form (4.3), then their intersections with $Z(\mathfrak{g})$ are not contained in each other and hence they cannot be compared with respect to the inclusion order. This proves the statement (ii) and completes the proof. $\square$

Corollary 4.13. *Every primitive ideal of $U(\mathfrak{g})$ is the annihilator of some simple highest weight module.*

Proof. The ideal $\mathcal{I}_\lambda$, $\lambda \in \mathbb{C}$, is the annihilator of $M(\lambda)$ by definition. If $\lambda \notin \mathbb{N}_0$, we have $M(\lambda) = L(\lambda)$ by Corollary 3.18. If $\lambda \in \mathbb{N}_0$, then $\mathcal{I}_\lambda = \mathcal{I}_{-\lambda-2}$ (Corollary 4.4(i)) and hence $\mathcal{I}_\lambda$ is the annihilator of $L(-\lambda-2)$ by Corollary 3.18.

The ideal $\mathcal{J}_n$, $n \in \mathbb{N}$, is the annihilator of $\mathbf{V}^{(n)}$ by definition and $\mathbf{V}^{(n)} \cong L(n-1)$ by Corollary 3.18. Now the statement follows from Theorem 4.11(i). This completes the proof. $\square$

Exercise 4.14. Show that the algebra $U(\mathfrak{g})$ has infinitely many two-sided ideals, which are not primitive.

4.4 Primitive quotients

For a primitive ideal I of $U(\mathfrak{g})$, the algebra $U(I) = U(\mathfrak{g})/I$ is called the *primitive quotient* of $U(\mathfrak{g})$. In this section we describe the structure of all primitive quotients of $U(\mathfrak{g})$.

Theorem 4.15.

(i) *For every $I \in \mathrm{Prim}(U(\mathfrak{g}))$ the algebra $U(I)$ is both left and right Noetherian.*

(ii) *For every $\lambda \in \mathbb{C}$ the algebra $U(\mathcal{I}_\lambda)$ is a domain, and is free over $\mathbb{C}[h]$ both a left and as a right module with basis $\mathbf{B}_1$.*

(iii) *For every $n \in \mathbb{N}$ we have $U(\mathcal{J}_n) \cong \mathrm{Mat}_{n \times n}(\mathbb{C})$.*

(iv) *For every $\lambda \in \mathbb{C} \setminus \mathbb{Z}$ the algebra $U(\mathcal{I}_\lambda)$ is a simple algebra.*

(v) *For every $n \in \mathbb{N}_0$ the algebra $U(\mathcal{I}_n)$ has a unique proper two-sided ideal, namely the image of $\mathcal{J}_{n+1}$ in $U(\mathcal{I}_n)$.*

Proof. Let $I \in \mathrm{Prim}(U(\mathfrak{g}))$. The algebra $U(I)$ is a quotient of the (both left and right) Noetherian algebra $U(\mathfrak{g})$ by Theorem 2.40. Hence $U(I)$ is both left and right Noetherian, proving the claim (i).

Consider the algebra $U(\mathcal{I}_\lambda)$ as a graded algebra with respect to the decomposition (4.2). By Theorem 2.33(ii), both the set $\mathbf{B}_2$ and the set $\mathbf{B}_3$ form a basis of $U(\mathcal{I}_\lambda)$. In particular, the algebra $U(\mathcal{I}_\lambda)$ is a free (both as a left and as a right) module over the domain $U(\mathcal{I}_\lambda)_0 = \mathbb{C}[h]$ with basis $\mathbf{B}_1$. For $x, y \in \mathbf{B}_1$ the element xy either belongs to $\mathbf{B}_1$ (and hence is non-zero), or can be written as $g(h)z$ for some non-zero $g(h) \in \mathbb{C}[h]$ and $z \in \mathbf{B}_1$, using the definition of the Casimir element. In either case $xy \neq 0$, which implies that the product of any two non-zero homogeneous elements from $U(\mathcal{I}_\lambda)$ is non-zero. As $U(\mathcal{I}_\lambda)$ is graded, it follows that the product of any two non-zero elements from $U(\mathcal{I}_\lambda)$ is non-zero. Thus $U(\mathcal{I}_\lambda)$ is a domain and the claim (ii) is proved.

For $n \in \mathbb{N}$, from the definition of $\mathcal{J}_n$ we have that the algebra $U(\mathcal{J}_n)$ is a subalgebra of the algebra $\mathrm{Mat}_{n \times n}(\mathbb{C})$, which is considered to be the algebra of all linear operators on $\mathbf{V}^{(n)}$. At the same time, the module $\mathbf{V}^{(n)}$ is a simple n-dimensional $U(\mathcal{J}_n)$-module. From the Artin–Wedderburn Theorem (Theorem 2.4.3 in [42]) it follows that the only simple finite-dimensional complex algebra having an n-dimensional simple module, is the algebra $\mathrm{Mat}_{n \times n}(\mathbb{C})$. Hence the algebra $\mathrm{Mat}_{n \times n}(\mathbb{C})$ is a quotient of $U(\mathcal{J}_n)$ and the claim (iii) follows.

Take now any $\lambda \in \mathbb{C} \setminus \mathbb{Z}$ and assume that I is a proper two-sided ideal of $U(\mathcal{I}_\lambda)$. As $U(\mathcal{I}_\lambda)$ is Noetherian by (i), the ideal I is contained in some maximal left ideal I', and hence the quotient $U(\mathcal{I}_\lambda)/I'$ is a simple $U(\mathfrak{g})$-module, whose annihilator contains I. However, none of the primitive ideals of $U(\mathfrak{g})$ properly contain $\mathcal{I}_\lambda$, $\lambda \in \mathbb{C} \setminus \mathbb{Z}$, by Theorem 4.11(ii). The obtained contradiction proves the claim (iv).

Now let $n \in \mathbb{N}_0$ and I be a proper two-sided ideal of the algebra $U(\mathcal{I}_n)$. We will need the following lemma:

Lemma 4.16. *The ideal I has finite codimension in $U(\mathcal{I}_n)$.*

Proof. By Lemma 4.12, the ideal I is a homogeneous ideal of $U(\mathcal{I}_n)$ with respect to the grading (4.2). Hence I contains a homogeneous non-zero element of the form $g(h)x$, where $x \in \mathbf{B}_1$ and $g(h) \in \mathbb{C}[h]$. We assume $x = e^i$ for some $i \in \mathbb{N}_0$ (the case $x = f^i$ is treated similarly). Note that $x = 1$ and $g(h) \in \mathbb{C}$ is not possible since the ideal I is supposed to be a proper ideal.

Let us prove, by induction on the degree of $g(h)$, that I contains some element of the form e^j, $j \in \mathbb{N}$. If the polynomial $g(h)$ is constant, we have nothing to prove. If the polynomial $g(h)$ is not constant, for the element $[e, g(h)e^i] \in I$ we have:

$$
\begin{aligned}
[e, g(h)e^i] &= eg(h)e^i - g(h)e^{i+1} \\
(\text{using } (2.1)) &= g(h-2)e^{i+1} - g(h)e^{i+1} \\
&= (g(h-2) - g(h))e^{i+1}.
\end{aligned}
\tag{4.4}
$$

The polynomial $g(h-2) - g(h)$ has degree $\deg(g(h)) - 1$ and hence, by induction, we conclude that I contains e^j for some $j \in \mathbb{N}$. It follows that I contains e^l for all $l \in \mathbb{N}$ such that $l \geq j$. Since e^l is a free $\mathbb{C}[h]$-basis of $U(\mathcal{I}_n)_{2l}$ (by Theorem 4.15(ii), proved above), the ideal I contains $U(\mathcal{I}_n)_{2l}$ for all $l \in \mathbb{N}$ such that $l \geq j$.

Since the algebra $U(\mathcal{I}_n)$ is graded and is a domain, the ideal I intersects every graded component $U(\mathcal{I}_n)_{2s}$. Taking some element from $U(\mathcal{I}_n)_{2s}$ for $s < 0$ and repeating the above arguments, we get a $j' \in \mathbb{N}$ such that the ideal I contains $U(\mathcal{I}_n)_{-2l}$ for all $l \in \mathbb{N}$, $l \geq j'$.

Finally, every graded component $U(\mathcal{I}_n)_{2s}$, $-j' < s < j$, is a free $\mathbb{C}[h]$-module of rank one. The intersection $U(\mathcal{I}_n)_{2s} \cap I$ is non-zero by above and hence contains some non-zero element $g_s(h)x$, where $x \in \mathbf{B}_1$ and $\deg(g_s(h)) = k_s$. Then the images (in the quotient algebra $U(\mathcal{I}_n)/I$) of the elements from the set $\{x, hx, h^2 x, \ldots, h^{k_s-1}x\}$ generate the graded component $(U(\mathcal{I}_n)/I)_{2s}$ over $\mathbb{C}$. Uniting all these finite generating sets over all intermediate degrees $s = -j'+1, -j'+2, \ldots, j-1$ we get a finite generating set for the algebra $U(\mathcal{I}_n)/I$ over $\mathbb{C}$. The claim of the lemma follows. $\square$

Consider the quotient $U' = U(\mathcal{I}_n)/I$. By Lemma 4.16, it is a finite-dimensional associative algebra. By Theorem 4.11(ii), the only simple $U(\mathfrak{g})$-module, whose annihilator contains $\mathcal{I}_n$, is $\mathbf{V}^{(n+1)}$. Hence $\mathbf{V}^{(n+1)}$ is the only simple U'-module. By Weyl's Theorem (Theorem 1.29), every finite-dimensional U'-module is semi-simple. Hence U' is a simple algebra and thus is isomorphic to the algebra $\mathrm{Mat}_{(n+1)\times(n+1)}(\mathbb{C})$ by the Artin–Wedderburn Theorem (Theorem 2.4.3 in [42]). This yields $I = \mathcal{J}_{n+1}$, which proves (v) and completes the proof of our theorem. $\square$

It is also possible to describe all primitive quotients of $U(\mathfrak{g})$ as $\mathfrak{g}$-modules with respect to the adjoint action.

Proposition 4.17.

(i) *For every $\lambda \in \mathbb{C}$ the adjoint $\mathfrak{g}$-module $U(\mathcal{I}_\lambda)$ is isomorphic to the module*

$$\bigoplus_{i \in \mathbb{N}} \mathbf{V}^{(2i-1)}.$$

(ii) *For every $n \in \mathbb{N}$ the adjoint $\mathfrak{g}$-module $U(\mathcal{J}_n)$ is isomorphic to the module*

$$\bigoplus_{i=1}^{n} \mathbf{V}^{(2i-1)}.$$

Proof. That the adjoint $\mathfrak{g}$-modules $U(\mathcal{I}_\lambda)$ and $U(\mathcal{J}_n)$ are direct sums of finite-dimensional modules follows from Exercise 2.49. To prove our proposition, therefore we have to determine, for every $m \in \mathbb{N}$, the multiplicity of the simple module $\mathbf{V}^{(m)}$ in the adjoint $\mathfrak{g}$-modules $U(\mathcal{I}_\lambda)$ and $U(\mathcal{J}_n)$.

We first observe that $\mathbf{V}^{(m)}$ is the only simple finite-dimensional $U(\mathfrak{g})$-module, which has a non-zero element v of weight $m-1$ satisfying $e(v) = 0$. This follows directly from (1.9) and Theorem 1.22. From the decomposition (4.2) we have that all weights of the adjoint $\mathfrak{g}$-modules $U(\mathcal{I}_\lambda)$ and $U(\mathcal{J}_n)$ are even. Hence, for $\mathbf{V}^{(m)}$ to occur in these modules with a non-zero multiplicity we thus must have that m is odd.

Fix now some $\lambda \in \mathbb{C}$ and let $m = 2i - 1$ for some $i \in \mathbb{N}$. Consider the weight space $U(\mathcal{I}_\lambda)_{m-1}$. By Theorem 4.15(ii), the space $U(\mathcal{I}_\lambda)_{m-1}$ is a free left $\mathbb{C}[h]$-module with basis e^{i-1}. From the computation (4.4) we have that the only (up to scalar) element of $U(\mathcal{I}_\lambda)_{m-1}$, annihilated by the adjoint action of the element e, is the element e^{i-1}. Hence the module $\mathbf{V}^{(m)}$ occurs in $U(\mathcal{I}_\lambda)$ with multiplicity one. The claim (i) follows.

If $n \in \mathbb{N}$, then from (1.9) it follows that the elements $1, e, e^2, \ldots, e^{n-1}$ do not annihilate $\mathbf{V}^{(n)}$. Hence their images in the adjoint $\mathfrak{g}$-module $U(\mathcal{J}_n)$ are non-zero. From the previous paragraph we thus get that the modules $\mathbf{V}^{(2i-1)}$, $i = 1, \ldots, n$, occur in the adjoint $\mathfrak{g}$-module $U(\mathcal{J}_n)$ with multiplicity one. At the same time, using Theorem 4.15(iii), we have

$$\dim U(\mathcal{J}_n) = n^2$$
$$= \sum_{i=1}^{n}(2i - 1)$$
$$= \sum_{i=1}^{n} \dim \mathbf{V}^{(2i-1)}$$

and the claim (ii) follows. This completes the proof. $\qquad\square$

Exercise 4.18. Let $\lambda \in \mathbb{C}$ and $n \in \mathbb{N}$. Show that the multiplicity of the simple module $\mathbf{V}^{(n)}$ in the adjoint $\mathfrak{g}$-module $U(\mathcal{I}_\lambda)$ equals the dimension $\dim \mathbf{V}_0^{(n)}$ of the zero weight space of $\mathbf{V}^{(n)}$.

4.5 Centralizers of elements in primitive quotients

The aim of this section is to prove the following theorem:

Theorem 4.19. *Let* $\lambda \in \mathbb{C}$. *Then for any non-scalar element* $u \in U(\mathcal{I}_\lambda)$ *the centralizer*

$$\mathcal{C}(u) = \{v \in U(\mathcal{I}_\lambda) : vu = uv\}$$

is a commutative algebra and a free left $\mathbb{C}[u]$-*module of finite rank.*

To prove this theorem we will need some preparation. For $u \in U(\mathcal{I}_\lambda)$ there exists a (unique) graded decomposition of u with respect to the grading (4.2), and this decomposition has the form

$$u = \sum_{i=m}^n u_i,$$

for some $m, n \in \mathbb{Z}$, $m \leq n$, and some $u_i \in U(\mathcal{I}_\lambda)_{2i}$, $i \in \{m, m+1, \ldots, n\}$. If $u \neq 0$, then we may assume $u_m \neq 0$ and $u_n \neq 0$. Similarly to Section 2.6, we set

$$d_+(u) = n \quad \text{and} \quad d_-(u) = m.$$

By Theorem 2.33(ii), the space $U(\mathcal{I}_\lambda)_{2i}$ is a free left $\mathbb{C}[h]$-module with the basis e^i if $i > 0$, $f^{|i|}$ if $i < 0$, and 1 if $i = 0$. If $u_{d_+(u)} = g(h)x$ and $u_{d_-(u)} = \tilde{g}(h)\tilde{x}$ for some $x, \tilde{x} \in \mathbf{B}_1$ and $g(h), \tilde{g}(h) \in \mathbb{C}[h]$, we set

$$k_+(u) = g(h) \quad \text{and} \quad k_-(u) = \tilde{g}(h).$$

Lemma 4.20. *Assume that $u \in U(\mathcal{I}_\lambda)$ is such that $d_+(u) = n \geq 0$. Assume further that $a, b \in \mathcal{C}(u)$ are non-zero and such that $m = d_+(a) = d_+(b) \geq 0$. Then $k_+(a)$ and $k_+(b)$ are linearly dependent over $\mathbb{C}$.*

Proof. Let $k_+(u) = g(h)$, $k_+(a) = x(h)$ and $k_+(b) = y(h)$. The equality $ua = au$ implies the equality

$$g(h)e^n x(h)e^m = x(h)e^m g(h)e^n.$$

Since $U(\mathcal{I}_\lambda)$ is a domain (Theorem 4.15(ii)), using (2.1) we get the equality

$$g(h)x(h - 2n) = x(h)g(h - 2m).$$

Similarly we get the equality

$$g(h)y(h - 2n) = y(h)g(h - 2m).$$

This yields

$$\frac{x(h - 2n)}{y(h - 2n)} = \frac{x(h)}{y(h)}$$

(note that both $x(h)$ and $y(h)$ are non-zero polynomials by our assumptions) and hence $\frac{x(h)}{y(h)} \in \mathbb{C}$. The claim follows. $\quad\square$

Now we are ready to prove Theorem 4.19.

Proof. Let $u \in U(\mathcal{I}_\lambda)$ be non-scalar. First of all we observe that $\mathbb{C}[u]$ is a subalgebra of $\mathcal{C}(u)$ by Exercise 2.59. Set $m = d_-(u)$, $n = d_+(u)$ and $g(h) = k_+(u)$. If $m = n = 0$, then $u \in \mathbb{C}[h]$. In this case, we obviously have $C(u) = \mathbb{C}[h]$, which is commutative and a free $\mathbb{C}[u]$-module of finite rank. Therefore for the rest of the proof we may assume $n > 0$ (the case $m < 0$ is considered similarly). Under this assumption we proceed with the following lemma:

Lemma 4.21. *For any non-zero $a \in C(u)$ we have $d_+(a) \geq 0$.*

Proof. Assume that $d_+(a) = -k < 0$ and $k_+(a) = x(h)$. Consider the elements $a^n u^{2k}$ and u^k. Since $U(\mathcal{I}_\lambda)$ is a domain, we have

$$d_+(a^n u^{2k}) = d_+(u^k) = nk > 0.$$

Note that both $a^n u^{2k}$ and u^k belong to $C(u)$. Hence, by Lemma 4.20, the polynomials $k_+(a^n u^{2k})$ and $k_+(u^k)$ should be linearly dependent over $\mathbb{C}$ and have the same degree. However, using (2.1) we get that the degree of the polynomial $k_+(u^k) \in \mathbb{C}[h]$ equals $k \deg(k_+(u))$. At the same time, we claim that the degree of the polynomial $k_+(a^n u^{2k}) \in \mathbb{C}[h]$ equals

$$n \deg(k_+(a)) + 2k \deg(k_+(u)) + 2nk > k \deg(k_+(u)). \tag{4.5}$$

Indeed, to compute $k_+(a^n u^{2k})$ we have to compute the following product:

$$\underbrace{x(h)f^k \cdot x(h)f^k \cdots x(h)f^k}_{n \text{ factors}} \cdot \underbrace{g(h)e^n \cdot g(h)e^n \cdots g(h)e^n}_{2k \text{ factors}}.$$

Using (2.1), we get that the total contribution of the factors $x(h)$ and $g(h)$ to the degree of $k_+(a^n u^{2k})$ will be $n \deg(k_+(a)) + 2k \deg(k_+(u))$. Thereafter we will also get nk cancellations of the form $fe = \frac{1}{4}((\lambda + 1)^2 - (h + 1)^2)$. This contributes $2nk$ to the total degree of $k_+(a^n u^{2k})$. This proves (4.5) and hence we obtain a contradiction. The claim of the lemma follows. $\square$

Denote by $\bar{d}_+$ the composition of the maps $d_+ : C(u) \setminus \{0\} \to \mathbb{Z}$ and the natural projection $\mathbb{Z} \to \mathbb{Z}_n$. As $U(\mathcal{I}_\lambda)$ is a domain, the image of $\bar{d}_+$ is a subgroup

$$G = \{0 = l_1 < l_2 < \cdots < l_p\} \subset \mathbb{Z}_n.$$

Set $a_1 = 1$ and for every $i = 2, \ldots, p$ choose some non-zero $a_i \in C(u)$ such that $\bar{d}_+(a_i) = l_i$ and $d_+(a_i)$ is the minimal possible.

Lemma 4.22. *The elements $a_1, \ldots, a_p$ form a free left $\mathbb{C}[u]$-basis of $C(u)$.*

Proof. Denote by N the left $\mathbb{C}[u]$-submodule of $U(\mathcal{I}_\lambda)$ generated by the elements $a_1, \ldots, a_p$. Let $g_i(u) \in \mathbb{C}[u]$, $i = 1, \ldots, p$, and assume that not all of them are zero. Assume that

$$g_1(u)a_1 + g_2(u)a_2 + \cdots + g_p(u)a_p = 0.$$

The above sum should contain at least two non-zero summands with the same highest degree in $U(\mathcal{I}_\lambda)$. That is, there must exist some different element $i \neq j \in \{1, \ldots, p\}$ such that

$$d_+(g_i(u)a_i) = n \deg g_i + d_+(a_i) = n \deg g_j + d_+(a_j) = d_+(g_j(u)a_j).$$

The latter is not possible, since $d_+(a_i)$ and $d_+(a_j)$ are not congruent modulo n by our choice of the a_t's. This means that $a_1, \ldots, a_p$ form a free left $\mathbb{C}[u]$-basis of N.

Now let us show that $N = \mathcal{C}(u)$. Let $a \in \mathcal{C}(u)$ be non-zero. We use induction on $d_+(a) = k$. If $k = 0$, then from Lemma 4.21 we get $d_-(a) = 0$ and $a \in \mathbb{C}[h]$. If $a \notin \mathbb{C}$, then from $au = ua$, we would derive that $u \in \mathbb{C}[h]$, contradicting $d_+(u) > 0$. Hence $a \in \mathbb{C}$ in this case, in particular, $a = \alpha a_1$ for some $\alpha \in \mathbb{C}$.

If $k > 0$, then $\bar{d}_+(a) = l_i$ for some i and we have $k \geq d_+(a_i)$ by our choice of a_i. Hence $k = l_i + nj$ for some $j \in \mathbb{N}_0$. Consider the elements a and $a_i u^j$. They both belong to $\mathcal{C}(u)$ and $d_+(a) = d_+(a_i u^j)$. Hence, by Lemma 4.20, there exists $\gamma \in \mathbb{C}$ such that $d_+(a - \gamma a_i u^j) < k$. The claim of the lemma now follows by induction. $\qquad \square$

The proof of Theorem 4.19 is now completed by the following lemma:

Lemma 4.23. *The algebra $\mathcal{C}(u)$ is commutative.*

Proof. For the element a_2 we have that $\bar{d}_+(a_2) = l_2$ generates the group G. Denote by A the subalgebra of $\mathcal{C}(u)$, generated by u and a_2. The algebra A is obviously commutative. Since $\bar{d}_+(a_2)$ generates G and A is a domain, for any $x \in \mathcal{C}(u)$ such that $d_+(x) > p \cdot d_+(a_2)$ there exists $y \in A$ such that $d_+(x) = d_+(y)$. From Lemma 4.20 it thus follows that there exist some $c \in \mathbb{C}$ such that $d_+(x - cy) < d_+(x)$. This yields that the $\mathbb{C}[u]$-module $\mathcal{C}(u)/A$ is finite-dimensional (over $\mathbb{C}$). Then there exists a non-zero polynomial $\alpha(u) \in \mathbb{C}[u]$ which annihilates the finite-dimensional $\mathbb{C}[u]$-module $\mathcal{C}(u)/A$. In other words, for any $z \in \mathcal{C}(u)$ we have $\alpha(u)z \in A$.

Now for any $z_1, z_2 \in \mathcal{C}(u)$ we have

$$\alpha(u)\alpha(u)z_1 z_2 \overset{z_1 \in \mathcal{C}(u)}{=} (\alpha(u)z_1)(\alpha(u)z_2)$$
$$(\text{as } A \text{ is commutative}) = (\alpha(u)z_2)(\alpha(u)z_1)$$
$$(\text{as } z_2 \in \mathcal{C}(u)) = \alpha(u)\alpha(u)z_2 z_1.$$

Since $U(\mathcal{I}_\lambda)$ is a domain, we get $z_1 z_2 = z_2 z_1$. This completes the proof of the lemma. $\qquad \square$

Theorem 4.19 follows from Lemmas 4.22 and 4.23. $\qquad \square$

Exercise 4.24. Let $\lambda \in \mathbb{C}$ and $u \in U(\mathcal{I}_\lambda)$ be a non-zero element such that $d_+(u) > 0$. Show that the rank of the free $\mathbb{C}[u]$-module $\mathcal{C}(u)$ divides $d_+(u)$.

Exercise 4.25. Let $\lambda \in \mathbb{C}$ and $u \in U(\mathcal{I}_\lambda)$ be a non-zero element such that $d_+(u) > 0$, $d_-(u) < 0$ and the numbers $d_+(u)$ and $d_-(u)$ are relatively prime. Show that $\mathcal{C}(u) = \mathbb{C}[u]$.

4.6 Addenda and comments

4.6.1

The material, presented in Sections 4.1–4.3 can be found in a much more general set up in several textbooks and monographs; for example [37, 64, 66]. The description of the annihilators of Verma modules for all semi-simple finite-dimensional complex Lie algebras, as well as the description of primitive ideals via annihilators of simple highest weight modules, is obtained by Duflo (see [43]). A complete exposition of this can be found in Chapter 8 of [37]. In Section 2.6 of [37] one finds an alternative approach (due to Quillen, see [100]) to the fact that every simple module has a central character.

4.6.2

To obtain an analogue of Proposition 4.17 for all semi-simple finite-dimensional complex Lie algebras is still an open problem (see [67]). Some recent progress in this direction in obtained in [112].

4.6.3

An analogue of Theorem 4.15 can be proved for a large class of algebras, which include $U(\mathfrak{g})$, in particular, for a large class of generalized Weyl algebras (see [107, 10]).

4.6.4

For every $\lambda \in \mathbb{C}$ one has the following property of the primitive quotients $U(\mathcal{I}_\lambda)$, which we will need later on in Chapter 6.

Theorem 4.26 ([4, 22, 10]). *Let* $I \subset U(\mathcal{I}_\lambda)$ *be a non-zero left ideal. Then the* $U(\mathcal{I}_\lambda)$-*module* $U(\mathcal{I}_\lambda)/I$ *has finite length.*

Proof. We will use notation similar to that used in Section 4.5. For $i \in \mathbb{Z}$ set

$$\mathsf{b}_i = \begin{cases} e^i, & i > 0; \\ 1, & i = 0; \\ f^{|i|}, & i < 0. \end{cases}$$

By Theorem 4.15(ii), every non-zero $u \in U(\mathcal{I}_\lambda)$ can be written in the form

$$u = \sum_{i=m}^n \mathsf{b}_i a_i$$

for some $m, n \in \mathbb{N}$, $m \leq n$, and $a_i \in \mathbb{C}[h]$ such that $a_m, a_n \neq 0$. We have $d_-(u) = m$, $d_+(u) = n$, $k_-(u) = a_m$, $k_+(u) = a_n$, $l(u) = 1 + n - m$. Let J be a non-zero left ideal of $U(\mathcal{I}_\lambda)$. For $i \in \mathbb{N}_0$ let J_i and J_i' be the ideals of $\mathbb{C}[h]$, defined by (2.12). Set $\overline{J} = \bigcup_{i \in \mathbb{N}_0} J_i$, $\overline{J}' = \bigcup_{i \in \mathbb{N}_0} J_i'$ and define

$$l(J) = \min\{l(u) : u \in J, u \neq 0\}.$$

If $J \subset Q$ are two non-zero left ideals of $U(\mathcal{I}_\lambda)$, we obviously have $l(J) \geq l(Q)$, $\overline{J} \subset \overline{Q}$ and $\overline{J}' \subset \overline{Q}'$.

Lemma 4.27. *Let $J \subset Q$ be two non-zero left ideals of $U(\mathcal{I}_\lambda)$. Assume that $l(J) = l(Q)$, $\overline{J} = \overline{Q}$ and $\overline{J}' = \overline{Q}'$. Then the space Q/J is finite-dimensional.*

Proof. First, we claim that the $\mathbb{C}[h]$-module Q/J is torsion, in the sense that for any $u \in Q$ there exists a non-zero $a \in \mathbb{C}[h]$ such that $a \cdot u \subset J$. To prove this, we proceed by induction on $l(u)$. First let us assume $l(u) = l(Q) = l(J)$. Let $v \in J$ be any element such that $l(v) = l(u)$. Multiplying, if necessary, the element v from the left with some power of e or some power of f, we may assume $d_+(v) = d_+(u)$. Let $k_+(u) = g(h)$ and $k_+(v) = p(h)$. Consider the element

$$w = g(h - 2d_+(v))v - p(h - 2d_+(u))u.$$

We have $w \in Q$ and $l(w) < l(u)$. Hence $w = 0$. This means that we have $p(h - 2d_+(u))u = g(h - 2d_+(v))v \in J$, proving the basis of the induction.

To prove the induction step we take some $u \in Q$ such that $l(u) > l(Q)$. As in the previous paragraph, we can find some $v \in J$ such that $l(v) = l(J)$ and $d_+(v) = d_+(u)$. Consider the element w as in the previous paragraph. We have $w \in Q$ and $l(w) < l(u)$. By induction, $xw \in J$ for some $x \in \mathbb{C}[h]$. Hence $xp(h - 2d_+(u))u = xg(h - 2d_+(v))v - xw \in J$. This proves the induction step and establishes the fact that the $\mathbb{C}[h]$-module Q/J is torsion.

Now we claim that Q/J is finitely generated over $\mathbb{C}[h]$. As $\mathbb{C}[h]$ is
Noetherian, both $\overline{J}$ and $\overline{J}'$ are finitely generated. Let $a_1, \ldots, a_s$ be a set
of generators for $\overline{J}$ and $b_1, \ldots, b_t$ be a set of generators for $\overline{J}'$. Choose
some $x_i \in J$, $i = 1, \ldots, s$, such that $k_+(x_i) = a_i$. Choose also some
$y_i \in J$, $i = 1, \ldots, t$, such that $k_-(y_i) = b_i$. As $\mathbb{C}[h]$ is Noetherian, there
exists $n \in \mathbb{N}$ such that $n > d_+(x_i)$ for all $i = 1, \ldots, s$; $n > d_+(y_i)$ for
all $i = 1, \ldots, t$; $n > -d_-(x_i)$ for all $i = 1, \ldots, s$; $n > -d_-(y_i)$ for all
$i = 1, \ldots, t$; $\overline{J} = J_n = Q_n$ and $\overline{J}' = J'_n = Q'_n$. Consider the linear subspace

$$\hat{Q} = \{u \in Q : d_+(u) < n, \, d_-(u) > -n\}.$$

The linear space $\hat{Q}$ is obviously a $\mathbb{C}[h]$-module and is a submodule of the
finitely generated $\mathbb{C}[h]$-module $\displaystyle\bigoplus_{s=-n}^{n} U(\mathcal{I}_\lambda)_{2s}$. Hence $\hat{Q}$ is a finitely generated $\mathbb{C}[h]$-module.

On the other hand, let us show that $Q = \hat{Q} + J$. Let $u \in Q$. We
proceed by induction on $N = \max\{d_+(u), -d_-(u)\}$. If $N < n$, we have
$u \in \hat{Q}$. Assume that $d_+(u) > n$ (in the case $d_-(u) < -n$ the arguments
are similar). Then we can write $k_+(u) = \displaystyle\sum_{i=1}^{s} p_i(h)a_i$ for some $p_i(h) \in \mathbb{C}[h]$.
Consider the element

$$v = \sum_{i=1}^{s} p_i(h - 2d_+(u))e^{d_+(u) - d_+(x_i)}x_i \in J.$$

From all definitions we then get $d_+(u - v) < d_+(u - v)$ and $d_-(u - v) \geq$
$d_-(u - v)$. The claim now follows by applying the inductive assumption to
the element $u - v \in Q$. Therefore Q/J is a quotient of a finitely generated
$\mathbb{C}[h]$-module $\hat{Q}$, hence finitely generated.

The ring $\mathbb{C}[h]$ is a (commutative) principal ideal domain. The above
shows that the $\mathbb{C}[h]$-module Q/J is finitely generated and torsion. Hence
it is finite-dimensional. This completes the proof of the lemma. $\qquad\square$

Let $I(1) \supset I(2) \supset \cdots \supset I$ be a descending chain of left ideals in
$U(\mathcal{I}_\lambda)$, all containing the ideal I. As both $\mathbb{C}[h]/\overline{I}$ and $\mathbb{C}[h]/\overline{I}'$ are finite-
dimensional, there exists $n \in \mathbb{N}$ such that $l(I(m)) = l(I(n))$, $\overline{I(m)} = \overline{I(n)}$
and $\overline{I(m)}' = \overline{I(n)}'$ for all $m \geq n$.

If $\lambda \in \mathbb{C} \setminus \mathbb{Z}$, then the infinite-dimensional algebra $U(\mathcal{I}_\lambda)$ is simple by
Theorem 4.15(iv). Hence it does not have any finite-dimensional represen-
tations. Therefore from Lemma 4.27 we obtain $I_m = I_n$ for all $m \geq n$.

If $\lambda \in \mathbb{Z}$, then from Theorem 4.15(v) it follows that each I_n/I_m is
annihilated by the image of $\mathcal{J}_{n+1}$ in $U(\mathcal{I}_\lambda)$, which is an ideal of finite

codimension, say l. The algebra $U(\mathcal{I}_\lambda)$ is Noetherian by Theorem 2.40 and hence the ideal I_n is finitely generated, say by k elements. But then the dimension of the $U(\mathcal{I}_\lambda)$-module I_n/I_m is bounded by kl, which does not depend on m. Hence the chain $I_1 \supset I_2 \supset \ldots$ of ideals stabilizes, which completes the proof of the theorem. $\qquad\square$

4.6.5

Description of the centralizers of elements for the first Weyl algebra A_1, similar to the one obtained in Theorem 4.19, is a result of Amitsur (see [2]). Theorem 4.19 is true (with the same proof) for a wide class of generalized Weyl algebras (see [10, 83]). We closely followed Bavula's paper [10] during our exposition in Section 4.5.

4.6.6

A natural question to ask is what is the intersection of the annihilators of all simple finite-dimensional $U(\mathfrak{g})$-modules. The answer is given by the following theorem (see [55]), which is true for all simple finite-dimensional complex Lie algebras (and which we prove for the algebra $\mathfrak{sl}_2$):

Theorem 4.28 (Harish-Chandra). *For any $u \in U(\mathfrak{g})$, $u \neq 0$, there exists $n \in \mathbb{N}$ such that $u \cdot \mathbf{V}^{(n)} \neq 0$.*

Proof. It is, of course, enough to show that there exists $n \in \mathbb{N}$ and some weight vector $v \in \mathbf{V}^{(n)}$ such that $u(v) \neq 0$. Since $\mathbf{V}^{(n)}$ is a weight module, weight vectors of $\mathbf{V}^{(n)}$ of different weights are linearly independent. Hence it is enough to prove the statement for every homogeneous u with respect to the decomposition (2.11). By Theorem 2.33(i), we can write $u = xg(c,h)$, where $x \in \mathbf{B}_1$ and $g(c,h) \in \mathbb{C}[c,h]$ is a non-zero polynomial. Assume $x = e^k$ for some $k \in \mathbb{N}_0$ (the case $x = f^k$ for some $k \in \mathbb{N}$ is considered similarly). Consider the set

$$X = \{(m^2, m-1-2i) \ : \ m \in \mathbb{N},\ m > k,\ i \in \{k, k+1, \ldots, m-1\}\}.$$

The set X is obviously Zariski dense in $\mathbb{C}^2$. Hence there exists $m \in \mathbb{N}$, $m > k$, and $i \in \{k, k+1, \ldots, m-1\}$ such that $g(m^2, m-1-2i) \neq 0$. Set

$\alpha = \prod_{j=0}^{k-1}(i-j)(m-i+j) \neq 0$. For the vector $v_i \in \mathbf{V}^{(m)}$ we have

$$u(v_i) = e^k g(c,h)(v_i)$$
$$= e^k(g(c,h)(v_i))$$
$$\text{(by (1.9) and Exercise 1.33)} = g(m^2, m-1-2i)e^k(v_i)$$
$$\text{(by (1.9))} = g(m^2, m-1-2i)\,\alpha\,v_{i-k}$$
$$\neq 0.$$

This completes the proof.   □

4.7   Additional exercises

Exercise 4.29. For every $\lambda \in \mathbb{C}$ prove the equality $\text{Ann}_{U(\mathfrak{g})}(M(\lambda)) = \text{Ann}_{U(\mathfrak{g})}(M(\lambda)^{\circledast})$.

Exercise 4.30. Let $\xi \in \mathbb{C}/2\mathbb{Z}$ and $\tau \in \mathbb{C}$. Show that

$$\text{Ann}_{U(\mathfrak{g})}(\mathbf{W}(\xi,\tau)) \subsetneq \text{Ann}_{U(\mathfrak{g})}(\mathbf{V}(\xi,\tau)).$$

Exercise 4.31. Let $X \subset \mathbb{N}$ be an infinite set. Show that for every $u \in U(\mathfrak{g})$, $u \neq 0$, there exists $n \in X$ such that $u \cdot \mathbf{V}^{(n)} \neq 0$.

Exercise 4.32. Show that the annihilator in $U(\mathfrak{g})$ of the adjoint $\mathfrak{g}$-module $U(\mathfrak{g})$ is trivial.

Exercise 4.33. Let A be a finitely generated associative algebra over $\mathbb{C}$ and $z \in A$ be a central element. Show that z acts as a scalar on every simple A-module.

Exercise 4.34. Show that for every $u \in U(\mathfrak{g})$ there is a primitive quotient U' of $U(\mathfrak{g})$ such that the image of u in U' is non-zero.

Exercise 4.35 ([10, 38]). Let $\lambda \in \mathbb{C}$ and $A \subset U(\mathcal{I}_\lambda)$ be a commutative subalgebra, different from $\mathbb{C}$. Show that A is a finitely-generated integral domain of transcendence degree one over $\mathbb{C}$.

Exercise 4.36 ([10, 38]). Let $\lambda \in \mathbb{C}$, $A \subset U(\mathcal{I}_\lambda)$ be a commutative subalgebra, and $B \subset A$ be a subalgebra, different from $\mathbb{C}$. Show that A is a finitely-generated B-module.

Exercise 4.37 ([10, 38]). Let $\lambda \in \mathbb{C}$ and $A \subset U(\mathcal{I}_\lambda)$ be a subalgebra. Show that the following conditions are equivalent:

(a) A is a maximal commutative subalgebra of $U(\mathcal{I}_\lambda)$.
(b) $A = \mathcal{C}(u)$ for some non-scalar $u \in U(\mathcal{I}_\lambda)$.
(c) $A \neq \mathbb{C}$ and for any non-scalar $u \in A$ we have $A = \mathcal{C}(u)$.

Exercise 4.38 ([10, 38]). Let $\lambda \in \mathbb{C}$ and $A \subset U(\mathcal{I}_\lambda)$ be a non-commutative subalgebra. Show that

$$\{u \in U(\mathcal{I}_\lambda) : ua = au \text{ for all } a \in A\} = \mathbb{C}.$$

Exercise 4.39 ([10, 38]). Let $\lambda \in \mathbb{C}$ and $A \subset U(\mathcal{I}_\lambda)$ be a commutative subalgebra, different from $\mathbb{C}$. Show that the set

$$\{u \in U(\mathcal{I}_\lambda) : ua = au \text{ for all } a \in A\}$$

is a maximal commutative subalgebra of $U(\mathcal{I}_\lambda)$ containing A.

Exercise 4.40 ([10, 38]). Let $\lambda \in \mathbb{C}$ and $u, v \in U(\mathcal{I}_\lambda)$ be two non-scalar commuting elements. Show that $\mathcal{C}(u) = \mathcal{C}(v)$.

Exercise 4.41 ([10, 38]). Let $\lambda \in \mathbb{C}$ and $A \subset U(\mathcal{I}_\lambda)$ be a maximal commutative subalgebra. Assume that $u \in U(\mathcal{I}_\lambda)$ is such that $g(u) \in A$ for some non-scalar polynomial $g(u) \in \mathbb{C}[u]$. Show that $u \in A$.

Exercise 4.42 ([10, 38]). Let $\lambda \in \mathbb{C}$ and $A, B \subset U(\mathcal{I}_\lambda)$ be two different maximal commutative subalgebras. Show that $A \cap B = \mathbb{C}$.

Chapter 5

Category $\mathcal{O}$

5.1 Definition and basic properties

In this chapter we will study one of the most important objects in the representation theory of Lie algebras, the Bernstein–Gelfand–Gelfand (BGG) *category* $\mathcal{O}$. This category is defined as the full subcategory of the category of all $\mathfrak{g}$-modules, which consists of all $\mathfrak{g}$-modules M satisfying the following three conditions:

   (I) M is finitely generated;
   (II) M is weight;
   (III) $\dim \mathbb{C}[e](v) < \infty$ for all $v \in M$.

The condition (III) says that the action of the algebra $\mathbb{C}[e]$ on $M \in \mathcal{O}$ is *locally finite*.

Example 5.1. All finite-dimensional $\mathfrak{g}$-modules belong to $\mathcal{O}$. Indeed, for such modules, conditions (I) and (III) are obvious, and condition (II) follows from Theorems 1.22 and 1.29.

As a result of Example 5.1, the category $\mathcal{O}$ can be viewed as a natural generalization of the category of finite-dimensional $\mathfrak{g}$-modules. Also from the definition we have that $\mathcal{O}$ is a full subcategory of $\mathfrak{W}$. In this section we will list some elementary properties of $\mathcal{O}$.

Exercise 5.2. Let $M, N \in \mathcal{O}$. Show that $M \oplus N \in \mathcal{O}$.

Proposition 5.3. *The category $\mathcal{O}$ is closed with respect to taking submodules, quotients and finite direct sums. In particular, the category $\mathcal{O}$ is an abelian Krull–Schmidt category with usual kernels and cokernels and every simple object in $\mathcal{O}$ is a simple $\mathfrak{g}$-module.*

Proof. That $\mathcal{O}$ is Krull–Schmidt follows from Proposition 3.55(vii). That $\mathcal{O}$ is closed with respect to taking finite direct sums follows from Exercise 5.2. Let $M \in \mathcal{O}$ and $N \subset M$ be a submodule. Then condition (III) for N is obviously satisfied. The module N is a weight module by Proposition 3.8(i). The module N is finitely generated as M is finitely generated and $U(\mathfrak{g})$ is Noetherian (Theorem 2.40). Hence $N \in \mathcal{O}$.

For the quotient M/N conditions (I) and (III) are obviously satisfied and condition (II) is satisfied by Proposition 3.8(ii). The claim of the proposition follows. $\square$

Proposition 5.4. *For every* $\lambda \in \mathbb{C}$ *the Verma module* $M(\lambda)$ *belongs to the category* $\mathcal{O}$.

Proof. Conditions (I) and (II) for $M(\lambda)$ are obviously satisfied. Assume that $M(\lambda)$ is given by (3.4) and that $v = \sum_{i=0}^{k} a_i v_i$ is some element of $M(\lambda)$ (here $a_i \in \mathbb{C}$ for all i). From (3.4) we have that for every $m \in \mathbb{N}$ the element $e^m(v)$ is a linear combination of the vectors $v_0, \ldots, v_k$. Hence $\dim \mathbb{C}[e](v) \leq k < \infty$. Thus condition (III) is satisfied as well, and we have $M(\lambda) \in \mathcal{O}$. This completes the proof. $\square$

Proposition 5.5. *Simple objects of the category* $\mathcal{O}$ *are simple highest weight modules* $L(\lambda)$, $\lambda \in \mathbb{C}$.

Proof. That $L(\lambda) \in \mathcal{O}$ for $\lambda \in \mathbb{C}$ follows from Propositions 5.3 and 5.4. On the other hand, let $L \in \mathcal{O}$ be simple. Since L is a weight module, it contains a non-zero weight vector v, say of weight λ. We claim that $e^m(v) = 0$ for some $m \in \mathbb{N}$.

Indeed, for every $m \in \mathbb{N}_0$ the vector $e^m(v)$ is a weight vector of weight $\lambda + 2m$ (Lemma 3.5). If all $e^m(v)$ were non-zero, the vectors $e^m(v)$, $m \in \mathbb{N}_0$, would be linearly independent as weight vectors corresponding to different weights. This would imply $\dim \mathbb{C}[e](v) = \infty$, which contradicts condition (III). Hence $e^m(v) = 0$ for some minimal $m \in \mathbb{N}$. In particular, for $w = e^{m-1}(v)$ we have $w \neq 0$ and $e(w) = 0$, that is, w is a highest weight vector.

Since L is a simple $\mathfrak{g}$-module by Proposition 5.3, it is generated by the highest weight vector w. Hence L is a simple highest weight module and the claim of the proposition follows from Corollary 3.18. $\square$

Exercise 5.6. Show that $\mathcal{O}$ is not closed with respect to extensions of modules.

Exercise 5.7. Show that for every $M \in \mathcal{O}$ and every $v \in M$ there exists $k \in \mathbb{N}$ such that $e^k(v) = 0$.

Proposition 5.8. *For every $M \in \mathcal{O}$ and $\lambda \in \mathbb{C}$ we have* $\dim M_\lambda < \infty$, *in particular, the category $\mathcal{O}$ is a subcategory of $\overline{\mathfrak{W}}$.*

Proof. Assume that M is generated by $v_1, \ldots, v_k$. As M is a weight module, every vector of M is a finite sum of weight vectors. In particular, we may assume that the vectors $v_1, \ldots, v_k$ are weight vectors. We prove the statement by induction on k.

For $k = 1$ assume that $v = v_1$ is a weight vector of weight λ. Then from Lemma 3.5 we have supp $M \subset \lambda + 2\mathbb{Z}$. By Exercise 5.7, we have $e^l(v) = 0$ for some $l \in \mathbb{N}$. Consider the PBW basis of $U(\mathfrak{g})$, given by standard monomials of the form $f^i e^j h^k$, $i, j, k \in \mathbb{N}_0$. As v is a weight vector of weight λ, we have $h^k(v) = \lambda^k v$ for all $k \in \mathbb{N}_0$. Using $e^l(v) = 0$, we see that $U(\mathfrak{g})v$ is a linear span of the elements $f^i e^j(v)$, where $i \in \mathbb{N}_0$ and $j = 0, \ldots, l - 1$. In particular, for every $\mu \in \text{supp}\, M$, the dimension of M_μ does not exceed the number
$$|\{(i, j) \in \mathbb{N}_0 \times \{0, 1, \ldots, l - 1\} : \mu = \lambda + 2j - 2i\}| \leq l.$$
This shows that M has finite-dimensional weight spaces for $k = 1$.

To prove the induction step, assume $k > 1$ and denote by N the submodule of M, generated by $v_1, \ldots, v_{k-1}$. By induction, the submodule N has finite-dimensional weight spaces. By the previous paragraph we have that the submodule $N' = U(\mathfrak{g})v_k$ of M also has finite-dimensional weight spaces. Since every element of M is a sum of an element from N and an element from N', for all $\mu \in \mathbb{C}$ we have
$$\dim M_\mu \leq \dim N_\mu + \dim N'_\mu.$$
Hence M also has finite-dimensional weight spaces. This completes the proof. $\square$

Exercise 5.9. Show that for every $M \in \mathcal{O}$ there exists $\lambda_1, \ldots, \lambda_k \in \mathbb{C}$ such that
$$\text{supp}\, M \subset \bigcup_{i=1}^{k} (\lambda_i - 2\mathbb{N}_0).$$

Exercise 5.10. Show that for every $M \in \mathcal{O}$ there exists $n \in \mathbb{N}$ such that $\dim M_\lambda < n$ for all $\lambda \in \mathbb{C}$.

For $\xi \in \mathbb{C}/2\mathbb{Z}$ and $\tau \in \mathbb{C}$ set $\mathcal{O}^{\xi, \tau} = \mathcal{O} \cap \overline{\mathfrak{W}}^{\xi, \tau}$.

Theorem 5.11. *We have the decomposition*
$$\mathcal{O} = \bigoplus_{\substack{\xi \in \mathbb{C}/2\mathbb{Z} \\ \tau \in \mathbb{C}}} \mathcal{O}^{\xi, \tau}. \tag{5.1}$$

Proof. This follows from Proposition 5.8 and Lemma 3.54(ii). $\quad\square$

Corollary 5.12. *Every object $M \in \mathcal{O}$ has finite length.*

Proof. For every $\xi \in \mathbb{C}/2\mathbb{Z}$ and $\tau \in \mathbb{C}$ each object of the category $\mathcal{O}^{\xi,\tau}$ has finite length by Proposition 3.55(v). At the same time, by Theorem 5.11 each $M \in \mathcal{O}$ may be written as a finite direct sum of some $M^\xi(\tau) \in \mathcal{O}^{\xi,\tau}$. The claim follows. $\quad\square$

Proposition 5.13. *The duality $\circledast$ restricts to an exact contravariant self-equivalence of $\mathcal{O}$, which preserves isomorphism classes of simple objects.*

Proof. Because of Theorem 3.84, we only have to check that $\circledast$ preserves the category $\mathcal{O}$. Let $M \in \mathcal{O}$. As $\circledast$ preserves $\mathfrak{W}$, we get that $M^\circledast$ is a weight module. As M has finite length (Corollary 5.12) and $\circledast$ is exact and preserves isomorphism classes of simple objects, the module $M^\circledast$ has finite length as well. In particular, $M^\circledast$ is finitely generated. By Exercise 5.9, there exists $\lambda_1, \ldots, \lambda_k \in \mathbb{C}$ such that

$$\operatorname{supp} M \subset \bigcup_{i=1}^{k}(\lambda_i - 2\mathbb{N}_0).$$

By Proposition 3.82(ii), we have $\operatorname{supp} M^\circledast = \operatorname{supp} M$ and hence

$$\operatorname{supp} M^\circledast \subset \bigcup_{i=1}^{k}(\lambda_i - 2\mathbb{N}_0).$$

This implies that condition (III) is satisfied and hence $M^\circledast \in \mathcal{O}$. The claim follows. $\quad\square$

Exercise 5.14. Show that the functor $\circledcirc$ from 3.12.6 does not preserve the category $\mathcal{O}$.

Exercise 5.15. Let $\xi \in \mathbb{C}/2\mathbb{Z}$ and $\tau \in \mathbb{C}$. Show that

(a) The category $\mathcal{O}^{\xi,\tau}$ is trivial if, and only if, $\tau \neq (\lambda+1)^2$ for all $\lambda \in \xi$.
(b) If there exists a unique $\lambda \in \xi$ such that $\tau = (\lambda+1)^2$, then the category $\mathcal{O}^{\xi,\tau}$ has exactly one simple object; namely $M(\lambda) = L(\lambda)$.
(c) If there exists $\lambda_1, \lambda_2 \in \xi$ such that $\lambda_1 \neq \lambda_2$ and $\tau = (\lambda_1+1)^2 = (\lambda_2+1)^2$, then $\lambda_1 \in \mathbb{Z}$ and the category $\mathcal{O}^{\xi,\tau}$ has two (non-isomorphic) simple objects, namely $L(\lambda_1)$ and $L(\lambda_2)$.

5.2 Projective modules

The properties of the category $\mathcal{O}$, established in the previous section, look very similar to the properties of the categories $\overline{\mathfrak{W}}^{\xi,\tau}$ obtained in Proposition 3.55. However, in this section we are going to establish one very important property of $\mathcal{O}$, which none of the categories $\overline{\mathfrak{W}}^{\xi,\tau}$ has: the existence of projective covers. Recall that an abelian category is said to have *enough projective objects* if every object in this category is a quotient of some projective object.

Theorem 5.16. *The category $\mathcal{O}$ has enough projective objects.*

To prove this theorem we will need some preparation.

Lemma 5.17. *Let $\lambda \in \mathbb{C} \setminus \{-2, -3, \dots\}$. Then the module $M(\lambda)$ is projective in $\mathcal{O}$.*

Proof. Let $\xi = \lambda + 2\mathbb{Z}$ and $\tau = (\lambda + 1)^2$. Then $M(\lambda) \in \mathcal{O}^{\xi,\lambda}$ by Proposition 5.4 and Proposition 3.13(iii). Because of the decomposition (5.1) we have to check that $M(\lambda)$ is projective in $\mathcal{O}^{\xi,\lambda}$.

By Exercise 5.15, the simple objects in $\mathcal{O}^{\xi,\lambda}$ are $L(\lambda)$ (always) and $L(-\lambda - 2)$ (if $\lambda \in \mathbb{N}_0$). Every object in $\mathcal{O}^{\xi,\lambda}$ is obtained from this (these) simple object(s) by extensions. Note that both $\operatorname{supp} L(\lambda)$ and $\operatorname{supp} L(-\lambda - 2)$ (if $\lambda \in \mathbb{N}_0$) are contained in the set $X = \lambda - 2\mathbb{N}_0$. Hence for every $M \in \mathcal{O}^{\xi,\lambda}$ we have $\operatorname{supp} M \subset X$.

This means that for every $M \in \mathcal{O}^{\xi,\lambda}$ and every $v \in M_\lambda$ we have $e(v) = 0$. From the universal property of Verma modules (Corollary 3.15) we get that for every $M \in \mathcal{O}^{\xi,\lambda}$ we have

$$\operatorname{Hom}_{\mathfrak{g}}(M(\lambda), M) = M_\lambda.$$

By Exercise 3.11, the functor $M \mapsto M_\lambda$ is an exact functor from $\mathcal{O}^{\xi,\lambda}$ to the category $\mathbb{C}$-mod of all complex vector spaces. Hence the functor $\operatorname{Hom}_{\mathfrak{g}}(M(\lambda), \_)$ from $\mathcal{O}^{\xi,\lambda}$ to $\mathbb{C}$-mod is exact. This means that the module $M(\lambda)$ is projective. $\square$

Exercise 5.18. Let $\lambda \in \mathbb{C} \setminus \{-2, -3, \dots\}$. Show that the module $M(\lambda)^{\circledast}$ is injective in $\mathcal{O}$.

Corollary 5.19. *Let $\lambda \in \mathbb{C} \setminus \{-2, -3, \dots\}$. Then the simple module $L(\lambda)$ has a projective cover in $\mathcal{O}$.*

Proof. The module $L(\lambda)$ is a quotient of $M(\lambda)$ by Corollaries 3.18 and 3.19. For $\lambda \in \mathbb{C} \setminus \{-2, -3, \dots\}$ the module $M(\lambda)$ is projective by Lemma 5.17. This completes the proof. $\qquad\qquad\qquad\qquad\qquad\qquad\square$

Lemma 5.20. *For every finite-dimensional* $\mathfrak{g}$*-module* V *and every* $M \in \mathcal{O}$ *we have* $V \otimes M \in \mathcal{O}$. *In particular, the endofunctor* $V \otimes \_$ *of* $\mathfrak{g}$*-mod restricts to an exact and self-adjoint endofunctor of* $\mathcal{O}$.

Proof. The module $V \otimes M$ is weight by Proposition 3.8(iv). By Exercise 5.9, there exists $\lambda_1, \dots, \lambda_k \in \mathbb{C}$ such that

$$\operatorname{supp} M \subset \bigcup_{i=1}^{k} (\lambda_i - 2\mathbb{N}_0).$$

By Exercise 3.10, we have

$$\operatorname{supp} V \otimes M \subset \bigcup_{i=1}^{k} (\operatorname{supp} V + \lambda_i - 2\mathbb{N}_0).$$

As V is finite-dimensional, the set $\operatorname{supp} V$ is finite. It follows that condition (III) for the module $V \otimes M$ is satisfied.

Since $V \otimes \_$ is exact, from Theorem 3.81 it follows that for every simple module $L \in \mathcal{O}$ the module $V \otimes L$ has finite length. From Corollary 5.12 we know that M has finite length. Hence, using the exactness of $V \otimes \_$ again, we conclude that $V \otimes M$ has finite length and so is finitely generated.

This means that $V \otimes M \in \mathcal{O}$ and thus $V \otimes \_$ restricts to an endofunctor of $\mathcal{O}$. That $V \otimes \_$ is exact follows from the fact that it is self-adjoint, which, in turn, follows from Exercise 3.72. $\qquad\qquad\qquad\qquad\qquad\square$

Corollary 5.21. *For every finite-dimensional* $\mathfrak{g}$*-module* V *the endofunctor* $V \otimes \_$ *of* $\mathcal{O}$ *sends projective modules to projective modules.*

Proof. Let $P \in \mathcal{O}$ be projective. By the self-adjointness of $V \otimes \_$ we have the natural isomorphism of functors (from $\mathcal{O}$ to $\mathbb{C}$-mod) as follows:

$$\operatorname{Hom}_{\mathfrak{g}}(V \otimes P, \_) = \operatorname{Hom}_{\mathfrak{g}}(P, V \otimes \_).$$

Now the functor $V \otimes \_$ is exact by Lemma 5.20, and the functor $\operatorname{Hom}_{\mathfrak{g}}(P, \_)$ is exact since the module P is projective. Hence the functor $\operatorname{Hom}_{\mathfrak{g}}(V \otimes P, \_)$ is exact as a composition of two exact functors. This means that the module $V \otimes P$ is projective and the claim follows. $\qquad\qquad\qquad\square$

Exercise 5.22. Show that for every finite-dimensional $\mathfrak{g}$-module V, the endofunctor $V \otimes \_$ of $\mathcal{O}$ sends injective modules to injective modules.

Corollary 5.23. *For every* $n \in \{2, 3, 4, \ldots\}$ *the simple module* $L(-n)$ *has a projective cover in* $\mathcal{O}$.

Proof. First we observe that $L(-n) = M(-n)$ by Theorem 3.16. The module $M(0)$ is projective in $\mathcal{O}$ by Lemma 5.17. Hence, by Corollary 5.21, the module $\mathbf{V}^{(n+1)} \otimes M(0)$ is projective in $\mathcal{O}$ as well. At the same time, from the self-adjointness of $\mathbf{V}^{(n+1)} \otimes \_$ we have

$$\mathrm{Hom}_{\mathfrak{g}}(\mathbf{V}^{(n+1)} \otimes M(0), L(-n)) = \mathrm{Hom}_{\mathfrak{g}}(M(0), \mathbf{V}^{(n+1)} \otimes M(-n)). \quad (5.2)$$

By Proposition 3.13(ii) we have $\mathrm{supp}\, M(-n) = -n - 2\mathrm{N}_0$. From (1.9) we also have $\mathrm{supp}\, \mathbf{V}^{(n+1)} = \{-n, -n+2, \ldots, n-2, n\}$. Hence from Exercise 3.10 we obtain

$$\mathrm{supp}\, \mathbf{V}^{(n+1)} \otimes M(-n) = -2\mathrm{N}_0.$$

In particular, there exists a non-zero $v \in (\mathbf{V}^{(n+1)} \otimes M(-n))_0$ and this vector satisfies $e(v) = 0$. By the universal property of Verma modules (Corollary 3.15), we thus get a non-zero homomorphism from the module $M(0)$ to $\mathbf{V}^{(n+1)} \otimes M(-n)$. From (5.2) it thus follows that

$$\mathrm{Hom}_{\mathfrak{g}}(\mathbf{V}^{(n+1)} \otimes M(0), L(-n)) \neq 0,$$

which means that $\mathbf{V}^{(n+1)} \otimes M(0)$ is a projective cover of $L(-n)$. $\qquad\square$

Now we are ready to prove Theorem 5.16.

Proof. Since every object in $\mathcal{O}$ has finite length (Corollary 5.12), we can prove the existence of a projective cover for $M \in \mathcal{O}$ by induction on the length of M. If the module M is simple, the statement follows from Proposition 5.5 and Corollaries 5.19 and 5.23. Now assume that M is not simple and consider any short exact sequence $L \hookrightarrow M \twoheadrightarrow N$, where L is simple. Then the length of N is strictly smaller than the length of M. Let $P \twoheadrightarrow L$ and $Q \twoheadrightarrow N$ be projective covers, which exist by the inductive assumption. Using the projectivity of Q we can lift the surjection $Q \twoheadrightarrow N$ to a homomorphism $Q \to M$ such that the following diagram commutes:

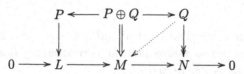

(here the maps $P \oplus Q \to P$ and $P \oplus Q \to Q$ are natural projections). The above diagram gives as a surjection from the projective module $P \oplus Q$ to M. Now the claim of the theorem follows by induction. $\qquad\square$

Exercise 5.24. Show that for any $n \in \mathbb{N}$ the module $\mathbf{V}^{(n)} \otimes M(-1)$ is projective in $\mathcal{O}$ and that it contains, as a direct summand, the projective cover of the module $L(-n)$.

Exercise 5.25. Show that for any $\lambda \in \mathbb{C} \setminus \{\ldots, -4, -3, -2, 0, 1, 2, 3, \ldots\}$ the module $M(\lambda)$ is both projective, injective and simple in $\mathcal{O}$.

Exercise 5.26. Show that $\mathcal{O}$ has *enough injectives*; that is, that every module in $\mathcal{O}$ has an injective envelope in $\mathcal{O}$.

As a result of Theorem 5.16 and Exercise 5.26, for every $\lambda \in \mathbb{C}$ we have the indecomposable projective cover $P(\lambda)$ of $L(\lambda)$ and the indecomposable injective envelope $I(\lambda)$ of $L(\lambda)$. By Lemma 5.17 and Exercise 5.18, for $\lambda \in \mathbb{C} \setminus \{-2, -3, \ldots\}$ we have $P(\lambda) = M(\lambda)$ and $I(\lambda) = M(\lambda)^{\circledast}$. For $\lambda \in \{-2, -3, \ldots\}$ the module $P(\lambda)$ is called the *big projective* module. For such λ the structure of $P(\lambda)$ and $I(\lambda)$ is described in the following statement:

Proposition 5.27. *For* $\lambda \in \{-2, -3, \ldots\}$ *we have the following:*

(i) $P(\lambda) \cong I(\lambda)$.
(ii) *The module* $P(\lambda)$ *has a basis*

$$\{v_\mu : \mu \in -\lambda - 2 - 2\mathbb{N}_0\} \cup \{w_\mu : \mu \in \lambda - 2\mathbb{N}_0\}$$

such that the action of $\mathfrak{g}$ *in this basis can be depicted as follows:*

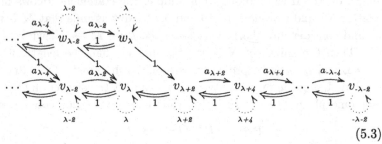

$$\tag{5.3}$$

(here $a_\mu = \frac{1}{4}(\mu + \lambda + 2)(\lambda - \mu)$ *for all* $\mu \in -\lambda - 2 - 2\mathbb{N}_0$*, and an absence of some arrow means that the corresponding linear operator acts on this basis vector as zero).*

(iii) *The module* $P(\lambda)$ *is uniserial of length 3. Its simple top and simple socle are isomorphic to* $L(\lambda)$ *and the intermediate subquotient is isomorphic to* $L(-\lambda - 2)$*, which may be depicted as follows:*

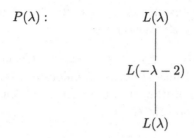

Proof. Let $\lambda = -n$ for $n \in \mathbb{N}$, $n \neq 1$. By our proof of Theorem 5.16, the module $P(-n)$ is a submodule of the module $\mathbf{V}^{(n+1)} \otimes M(0)$. As the module $M(0)$ is a submodule of $\mathbf{V}(2\mathbb{Z}, 1^2)$, using the exactness of the functor $\mathbf{V}^{(n+1)} \otimes \_$ we have that $P(-n)$ is a submodule of $\mathbf{V}^{(n+1)} \otimes \mathbf{V}(2\mathbb{Z}, 1^2)$. Observe that $P(-n) \in \overline{\mathfrak{W}}^{-n+2\mathbb{Z},(n-1)^2}$ and hence from Theorem 3.81 we obtain that $P(-n)$ is a submodule of $\mathbf{W}(-n+2\mathbb{Z}, (n-1)^2)$. If the module $\mathbf{W}(-n+2\mathbb{Z}, (n-1)^2)$ is given by (3.32), then a direct calculation shows that the maximal submodule of $\mathbf{W}(-n+2\mathbb{Z}, (n-1)^2)$, which belongs to $\mathcal{O}$, is given by (5.3). Call this module N.

A direct calculation shows that the module N, given by (5.3), is a uniserial module satisfying the assertion of (iii). By the above paragraph, the module $P(-n)$ is a submodule of N with simple top $L(-n)$. Hence we either have $P(-n) = N$ or $P(-n) = L(-n)$. On the other hand, the module N has simple top $L(-n)$ and belongs to $\mathcal{O}$, hence N is a quotient of the projective module $P(-n)$. This yields $P(-n) = N$, proving both (ii) and (iii).

The module $M(-1)$ is projective in $\mathcal{O}$ by Lemma 5.17 and is injective in $\mathcal{O}$ by Exercise 5.18. The functor $\mathbf{V}^{(n)} \otimes \_$ sends projective modules to projective (Corollary 5.21) and injective modules to injective (Exercise 5.22). Hence every direct summand of the projective module $\mathbf{V}^{(n)} \otimes M(-1)$ is also injective. Using Exercise 5.24, we thus get that the module $P(-n)$ is injective. By (iii) it has simple socle $L(\lambda)$, which yields (i). This completes the proof. $\qquad\square$

Corollary 5.28. *Let $\lambda \in \mathbb{C}$. The indecomposable projective module $P(\lambda)$ is injective if, and only if, $\lambda \notin \mathbb{N}_0$. Moreover, for $\lambda \notin \mathbb{N}_0$ we have $P(\lambda) = I(\lambda)$.*

Proof. This follows directly from Proposition 5.27, Lemma 5.17, Theorem 3.16 and Exercise 5.18. $\qquad\square$

Exercise 5.29. Show that for any $\lambda \in \{-2, -3, \dots\}$ the module $P(\lambda)$ is isomorphic to $\mathcal{E}(\mathbf{W}(\lambda + 2\mathbb{Z}, (\lambda + 1)^2))$.

Exercise 5.30. Let $\lambda \in \{-2, -3, \dots\}$. Show that multiplication with the element $c - (\lambda + 1)^2$ defines a non-zero endomorphism φ of $P(\lambda)$, which satisfies $\varphi^2 = 0$. Show further that the endomorphism algebra of $P(\lambda)$ is isomorphic to the algebra $\mathbb{C}[x]/(x^2)$.

5.3   Blocks via quiver and relation

The summands $\mathcal{O}^{\xi, \tau}$ from the decomposition (5.1) are called *blocks* of $\mathcal{O}$. From previous sections we have that every block of $\mathcal{O}$ is an abelian category with enough projective objects, moreover, all objects of $\mathcal{O}$ have finite length. From Exercise 5.15 we also have that every block $\mathcal{O}^{\xi, \tau}$ has only finitely many simple objects (up to isomorphism). This suggests that blocks of $\mathcal{O}$ can be described using finite-dimensional associative algebras.

Theorem 5.31 (Description of blocks of $\mathcal{O}$). *Let $\xi \in \mathbb{C}/2\mathbb{Z}$, $\tau \in \mathbb{C}$.*

(i) *If $(\lambda + 1)^2 \neq \tau$ for all $\lambda \in \xi$, then the block $\mathcal{O}^{\xi, \tau}$ is zero.*

(ii) *If $(\lambda + 1)^2 = \tau$ for a unique $\lambda \in \xi$, then the block $\mathcal{O}^{\xi, \tau}$ is semi-simple and equivalent to the category $\mathbb{C}$-mod of complex vector spaces (or, equivalently, $\mathbb{C}$-modules).*

(iii) *If $(\lambda_1 + 1)^2 = (\lambda_2 + 1)^2 = \tau$ for $\lambda_1, \lambda_2 \in \xi$, $\lambda_1 \neq \lambda_2$, then $\tau = n^2$ for some $n \in \mathbb{N}$ and the block $\mathcal{O}^{\xi, \tau}$ is equivalent to the category of modules over the following $\mathbb{C}$-category, whose path algebra is finite-dimensional:*

$$\mathfrak{D}: \qquad \mathfrak{p} \underset{b}{\overset{a}{\rightleftarrows}} \mathfrak{q}, \qquad ab = 0.$$

The blocks $\mathcal{O}^{\xi, \tau}$ from Theorem 5.31(iii) are called *regular* blocks. This terminology stems from the observation that simple objects in such blocks are indexed by regular orbits of the dot-action of the Weyl group on $\mathfrak{h}^*$, described in Section 2.5. For similar reasons, the block $\mathcal{O}^{-1+2\mathbb{Z}, 0}$ is called *singular*.

Proof. As the module $\mathbf{V}(\xi, \tau)$ never belongs to $\mathcal{O}$ (since for the module $\mathbf{V}(\xi, \tau)$ condition (III) obviously fails), the claim (i) follows directly from Proposition 3.55(ii).

If $(\lambda + 1)^2 = \tau$ for a unique $\lambda \in \xi$, then, comparing Proposition 3.55(iii) and Proposition 5.5, we obtain that $\mathcal{O}^{\xi,\tau}$ contains a unique simple object, namely $M(\lambda)$. In this case we also have $\tau \neq n^2$ for any $n \in \mathbb{N}$ (otherwise we would be in the situation of claim (iii)). Hence $\lambda \notin \mathbb{Z} \setminus \{-1\}$ and from Lemma 5.17 we get that the simple object $M(\lambda)$ is also projective. Hence the category $\mathcal{O}^{\xi,\tau}$ is a semi-simple category with one simple object, whose endomorphism algebra is $\mathbb{C}$ (Exercise 3.17). This means that the category $\mathcal{O}^{\xi,\tau}$ is equivalent to $\mathbb{C}$-mod, completing the proof of (ii).

Finally, assume that $\tau = n^2$ for some $n \in \mathbb{N}$. Comparing Proposition 3.55(iv) and Proposition 5.5, we obtain that $\mathcal{O}^{\xi,\tau}$ has two simple objects, namely $L(-n - 1)$ and $L(n - 1)$. Consider the corresponding indecomposable projective covers $P(-n-1)$ and $P(n-1)$. We have $P(n-1) \cong M(n-1)$, which is given by (3.4), and the big projective module $P(-n-1)$ is given by (5.3). As $L(-n - 1) \cong M(-n - 1) \subset M(n - 1)$ is a unique occurrence of the subquotient $L(-n-1)$ in $M(n-1)$ (by Theorem 3.16(ii)) and $P(-n - 1)$ is projective with simple top $L(-n - 1)$, we have a unique (up to a non-zero scalar) non-zero morphism $\varphi : P(-n - 1) \to M(n - 1)$. As $L(n - 1)$ occurs with multiplicity one as a subquotient of $P(-n - 1)$ by Proposition 5.27(iii) and $M(n - 1)$ is projective, we have a unique (up to a non-zero scalar) non-zero morphism $\psi : M(n - 1) \to P(-n - 1)$. Denote by $\mathcal{D}$ the full subcategory of $\mathcal{O}^{\xi,\tau}$, which has two objects $P(-n - 1)$ and $P(n - 1)$. The path algebra of $\mathcal{D}$ is finite-dimensional by Proposition 3.55(vi).

Lemma 5.32. *The map*
$$\mathbf{p} \mapsto P(-n - 1), \quad \mathbf{q} \mapsto P(n - 1), \quad \mathbf{a} \mapsto \varphi, \quad \mathbf{b} \mapsto \psi.$$
extends uniquely to an isomorphism $\mathfrak{f}$ from the $\mathbb{C}$-category $\mathfrak{D}$ to the $\mathbb{C}$-category $\mathcal{D}$.

Proof. A direct computation using (3.4), (5.3) and the definitions of φ and ψ gives $\varphi\psi = 0$. As $\mathbf{ab} = 0$ is the only defining relation of $\mathfrak{D}$, using the standard universal construction (quotient of a free category modulo relations) we get that our map extends to a functor $\mathfrak{f}$ from $\mathfrak{D}$ to $\mathcal{D}$. To prove that this functor is, in fact, an isomorphism, we have to see how it acts on the spaces of homomorphisms between objects. Set
$$V_1 = \mathfrak{D}(\mathbf{p}, \mathbf{p}), \quad W_1 = \operatorname{Hom}_{\mathfrak{g}}(P(-n - 1), P(-n - 1)),$$
$$V_2 = \mathfrak{D}(\mathbf{q}, \mathbf{q}), \quad W_2 = \operatorname{Hom}_{\mathfrak{g}}(P(n - 1), P(n - 1)),$$
$$V_3 = \mathfrak{D}(\mathbf{p}, \mathbf{q}), \quad W_3 = \operatorname{Hom}_{\mathfrak{g}}(P(-n - 1), P(n - 1)),$$
$$V_4 = \mathfrak{D}(\mathbf{q}, \mathbf{p}), \quad W_4 = \operatorname{Hom}_{\mathfrak{g}}(P(n - 1), P(-n - 1)).$$

First we claim that $\mathfrak{f}$ is full. Indeed, from the above we have that $\dim W_3 = \dim W_4 = 1$ and also $\dim W_2 = 1$ by Exercise 3.17. Hence the surjectivity of $\mathfrak{f}$ on these spaces follows immediately from the definition. By Exercise 5.30 we have $\dim W_1 = 2$. The space $\mathfrak{f}(V_1)$ contains the identity map and the homomorphism $\mathfrak{f}(\mathbf{ba}) = \psi\varphi$. From the definitions it follows that $\psi\varphi \neq 0$, while $(\psi\varphi)^2 = 0$. Hence the identity and $\psi\varphi$ are linearly independent and thus the functor $\mathfrak{f}$ is full.

To complete the proof, it is enough to check that $\dim V_i = \dim W_i$ for all $i = 1, 2, 3, 4$. From the previous paragraph we know that $\dim W_1 = 2$ and $\dim W_i = 1$, $i = 2, 3, 4$. To check that $\dim V_1 = 2$ and that $\dim V_i = 1$, $i = 2, 3, 4$, is a direct calculation using the definition of $\mathfrak{D}$ and is left to the reader. This completes the proof. $\qquad\square$

Lemma 5.33. *The category $\mathfrak{D}$ is isomorphic to the opposite category $\mathfrak{D}^{op}$ via an isomorphism $\mathfrak{t}$, which swaps $\mathbf{a}$ and $\mathbf{b}$.*

Proof. We have $\mathfrak{t}(\mathbf{ab}) = \mathfrak{t}(\mathbf{b})\mathfrak{t}(\mathbf{a}) = \mathbf{ab} = 0$, and hence $\mathfrak{t}$ extends to a functor from $\mathfrak{D}$ to $\mathfrak{D}^{op}$. This functor is full since its image contains both generators $\mathbf{a}$ and $\mathbf{b}$. The claim now follows by the obvious comparison of the dimensions of the corresponding homomorphism spaces in the categories $\mathfrak{D}$ and $\mathfrak{D}^{op}$. $\qquad\square$

For $M \in \mathcal{O}^{\xi,\tau}$ we consider the following diagram $\mathrm{F}\,M$:

$$\mathrm{Hom}_{\mathfrak{g}}(P(-n-1), M) \underset{-\circ\varphi}{\overset{-\circ\psi}{\rightleftarrows}} \mathrm{Hom}_{\mathfrak{g}}(P(n-1), M) \qquad (5.4)$$

Exercise 5.34.

(a) Check that the diagram (5.4) defines a $\mathcal{D}^{op}$-module.
(b) Check that F extends to a functor from $\mathcal{O}^{\xi,\tau}$ to $\mathcal{D}^{op}$-mod by defining it on morphisms in the natural way.
(c) Check that the functor F sends simple objects from $\mathcal{O}^{\xi,\tau}$ to one-dimensional (and hence simple) objects in $\mathcal{D}^{op}$-mod.

The category $\mathcal{D}$ itself has the natural structure of a left $U(\mathfrak{g})$-module $P(-n-1) \oplus P(n-1)$, and also of a left $\mathcal{D}$-module (or, equivalently, right $\mathcal{D}^{op}$-module). These two structures commute giving us on $\mathcal{D}$ the structure of a $U(\mathfrak{g}) - \mathcal{D}^{op}$ bimodule. Hence, for any $\mathcal{D}^{op}$-module V we can consider the module

$$\mathrm{G}\,V = \mathcal{D} \bigotimes_{\mathcal{D}^{op}} V.$$

Exercise 5.35.

(a) Check that $GV \in \mathcal{O}^{\xi,\tau}$.

(b) Check that G extends to a functor from $\mathcal{D}^{op}$-mod to $\mathcal{O}^{\xi,\tau}$ by defining it on morphisms in the natural way.

(c) Check that G sends simple objects from $\mathcal{D}^{op}$-mod to simple objects in $\mathcal{O}^{\xi,\tau}$.

(d) Check that (G, F) is an adjoint pair of functors.

Lemma 5.36. *The functors F and G are mutually inverse equivalences of categories between $\mathcal{O}^{\xi,\tau}$ to the category $\mathcal{D}^{op}$-mod.*

Proof. From Exercise 5.35(d) we have adjunction morphisms

$$\mathrm{ID}_{\mathcal{D}^{op}\text{-mod}} \to FG, \quad GF \to \mathrm{ID}_{\mathcal{O}^{\xi,\tau}}.$$

By Exercise 5.35(c) and Exercise 5.34(c), the adjunction morphisms are isomorphisms when applied to simple modules. Since every module in both $\mathcal{O}^{\xi,\tau}$ and $\mathcal{D}^{op}$-mod has finite length, using the Five Lemma and induction on the length of a module, we can see that adjunction morphisms are isomorphisms for all modules (see the proof of Theorem 3.58). The claim of the lemma follows. $\square$

The statement Theorem 5.31(iii) follows directly from Lemmas 5.32, 5.33 and 5.36. $\square$

Remark 5.37. An alternative way to prove Theorem 5.31 is to use the results of Sections 3.8 and 3.9. The only nontrivial case is that of Theorem 5.31(iii). In this case the claim reduces to determination of the quotient category of the category $\mathfrak{B}$ modulo the ideal, generated by the maps c, d and 1_r.

Corollary 5.38. *All regular blocks of $\mathcal{O}$ are equivalent.*

Proof. This follows directly from Theorem 5.31(iii). $\square$

Corollary 5.39. *All blocks of $\mathcal{O}$ are indecomposable (as categories).*

Proof. This also follows directly from Theorem 5.31(iii). $\square$

From Theorem 5.31(ii) we see that those blocks $\mathcal{O}^{\xi,\tau}$, which contain only one simple object, are semi-simple. In particular, every indecomposable object in such block is simple. Using the very nice description of the regular blocks obtained in Theorem 5.31(iii), we can classify all indecomposable objects in such blocks as well.

Theorem 5.40. *Let* $n \in \mathbb{N}$, $\xi = n - 1 + 2\mathbb{Z}$ *and* $\tau = n^2$. *Then the block* $\mathcal{O}^{\xi,\tau}$ *contains (up to isomorphism) exactly five pairwise non-isomorphic indecomposable objects, namely,*

$$L(-n-1), \quad L(n-1), \quad M(n-1), \quad P(-n-1) \quad and \quad M(n-1)^{\circledast}. \quad (5.5)$$

Proof. Any $\mathfrak{D}$-module V has the form

$$V(\mathsf{p}) \underset{V(\mathsf{b})}{\overset{V(\mathsf{a})}{\rightleftarrows}} V(\mathsf{q}), \qquad V(\mathsf{a})V(\mathsf{b}) = 0.$$

Choosing bases in $V(\mathsf{p})$ and $V(\mathsf{q})$ we can consider the matrices A and B of the linear operators $V(\mathsf{a})$ and $V(\mathsf{b})$, respectively. Let $\dim V(\mathsf{p}) = m$ and $\dim V(\mathsf{q}) = n$. Using the Gaussian elimination method, we obtain that the bases in $V(\mathsf{p})$ and $V(\mathsf{q})$ can be chosen such that the matrix B has has the form

$$B = \left(\begin{array}{c|c} E & 0 \\ \hline 0 & 0 \end{array}\right), \qquad (5.6)$$

where E is the identity matrix of size $k \le \min(m,n)$. From $AB = 0$ it thus follows that the matrix A has the form

$$A = \left(\begin{array}{c|c} 0 & A' \\ \hline 0 & A'' \end{array}\right),$$

where again the zero matrix in the left upper corner has size $k \times k$. Using elementary matrix transformations we can reduce A' to the form (5.6). The elementary transformations of the first k rows of A correspond to the elementary transformations of the first k columns of B, which affects the identity part of B. However, this can be compensated by elementary row transformations of the identity part of B, which correspond to elementary transformations of the first k columns of A, and the latter do not affect A at all. This means that the matrix A can be reduced to the form

$$A = \left(\begin{array}{c|cc} 0 & E & 0 \\ 0 & 0 & 0 \\ \hline 0 & A_1'' & A_2'' \end{array}\right).$$

Adding one of the first k rows of A to some row $l > k$ does not affect the matrix B (as it corresponds to the adding of the zero column l of B to one of the first k columns). Hence we can use such elementary transformations to make $A_1'' = 0$, reducing A to the form

$$A = \left(\begin{array}{c|cc} 0 & E & 0 \\ 0 & 0 & 0 \\ \hline 0 & 0 & A_2'' \end{array}\right).$$

Finally, any elementary transformation of rows or columns of the matrix A_2'' does not affect the matrix B and hence we can reduce the matrix A to the form

$$A = \begin{pmatrix} 0 & E & 0 & 0 \\ 0 & 0 & 0 & 0 \\ \hline 0 & 0 & E & 0 \\ 0 & 0 & 0 & 0 \end{pmatrix}. \tag{5.7}$$

This means that in any $\mathfrak{D}$-module V, we can choose some bases in $V(\mathbf{p})$ and $V(\mathbf{q})$ such that the matrix $A = V(\mathbf{a})$ has the form (5.7) and that the matrix $B = V(\mathbf{b})$ has the form (5.6). It is clear that there are only five pairs (A, B) of the matrices, given by (5.7) and (5.6), such that the module V is indecomposable, namely, the following pairs:

(a) $m = 1,\ n = 0$;
(b) $m = 0,\ n = 1$;
(c) $A = (1),\ B = (0)$;
(d) $A = (0),\ B = (1)$;
(e) $A = \begin{pmatrix} 0 & 1 \end{pmatrix},\ B = \begin{pmatrix} 1 \\ 0 \end{pmatrix}$.

It follows that $\mathfrak{D}$-mod contains (up to isomorphism) exactly five pairwise nonisomorphic indecomposable objects.

We leave it to the reader to check that all five objects in our list (5.5) are indeed indecomposable and pairwise nonisomorphic. Application of Theorem 5.31(iii) completes the proof. □

Exercise 5.41. Check that under the equivalence, constructed during the proof of Theorem 5.31(iii), we have the following correspondence between the indecomposable $\mathfrak{D}$-modules, given by (a)–(e) above, and the modules from the list (5.5):

$$\text{(a)} \leftrightarrow L(-n-1), \quad \text{(b)} \leftrightarrow L(n-1), \quad \text{(c)} \leftrightarrow M(n-1)^{\circledast},$$
$$\text{(d)} \leftrightarrow M(n-1), \quad \text{(e)} \leftrightarrow P(-n-1).$$

5.4 Structure of a highest weight category

The description of the blocks of the category $\mathcal{O}$, in particular, of all indecomposable objects in these blocks, obtained in the previous section, is so complete and explicit that it can easily be used to obtain structural information about this category and the finite-dimensional associative algebras describing its blocks. As the study of semi-simple blocks does not really sound as an intriguing problem, we will concentrate our structural study on regular blocks. By Corollary 5.38 we can actually pick one of them. We consider the block $\mathcal{O}^{2\mathbb{Z},1}$, which is called the *principal* block of $\mathcal{O}$. We will denote this block by $\mathcal{O}_0$. The block $\mathcal{O}_0$ is characterized as the indecomposable direct summand of $\mathcal{O}$, containing the trivial (one-dimensional) $\mathfrak{g}$-module. We also denote by D the path algebra of the category $\mathfrak{D}$.

Exercise 5.42. Show that the algebra D is isomorphic to the following matrix algebra:

$$\left\{ \begin{pmatrix} c & b & d \\ 0 & x & a \\ 0 & 0 & c \end{pmatrix} : a, b, c, d, x \in \mathbb{C} \right\}.$$

The simple objects in $\mathcal{O}_0$ are the modules $L(-2)$ and $L(0)$. These modules are indexed by the corresponding highest weights -2 and 0. On the set $\mathrm{I} = \{-2, 0\}$ we have the linear order, coming from the natural linear order on $\mathbb{Z}$. Our first observation is the following:

Proposition 5.43.

(i) *For every* $\lambda \in \mathrm{I}$ *all composition subfactors of the Verma module* $M(\lambda)$ *have the form* $L(\mu)$, $\mu \leq \lambda$. *Moreover, the module* $L(\lambda)$ *occurs with multiplicity one.*

(ii) *For every* $\lambda \in \mathrm{I}$ *the indecomposable projective module* $P(\lambda)$ *has a filtration, whose subquotients have the form* $M(\mu)$, $\mu \geq \lambda$. *Moreover, the module* $M(\lambda)$ *occurs exactly once.*

Proof. The claim (i) follows from Theorem 3.16.

The module $P(0) = M(0)$ is a Verma module and has a unique filtration $0 \subset M(0)$, whose subquotients are isomorphic to Verma modules. This filtration obviously satisfies the assertion of the claim (ii).

From the explicit description of $P(-2)$, given by Proposition 5.27(iii), we have that $P(-2)$ has a unique filtration, whose subquotients are isomorphic to Verma modules, namely, the filtration $0 \subset N \subset P(-2)$, where N is the image of any non-zero homomorphism from $M(0)$ to $P(-2)$. The subquotients of this filtration are isomorphic to $M(0)$ and $M(-2)$, respectively, and hence this filtration satisfies the assertion of the claim (ii). This completes the proof. $\square$

Exercise 5.44 (BGG-reciprocity). For any $\lambda, \mu \in I$ we denote by $[P(\lambda) : M(\mu)]$ the multiplicity of $M(\mu)$ as a subquotient of the filtration of $P(\lambda)$, given by Proposition 5.43(ii). Show that

$$[P(\lambda) : M(\mu)] = [M(\mu) : L(\lambda)].$$

Proposition 5.43 says that the category $\mathcal{O}_0$ is a *highest weight* category. Equivalently, the algebra D is a *quasi-hereditary* algebra. The Verma modules $M(-2)$ and $M(0)$, which play an important role in the definition of the highest weight structure, are called *standard* modules, and are usually denoted as follows: $\Delta(-2) = M(-2)$ and $\Delta(0) = M(0)$.

Dually, the modules $\nabla(-2) = M(-2)^{\circledast} \cong M(-2)$ and $\nabla(0) = M(0)^{\circledast}$ are called *costandard* modules. For these modules we have the following dual version of Proposition 5.43:

Exercise 5.45.

(a) Show that for every $\lambda \in I$ all composition subfactors of the module $\nabla(\lambda)$ have the form $L(\mu)$, $\mu \leq \lambda$. Moreover, the module $L(\lambda)$ occurs with multiplicity one.

(b) Show that for every $\lambda \in I$, the indecomposable injective module $I(\lambda)$ has a filtration whose subquotients have the form $\nabla(\mu)$, $\mu \geq \lambda$. Moreover, the module $\nabla(\lambda)$ occurs exactly once.

Denote by $\mathcal{F}(\Delta)$ and $\mathcal{F}(\nabla)$ the full subcategories of $\mathcal{O}_0$, consisting of all modules M, which admit a filtration whose subquotients are standard or costandard modules, respectively. The modules from $\mathcal{F}(\Delta)$ are called Δ-*filtered* modules or modules with a *standard* filtration. The modules from $\mathcal{F}(\nabla)$ are called ∇-*filtered* modules or modules with a *costandard* filtration.

Exercise 5.46. Show that $M, N \in \mathcal{F}(\Delta)$ implies $M \oplus N \in \mathcal{F}(\Delta)$, and that $M, N \in \mathcal{F}(\nabla)$ implies $M \oplus N \in \mathcal{F}(\nabla)$.

Proposition 5.47. *For $M \in \mathcal{O}_0$ the following conditions are equivalent:*

(i) $M \in \mathcal{F}(\Delta)$.

(ii) *The socle* $\mathrm{soc}(M)$ *does not contain* $L(0)$ *as a direct summand.*

(iii) $\mathrm{Hom}_{\mathfrak{g}}(L(0), M) = 0$.

Proof. The equivalence of conditions (ii) and (iii) follows from definitions. By Theorem 3.16, we have $\mathrm{soc}(\Delta(0)) = \mathrm{soc}(\Delta(-2)) = L(-2)$. Hence (i) implies (ii). On the other hand, assume that M satisfies the condition

(ii). Then every direct summand of M satisfies (ii) as well. The only indecomposable modules from the list (5.5) that satisfy (ii) are $L(-2) = \Delta(-2)$,
$\Delta(0)$ and $P(-2)$. The first two of these are standard modules, and the last
one has a standard filtration by Proposition 5.43(ii). This means that every
indecomposable direct summand of M has a standard filtration. Thus M
also has a standard filtration by Exercise 5.46, meaning that M satisfies
(i). This completes the proof.   □

Exercise 5.48. Show that the following conditions are equivalent for any
module $M \in \mathcal{O}_0$:

(a) $M \in \mathcal{F}(\nabla)$.
(b) The quotient of M modulo the radical $\mathrm{rad}(M)$ does not contain $L(0)$
    as a direct summand.
(c) $\mathrm{Hom}_\mathfrak{g}(M, L(0)) = 0$.

Corollary 5.49. *Let $M, N \in \mathcal{O}$. Then we have*

 (i) $M \oplus N \in \mathcal{F}(\Delta)$ *if, and only if,* $M, N \in \mathcal{F}(\Delta)$.
 (ii) $M \oplus N \in \mathcal{F}(\nabla)$ *if, and only if,* $M, N \in \mathcal{F}(\nabla)$.
 (iii) $M \oplus N \in \mathcal{F}(\Delta) \cap \mathcal{F}(\nabla)$ *if, and only if,* $M, N \in \mathcal{F}(\Delta) \cap \mathcal{F}(\nabla)$.

Proof. For claims (i) and (ii) the "if" parts follow from Exercise 5.46.
The "only if" parts follow from Proposition 5.47 and Exercise 5.48 because of the the additivity of the conditions Proposition 5.47(iii) and Exercise 5.48(c). The claim (iii) follows from (i) and (ii).   □

The modules in the category $\mathcal{F}(\Delta) \cap \mathcal{F}(\nabla)$ are called *tilting modules*.
Tilting modules have both a standard and a costandard filtration. By
Corollary 5.49(iii), every tilting module is a direct sum of indecomposable
tilting modules. From the list (5.5) we see that the only indecomposable
tilting modules are $L(-2)$ and $P(-2)$. These are indexed using their highest weights as follows: $T(-2) = L(-2)$ and $T(0) = P(-2)$. The module
$T(-2) \oplus T(0)$ is called the *characteristic* tilting module. Later we will see
that these modules have nice homological properties.

Exercise 5.50. Show that the endomorphism algebra of the characteristic
tilting module in $\mathcal{O}_0$ is isomorphic to D; in particular, that it is quasi-
hereditary.

5.5   Grading

Although the categories $\mathcal{O}_0$ and D-mod are equivalent, in this section we will need to distinguish their objects. For this we will use indices -2 and 0 to denote objects in $\mathcal{O}_0$, while we will use indices p and q to denote objects in D-mod. In particular, the equivalence from Theorem 5.31(iii) induces the following correspondence between the indecomposable objects in $\mathcal{O}_0$ and D-mod:

$\mathcal{O}_0$	$L(-2)$	$L(0)$	$P(-2)$	$P(0)$	$I(0)$
D-mod	$L(\mathrm{p})$	$L(\mathrm{q})$	$P(\mathrm{p})$	$P(\mathrm{q})$	$I(\mathrm{q})$

The principal observation motivating the results of this section is that the category $\mathfrak{D}$ is defined using one *homogeneous* relation ab. This automatically equips $\mathfrak{D}$ and the corresponding path algebra D with the structure of a graded category and algebra, respectively. Let us start by recalling the definitions.

Let $\mathcal{X}$ be a $\mathbb{C}$-category. The category $\mathcal{X}$ is called $\mathbb{Z}$-*graded* or, simply, *graded* provided that for all objects $x, y \in \mathcal{X}$ the homomorphism set $\mathcal{X}(x, y)$ is equipped with the structure of a $\mathbb{Z}$-graded vector space. That is

$$\mathcal{X}(x, y) = \bigoplus_{i \in \mathbb{Z}} \mathcal{X}(x, y)_i,$$

such that the multiplication of morphisms in $\mathcal{X}$ preserves this structure in the following sense:

$$\mathcal{X}(y, z)_j \circ \mathcal{X}(x, y)_i \subset \mathcal{X}(x, z)_{i+j}$$

for all $x, y, z \in \mathcal{X}$ and $i, j \in \mathbb{Z}$. An algebra is just a category with one object, so from the above definition we get the definition of a graded algebra as a special case.

A graded category (algebra) $\mathcal{X}$ is said to be *positively* graded provided that the following conditions are satisfied:

- $\mathcal{X}(x, y)_i \neq 0$ implies $i \geq 0$ for all $x, y \in \mathcal{X}$ and $i \in \mathbb{Z}$;
- $\mathcal{X}(x, y)_0 \neq 0$ implies $x = y$ for all $x, y \in \mathcal{X}$;
- $\mathcal{X}(x, x)_0 \neq 0$ is a semi-simple algebra for all $x \in \mathcal{X}$.

Example 5.51. Each $\mathbb{C}$-category $\mathcal{X}$ carries the *trivial* structure of the graded category given by $\mathcal{X}(x, y) = \mathcal{X}(x, y)_0$ for all $x, y \in \mathcal{X}$.

Example 5.52. The category $\mathbb{C}$-gMod of graded vector spaces consists of all graded complex vector spaces; if $V = \oplus_{i \in \mathbb{Z}} V_i$ and $W = \oplus_{i \in \mathbb{Z}} W_i$ are two graded vector spaces, then for every $j \in \mathbb{Z}$ the vector space $\mathbb{C}$-gMod$(V,W)_j$ is defined as the set of all linear maps $\varphi : V \to W$ such that $\varphi(V_i) \subset W_{i+j}$. This grading is not positive. The category $\mathbb{C}$-gMod has two classical full subcategories, the category $\mathbb{C}$-gmod of *finite-dimensional* graded vector spaces (in which case V is finite-dimensional), and the category $\mathbb{C}$-fgmod of *locally finite-dimensional* graded vector spaces (in which case each V_i is finite-dimensional).

Example 5.53. The free $\mathbb{C}$-category $\mathcal{X}$ with generators $a_1, a_2, \ldots$ carries the natural structure of a positively graded category given for $x, y \in \mathcal{X}$ and $i \in \mathbb{Z}$ as follows:

$$\mathcal{X}(x,y)_i = \begin{cases} \mathbb{C}1_x, & x = y, i = 0; \\ X(x,y,i), & i > 0; \\ 0, & \text{otherwise;} \end{cases}$$

where $X(x,y,i)$ denotes the linear span of all paths from x to y in $\mathcal{X}$ of length i, that is of the form $a_{j_1} a_{j_1} \ldots a_{j_i}$.

Exercise 5.54. Show that each quotient of the free category $\mathcal{X}$ from Example 5.53 modulo an ideal, generated by homogeneous elements, inherits the natural structure of a positively graded category.

By Exercise 5.54, the category $\mathfrak{D}$ (and hence the algebra D as well) is equipped with the natural structure of a positively graded category (respectively algebra). The bases of the graded components of the category $\mathfrak{D}$ and the algebra D are given in the following table (it will be clear later on why it is more natural for the grading to increase downwards and not upwards):

Degree	$\mathfrak{D}(p,p)$	$\mathfrak{D}(p,q)$	$\mathfrak{D}(q,p)$	$\mathfrak{D}(q,q)$	D
$\vdots$	$\vdots$	$\vdots$	$\vdots$	$\vdots$	$\vdots$
-1	$\varnothing$	$\varnothing$	$\varnothing$	$\varnothing$	$\varnothing$
0	1_p	$\varnothing$	$\varnothing$	1_q	$1_p, 1_q$
1	$\varnothing$	a	b	$\varnothing$	a, b
2	ba	$\varnothing$	$\varnothing$	$\varnothing$	ba
3	$\varnothing$	$\varnothing$	$\varnothing$	$\varnothing$	$\varnothing$
$\vdots$	$\vdots$	$\vdots$	$\vdots$	$\vdots$	$\vdots$

Exercise 5.55. Show that the radical of D is a homogeneous ideal of D.

A *graded* module V over a graded category $\mathfrak{X}$ is a functor from $\mathfrak{X}$ to $\mathbb{C}$-gMod. A *finite-dimensional graded* module over a graded category $\mathfrak{X}$ is a functor from $\mathfrak{X}$ to $\mathbb{C}$-gmod. Here we will always work with finite-dimensional modules. Such module can be realized by assigning to every $x \in \mathfrak{X}$ a (finite-dimensional) graded vector space $V(x) = \oplus_{i \in \mathbb{Z}} V(x)_i$, and to every element $\alpha \in \mathfrak{X}(x, y)_j$, where $x, y \in \mathfrak{X}$ and $j \in \mathbb{Z}$, a linear operator $V(\alpha) : V(x) \to V(y)$ such that $V(\alpha) V(x)_i \subset V(y)_{i+j}$ for all $i \in \mathbb{Z}$. The assignment $\alpha \mapsto V(\alpha)$ should respect the multiplication in $\mathfrak{X}$.

Now comes the most important and nontrivial notion, the notion of homomorphisms between graded modules. If V and W are two graded modules over a graded category $\mathfrak{X}$, then a *homomorphism* φ from V to W is a homomorphism of the underlying ungraded $\mathfrak{X}$-modules (a collection of linear maps $\varphi_x : V(x) \to W(x)$ for all $x \in \mathfrak{X}$, which intertwine the action of $\mathfrak{X}$ on V and W), satisfying the condition $\varphi_x(V(x)_i) \subset W(x)_i$ for all $x \in \mathfrak{X}$ and $i \in \mathbb{Z}$. In other words, every map φ_x should be homogeneous of degree zero. Under such definition of morphisms, all graded $\mathfrak{X}$-modules form the category $\mathfrak{X}$-gMod, which has naturally defined subcategories $\mathfrak{X}$-gmod and $\mathfrak{X}$-fgmod, consisting of all finite-dimensional and locally finite-dimensional modules, respectively.

Exercise 5.56. Show that the categories $\mathfrak{X}$-gMod, $\mathfrak{X}$-gmod and $\mathfrak{X}$-fgmod, defined in the previous paragraph, are abelian categories with the usual kernels and cokernels.

Exercise 5.57. Check that one will not get an abelian category in the general case if one redefines the notion of a morphism between graded modules to include all homogeneous maps (and not only those of degree zero, as defined above).

The group $\mathbb{Z}$ acts on the category $\mathfrak{X}$-gMod (and also on the categories $\mathfrak{X}$-gmod and $\mathfrak{X}$-fgmod) functorially by shifting the gradings on these categories as follows: for any $k \in \mathbb{Z}$ and any $V \in \mathfrak{X}$-gMod the module $V\langle k \rangle$ is defined as follows: $(V\langle k \rangle)_i = V_{i+k}$ and $V\langle k \rangle(\alpha) = V(\alpha)$ for any morphism α. The above action of the group $\mathbb{Z}$ on graded $\mathfrak{X}$-modules naturally extends to an action on the category $\mathfrak{X}$-gMod (which preserves both subcategories $\mathfrak{X}$-gmod and $\mathfrak{X}$-fgmod). Graphically, the action of the functors $\langle \pm 1 \rangle$ can be described as follows:

Degree	$V(x)$	$V\langle 1\rangle(x)$	$V\langle -1\rangle(x)$
$\vdots$	$\vdots$	$\vdots$	$\vdots$
-2	$V(x)_{-2}$	$V(x)_{-1}$	$V(x)_{-3}$
-1	$V(x)_{-1}$	$V(x)_0$	$V(x)_{-2}$
0	$V(x)_0$	$V(x)_1$	$V(x)_{-1}$
1	$V(x)_1$	$V(x)_2$	$V(x)_0$
2	$V(x)_2$	$V(x)_3$	$V(x)_1$
$\vdots$	$\vdots$	$\vdots$	$\vdots$

There is an obvious functor F from $\mathfrak{X}$-gmod to $\mathfrak{X}$-mod; namely, the functor of forgetting the grading. An $\mathfrak{X}$-module M is called *gradable* if $M = \mathrm{F}\, M'$ for some graded $\mathfrak{X}$-module M'. The module M' is called a *graded lift* of M. As it follows from the following exercise, the image of the functor F usually is not dense in $\mathfrak{X}$-mod:

Exercise 5.58. The polynomial algebra $\mathbb{C}[x]$ has the natural structure of a positively graded algebra by assigning the element x degree 1. Show that a finite-dimensional $\mathbb{C}[x]$-module M is gradable if and only if x acts nilpotently on M.

Exercise 5.59. Show that if both $\mathfrak{X}$-modules M and N are gradable, then the module $M \oplus N$ is gradable as well.

Taking Exercise 5.58 into account, the following statement looks quite remarkable.

Theorem 5.60.

(i) *Every finite-dimensional D-module is gradable.*

(ii) *Graded lifts of indecomposable D-module are unique up to isomorphism and shift of gradings.*

Proof. Because of Exercise 5.59 and Theorem 5.40, to prove the claim (i) it is enough to construct graded lifts of all indecomposable modules from the list (5.5) (considered as D-modules via the equivalence established in Theorem 5.31(iii), see also Exercise 5.41). We present *standard lifts* of these modules in Figure 5.1 (in fact, in Figure 5.1 we even give standard lifts of all indecomposable structural D-modules). We use the convention that $\mathbb{C}_p$ and $\mathbb{C}_q$ denote one-dimensional spaces on which the elements 1_p and 1_q act as the identity, respectively. If the action of a generator on some element is specified, it is assumed to be given by the identity linear transformation; if the action is not specified, it is assumed to be zero. We leave it to the reader to check that Figure 5.1 does provide lifts of the corresponding indecomposable modules. This proves the claim (i).

To prove the claim (ii) we first note that for one-dimensional modules (in particular, for modules $L(\mathsf{p})$ and $L(\mathsf{q})$) the claim is obvious. The indecomposable modules which are left are either projective or injective. We will prove the claim (ii) for graded lifts of indecomposable projectives (for injectives the arguments are similar).

Let P and P' be two graded lifts of the same indecomposable projective module and let L be the simple top of both P and P'. Consider the standard graded lift of L, concentrated in degree zero. Since the radical $\mathrm{rad}(\mathsf{D})$ of D is a homogeneous ideal (Exercise 5.55), we have that both

$$\mathrm{rad}(P) = \mathrm{rad}(\mathsf{D})P \quad \text{and} \quad \mathrm{rad}(P') = \mathrm{rad}(\mathsf{D})P',$$

are graded submodules of P and P', respectively. In particular, the corresponding simple quotients are also graded. Due to the uniqueness of graded lifts for simple modules (established above), shifting P and P' in grading, if necessary, we may assume that the projections $P \twoheadrightarrow L$ and $P' \twoheadrightarrow L$ are homogeneous maps of degree zero (that is belong to D-gmod). Using the projectivity of P in D-mod we thus can lift the projection $P \twoheadrightarrow L$ to a map $P \to P'$, which might be non-homogeneous. Write this lift as a sum of homogeneous maps and take the homogeneous component of degree zero. This gives us a homogeneous lift of $P \twoheadrightarrow L$, that is a lift in D-gmod. Since both P and P' are projective and the image of our homogeneous lift does not belong to the radical of P', this lift must be an isomorphism. This completes the proof. $\qquad\square$

The following corollary is also rather interesting:

Corollary 5.61. *The endomorphism algebra of every indecomposable object in D-gmod is $\mathbb{C}$ and consists of scalar maps.*

Proof. That all scalar maps belong to the endomorphism algebra of any object in D-gmod is obvious. For the modules $L(\mathsf{p})$, $L(\mathsf{q})$, $P(\mathsf{q})$ and $I(\mathsf{q})$, we know that their endomorphism algebras in D-mod reduce to scalars (this follows from Theorem 1.26, Exercise 3.17 and Theorem 3.84).

For the module $P(\mathsf{p})$, we know that its endomorphism algebra in D-mod is of dimension two and is isomorphic to $\mathbb{C}[x]/(x^2)$ (see Exercise 5.30). Furthermore, the nontrivial nilpotent endomorphism of $P(\mathsf{p})$ corresponds to the multiplication by the element ba, which has degree two. Hence every

Degree	$L(\mathbf{p})$	$L(\mathbf{q})$	$P(\mathbf{p})$	$P(\mathbf{q})$	$I(\mathbf{p})$	$I(\mathbf{q})$
$\vdots$	$\vdots$	$\vdots$	$\vdots$	$\vdots$	$\vdots$	$\vdots$
-3	0	0	0	0	0	0
-2	0	0	0	0	$\mathbb{C}_p$ $\downarrow a$	0
-1	0	0	0	0	$\mathbb{C}_q$ $\downarrow b$	$\mathbb{C}_p$ $\downarrow a$
0	$\mathbb{C}_p$	$\mathbb{C}_q$	$\mathbb{C}_p$ $\downarrow a$	$\mathbb{C}_q$ $\downarrow b$	$\mathbb{C}_p$	$\mathbb{C}_q$
1	0	0	$\mathbb{C}_q$ $\downarrow b$	$\mathbb{C}_p$	0	0
2	0	0	$\mathbb{C}_p$	0	0	0
3	0	0	0	0	0	0
$\vdots$	$\vdots$	$\vdots$	$\vdots$	$\vdots$	$\vdots$	$\vdots$

Degree	$\Delta(\mathbf{p})$	$\Delta(\mathbf{q})$	$\nabla(\mathbf{p})$	$\nabla(\mathbf{q})$	$T(\mathbf{p})$	$T(\mathbf{q})$
$\vdots$	$\vdots$	$\vdots$	$\vdots$	$\vdots$	$\vdots$	$\vdots$
-2	0	0	0	0	0	0
-1	0	0	0	$\mathbb{C}_p$ $\downarrow a$	0	$\mathbb{C}_p$ $\downarrow a$
0	$\mathbb{C}_p$	$\mathbb{C}_q$ $\downarrow b$	$\mathbb{C}_p$	$\mathbb{C}_q$	$\mathbb{C}_p$	$\mathbb{C}_q$ $\downarrow b$
1	0	$\mathbb{C}_p$	0	0	0	$\mathbb{C}_p$
2	0	0	0	0	0	0
$\vdots$	$\vdots$	$\vdots$	$\vdots$	$\vdots$	$\vdots$	$\vdots$

Fig. 5.1   Standard graded lifts of all indecomposable structural D-modules.

endomorphism of $P(\mathrm{p})$ in D-mod is a linear combination of the identity map (homogeneous of degree zero) and the multiplication by ba (homogeneous of degree two). In particular, the degree zero homogeneous part reduces to scalars. The claim follows.   □

From the proof of Theorem 5.60 it becomes clear why the grading on our pictures increases downwards. The unique simple quotient of an indecomposable projective module is usually called the *top* of this module. It is thus natural to depict the top of a module as its *highest* component on the graded picture. At the same time, the rest of the module consists of the radical, which is given by multiplication with the radical of a positively graded algebra. In particular, the rest of the module should be of larger degree and hence the grading must increase downwards.

Exercise 5.62. Show that the category of graded finite-dimensional $\mathfrak{D}$-modules is equivalent to the category of modules over the following $\mathbb{C}$-category

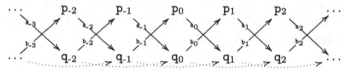

with relations $a_i b_{i-1} = 0$, $i \in \mathbb{Z}$ (which are depicted by dotted arrows).

We refer the reader to Figure 5.1, where the standard graded lifts of all indecomposable structural D-modules are given. Our convention for these standard graded lifts has a natural explanation. For the indecomposable projective modules, corresponding to the idempotents 1_p and 1_q, the standard graded lifts are simply $D1_p$ and $D1_q$, respectively, with the grading induced from the grading on D. Standard graded lifts of simple modules are, then, quotients of these graded projective modules and are automatically concentrated in degree zero (since the idempotents 1_p and 1_q have degree zero as elements of D). Standard modules are also quotients of projective modules and so their standard graded lifts can be defined similarly. Analogously, for an indecomposable injective module (and for the corresponding costandard module) we require that its socle coincides with the standard lift of the corresponding graded simple module. For the indecomposable tilting module $T(\lambda)$ we require that the unique (up to scalar) non-zero morphism $\Delta(\lambda) \to T(\lambda)$ is homogeneous of degree zero. Abusing notation, we will denote modules and their standard graded lifts in the same way.

Exercise 5.63. Prove that the standard graded lift of the injective module $I(\mathbf{p})$ is isomorphic to $P(\mathbf{p})\langle 2\rangle$; that the standard graded lift of the tilting module $T(\mathbf{p})$ is isomorphic to $L(\mathbf{p})$; and that the standard graded lift of the tilting module $T(\mathbf{q})$ is isomorphic to $P(\mathbf{p})\langle 1\rangle$.

Exercise 5.64. Check that for $\lambda = \mathbf{p}, \mathbf{q}$ the unique (up to scalar) non-zero map $T(\lambda) \to \nabla(\lambda)$ is homogeneous of degree zero.

5.6 Homological properties

In this chapter we will describe some homological properties of the principal block $\mathcal{O}_0$. As before, we identify $\mathcal{O}_0$ with the category of all finite-dimensional D-modules via the equivalence established in Theorem 5.31(iii). We start with a construction of projective resolutions of all indecomposable modules.

Proposition 5.65. *The following exact sequences give minimal (graded) projective resolutions of all (standard lifts of) indecomposable (graded) D-modules, the latter being written in bold:*

$$
\begin{aligned}
0 \quad &\to \quad P(\mathbf{p}) \to \mathbf{P(p)} \to 0, \\
0 \quad &\to \quad P(\mathbf{q}) \to \mathbf{P(q)} \to 0, \\
0 \quad \to P(\mathbf{q})\langle -1\rangle \to \quad &P(\mathbf{p}) \quad \to \mathbf{L(p)} \to 0, \\
0 \to P(\mathbf{q})\langle -2\rangle \to P(\mathbf{p})\langle -1\rangle \to \quad &P(\mathbf{q}) \quad \to \mathbf{L(q)} \to 0, \\
0 \to P(\mathbf{q})\langle -2\rangle \to P(\mathbf{p})\langle -1\rangle \to &P(\mathbf{p})\langle 1\rangle \to \mathbf{I(q)} \to 0.
\end{aligned}
$$

Proof. The first two resolutions are obvious. In Figure 5.2 we give a complete picture for the graded projective resolution of the module $L(\mathbf{p})$. All dotted arrows on this picture represent zero maps, and all solid and dashed arrows represent identity maps. In Figure 5.3 we give the non-zero part of the graded projective resolution of the module $L(\mathbf{q})$. In Figure 5.4 we give the non-zero part of the graded projective resolution of the module $I(\mathbf{q})$. It is straightforward to verify that these diagrams commute and are exact, giving required projective resolutions. The minimality of all resolutions follows from the fact that all maps between projective modules end up in the radical of the target module. This completes the proof. $\square$

Recall that the *global dimension* of an algebra A is the length of the longest projective resolution of an A-module. From Proposition 5.65 we immediately get the following:

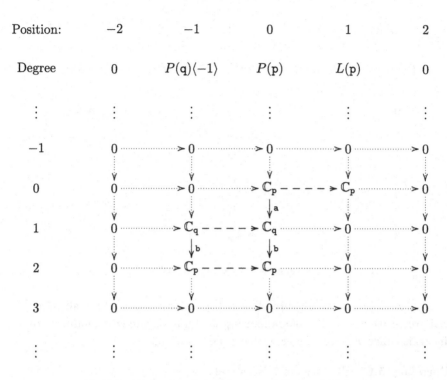

Fig. 5.2   Minimal graded projective resolution of $L(\mathsf{p})$.

Corollary 5.66. *The global dimension of the algebra* D *equals two.*

Proof. From Proposition 5.65 we see that the longest projective resolution for D-modules has length two. □

Let us look closer at the minimal projective resolutions of the simple modules $L(\mathsf{p})$ and $L(\mathsf{q})$, given by Proposition 5.65 (see Figures 5.2 and 5.3, respectively). One observes that both these resolutions are *linear* in the following sense: the top of the projective module staying in position $-i$ of this resolution (where $i \in \mathbb{N}_0$) is concentrated in degree i (that is, these tops belong to the diagonal "position= $-$degree" on Figures 5.2 and 5.3, respectively). One can also say that the projective module in position $-i$ is obtained from the standard lift of the projective module using the shift $\langle -i \rangle$. Note that the projective resolution of the module $I(\mathsf{q})$, presented on Figure 5.4, is not linear.

Position:	-2	-1	0	1
Degree	$P(\mathsf{q})\langle -2\rangle$	$P(\mathsf{p})\langle -1\rangle$	$P(\mathsf{q})$	$L(\mathsf{q})$

$$
\begin{array}{ccc}
0 & & \mathbb{C}_\mathsf{q} \dashrightarrow \mathbb{C}_\mathsf{q} \\
& & \downarrow{\scriptstyle b} \\
1 & \mathbb{C}_\mathsf{p} \dashrightarrow \mathbb{C}_\mathsf{p} \\
& & \downarrow{\scriptstyle a} \\
2 & \mathbb{C}_\mathsf{q} \dashrightarrow \mathbb{C}_\mathsf{q} \\
& \downarrow{\scriptstyle b} \quad\; \downarrow{\scriptstyle b} \\
3 & \mathbb{C}_\mathsf{p} \dashrightarrow \mathbb{C}_\mathsf{p}
\end{array}
$$

Fig. 5.3   Minimal graded projective resolution of $L(\mathsf{q})$.

A positively graded algebra A, which has the property that all minimal projective resolutions of standard graded lifts of simple A-modules are linear, is called *Koszul*. Hence we have the following:

Corollary 5.67. *The algebra* D *is Koszul.*

Exercise 5.68. Construct minimal injective resolutions of all indecomposable D-modules.

Exercise 5.69. Show that all minimal graded injective resolutions of the standard graded lifts of simple D-modules are linear in the sense that the socles of all graded injective modules in these resolutions belong to the diagonal "position= −degree".

Exercise 5.70. Verify that the quasi-hereditary algebra D is *standard Koszul* in the sense that all minimal graded projective (resp. injective) resolutions of the standard graded lifts of standard (resp. costandard) D-modules are linear.

Corollary 5.71. *For any $i \in \mathbb{N}_0$ and $j \in \mathbb{Z}$ we have*

$$
\mathrm{Ext}^i_{\text{D-gmod}}(L(\mathsf{q}), L(\mathsf{q})\langle j\rangle) = \begin{cases} \mathbb{C}, & i = j = 0; \\ \mathbb{C}, & i = 2, j = -2; \\ 0, & \text{otherwise}; \end{cases}
$$

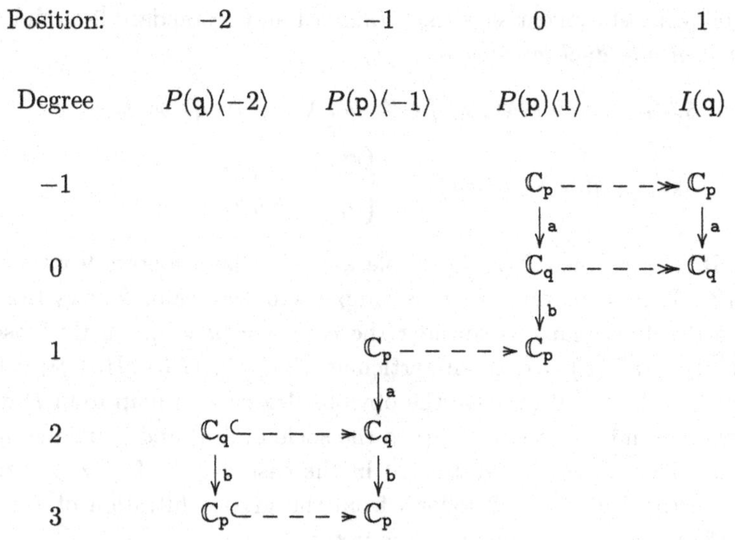

Position: −2   −1   0   1

Degree $P(\mathsf{q})\langle -2\rangle$ $P(\mathsf{p})\langle -1\rangle$ $P(\mathsf{p})\langle 1\rangle$ $I(\mathsf{q})$

Fig. 5.4   Minimal graded projective resolution of $I(\mathsf{q})$.

$$\mathrm{Ext}^i_{D\text{-gmod}}(L(\mathsf{q}), L(\mathsf{p})\langle j\rangle) = \begin{cases} \mathbb{C}, & i = 1, j = -1; \\ 0, & \textit{otherwise}; \end{cases}$$

$$\mathrm{Ext}^i_{D\text{-gmod}}(L(\mathsf{p}), L(\mathsf{q})\langle j\rangle) = \begin{cases} \mathbb{C}, & i = 1, j = -1; \\ 0, & \textit{otherwise}; \end{cases}$$

$$\mathrm{Ext}^i_{D\text{-gmod}}(L(\mathsf{p}), L(\mathsf{p})\langle j\rangle) = \begin{cases} \mathbb{C}, & i = j = 0; \\ 0, & \textit{otherwise}. \end{cases}$$

Proof.   This follows directly from Figures 5.2 and 5.3.   □

Exercise 5.72. Show that

$$\mathrm{Ext}^i_{D\text{-gmod}}(\Delta(\lambda), L(\mu)\langle j\rangle) = \begin{cases} \mathbb{C}, & \lambda = \mu, i = 0, j = 0; \\ \mathbb{C}, & \lambda = \mathsf{p}, \mu = \mathsf{q}, i = 1, j = -1; \\ 0, & \textit{otherwise}; \end{cases}$$

$$\mathrm{Ext}^i_{D\text{-gmod}}(L(\mu)\langle j\rangle, \nabla(\lambda)) = \begin{cases} \mathbb{C}, & \lambda = \mu, i = 0, j = 0; \\ \mathbb{C}, & \lambda = \mathsf{p}, \mu = \mathsf{q}, i = 1, j = 1; \\ 0, & \textit{otherwise}. \end{cases}$$

The following statement says that standard and costandard D-modules form *homologically dual families*:

Corollary 5.73. *For any* $i \in \mathbb{N}_0$, $j \in \mathbb{Z}$ *and* $\lambda, \mu \in \{\mathsf{p}, \mathsf{q}\}$ *we have*

$$\mathrm{Ext}^i_{\mathrm{D\text{-}gmod}}(\Delta(\lambda), \nabla(\mu)\langle j \rangle) = \begin{cases} \mathbb{C}, & \lambda = \mu, i = j = 0; \\ 0, & \textit{otherwise.} \end{cases}$$

Proof. If $\lambda = \mathsf{p}$, then $\Delta(\lambda)$ is simple and the claim follows from Exercise 5.72. If $\mu = \mathsf{p}$, then $\nabla(\mu)$ is simple and the claim follows from Exercise 5.72. It remains to consider the case $\lambda = \mu = \mathsf{q}$. In this case $\Delta(\lambda) = P(\mathsf{q})$ and $\nabla(\mu) = I(\mathsf{q})$. In particular, $\mathrm{Ext}^i_{\mathrm{D\text{-}gmod}}(P(\mathsf{q}), I(\mathsf{q})\langle j \rangle) \neq 0$ implies $i = 0$. For $i = 0$ we have the obvious degree zero map from $P(\mathsf{q})$ to $I(\mathsf{q})$, which sends the top of $P(\mathsf{q})$ to the socle of $I(\mathsf{q})$ and is unique up to a scalar. This proves the statement in the case $j = 0$. If $j \neq 0$ then $\mathrm{Hom}_{\mathrm{D\text{-}gmod}}(P(\mathsf{q}), I(\mathsf{q})\langle j \rangle) = 0$ follows from the graded filtration of $I(\mathsf{q})$, given by Figure 5.1. This completes the proof. $\square$

Exercise 5.74. Show that for any $i \in \mathbb{N}_0$, $j \in \mathbb{Z}$ and $\lambda, \mu \in \{\mathsf{p}, \mathsf{q}\}$ we have

$$\mathrm{Ext}^i_{\mathrm{D\text{-}gmod}}(P(\lambda), L(\mu)\langle j \rangle) = \begin{cases} \mathbb{C}, & \lambda = \mu, i = j = 0; \\ 0, & \text{otherwise.} \end{cases}$$

Exercise 5.75. Show that for any $i \in \mathbb{N}_0$, $j \in \mathbb{Z}$ and $\lambda, \mu \in \{\mathsf{p}, \mathsf{q}\}$ we have

$$\mathrm{Ext}^i_{\mathrm{D\text{-}gmod}}(L(\lambda), I(\mu)\langle j \rangle) = \begin{cases} \mathbb{C}, & \lambda = \mu, i = j = 0; \\ 0, & \text{otherwise.} \end{cases}$$

5.7 Category of bounded linear complexes of projective graded D-modules

Recall that for a (graded) algebra A and a (graded) A-module M, the *additive closure* of M is a full subcategory $\mathrm{add}(M)$ of the category of all (graded) A-modules. This consists of all (graded) modules, isomorphic to finite direct sums of direct summands of M.

We will use the standard notation $(\mathcal{X}^\bullet, d_\bullet)$ (or, simply $\mathcal{X}^\bullet$) for a complex

$$\cdots \xrightarrow{d_{i-2}} \mathcal{X}^{i-1} \xrightarrow{d_{i-1}} \mathcal{X}^i \xrightarrow{d_i} \mathcal{X}^{i+1} \xrightarrow{d_{i+1}} \cdots$$

of (graded) D-modules. Denote by $\mathfrak{LP}$ the category, whose objects are all bounded linear complexes of graded finite-dimensional projective D-modules, and morphisms are all possible morphisms (chain maps) between complexes of graded D-modules. Then a finite complex $\mathcal{X}^\bullet$ of graded projective D-modules belongs to $\mathfrak{LP}$ if, and only if, for every $i \in \mathbb{Z}$ the top of the graded projective module $\mathcal{X}^i$ is concentrated in degree $-i$ (if $\mathcal{X}^i \neq 0$). Equivalently, we may require $\mathcal{X}^i \in \mathrm{add}(D\langle i \rangle)$ for all $i \in \mathbb{Z}$. For example, from the previous section we have that minimal projective resolutions of both simple D-modules belong to $\mathfrak{LP}$ (here we mean the genuine projective resolutions, which are obtained from the exact sequences mentioned in Proposition 5.65 by deleting the bold elements). These and some other objects of $\mathfrak{LP}$ are presented on Figures 5.5–5.9 (only non-zero maps and components are shown).

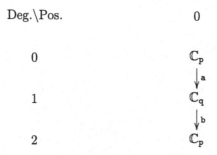

Fig. 5.5   Linear complex $\mathcal{C}(1)^\bullet$ of projective D-modules.

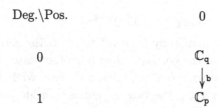

Fig. 5.6   Linear complex $\mathcal{C}(2)^\bullet$ of projective D-modules.

Let $\mathcal{X}^\bullet$ and $\mathcal{Y}^\bullet$ be two complexes from $\mathfrak{LP}$ and $i \in \mathbb{Z}$. By the definition of $\mathfrak{LP}$, the top of $\mathcal{X}^i$ is concentrated in degree $-i$. At the same time, from the definition of $\mathfrak{LP}$ we also have that the graded component $\mathcal{Y}^{i-1}_{-i}$ is zero.

Deg.\Pos.	-1	0
0		$\mathbb{C}_q$
		$\downarrow$ b
1	$\mathbb{C}_p$ - - - - $\to$	$\mathbb{C}_p$
	$\downarrow$ a	
2	$\mathbb{C}_q$	
	$\downarrow$ b	
3	$\mathbb{C}_p$	

Fig. 5.7   Linear complex $C(3)^\bullet$ of projective D-modules.

Deg.\Pos.	-1	0
0		$\mathbb{C}_p$
		$\downarrow$ a
1	$\mathbb{C}_q$ - - - $\to$	$\mathbb{C}_q$
	$\downarrow$ b	$\downarrow$ b
2	$\mathbb{C}_p$ - - - $\to$	$\mathbb{C}_p$

Fig. 5.8   Linear complex $C(4)^\bullet$ of projective D-modules.

In particular,

$$\mathrm{Hom}_{\text{D-gmod}}(\mathcal{X}^i, \mathcal{Y}^{i-1}) = 0,$$

implying that the only homotopy from $\mathcal{X}^\bullet$ to $\mathcal{Y}^\bullet$ is the zero map. It follows that $\mathfrak{LP}$ is equal to the corresponding homotopy category. Note that the category of all complexes of D-modules is abelian, while the homotopy category is not. Since for $\mathfrak{LP}$ the two categories coincide, one might hope that $\mathfrak{LP}$ should be abelian. This turns out to be the case.

Proposition 5.76. *The category* $\mathfrak{LP}$ *is abelian.*

Proof.   Since the category of all complexes is abelian, it is enough to show that for any two complexes $\mathcal{X}^\bullet$ and $\mathcal{Y}^\bullet$ from $\mathfrak{LP}$ and any homomorphism $\varphi_\bullet : \mathcal{X}^\bullet \to \mathcal{Y}^\bullet$ both the kernel and the cokernel of $\varphi_\bullet$ belong to $\mathfrak{LP}$.

Deg.\Pos.	−2	−1	0

$$
\begin{array}{ccc}
0 & & & \mathbb{C}_{\mathsf{q}} \\
 & & & \downarrow b \\
1 & & \mathbb{C}_{\mathsf{p}} \dashrightarrow \mathbb{C}_{\mathsf{p}} \\
 & & & \downarrow a \\
2 & \mathbb{C}_{\mathsf{q}} \dashrightarrow \mathbb{C}_{\mathsf{q}} \\
 & \downarrow b & \downarrow b \\
3 & \mathbb{C}_{\mathsf{p}} \dashrightarrow \mathbb{C}_{\mathsf{p}}
\end{array}
$$

Fig. 5.9   Linear complex $\mathcal{C}(5)^{\bullet}$ of projective D-modules.

Both the kernel and the cokernel of $\varphi_{\bullet}$ are, obviously, finite complexes. Hence it is left to show that the category $\mathrm{add}(D\langle i \rangle)$ is an abelian category for every $i \in \mathbb{Z}$.

Lemma 5.77. *The category* $\mathrm{add}(D)$ *is an abelian subcategory of* D-gmod.

Proof. Let $P, Q \in \mathrm{add}(D)$ and $\psi : P \to Q$ be a homomorphism of graded modules. Consider the component of degree zero, $\psi_0 : P_0 \to Q_0$, which splits into a direct sum of the following two maps, $\psi_0^{\mathsf{p}} : P_0(\mathsf{p}) \to Q_0(\mathsf{p})$ and $\psi_0^{\mathsf{q}} : P_0(\mathsf{q}) \to Q_0(\mathsf{q})$.

Let $\{v_i\}$ be some basis of $\mathrm{Ker}(\psi_0^{\mathsf{p}})$, $\{w_j\}$ be some extension of $\{v_i\}$ to a basis of $P_0(\mathsf{p})$, and $\{z_l\}$ be some extension of $\{\varphi_0(w_j)\}$ to a basis of $Q_0(\mathsf{p})$. Similarly, let $\{v_i'\}$ be some basis of $\mathrm{Ker}(\psi_0^{\mathsf{q}})$, $\{w_j'\}$ be some extension of $\{v_i'\}$ to a basis of $P_0(\mathsf{q})$, and $\{z_l'\}$ be some extension of $\{\varphi_0(w_j')\}$ to a basis of $Q_0(\mathsf{q})$.

Since $P \in \mathrm{add}(D)$, the D-submodule K of P, generated by $\{v_i, v_i'\}$, belongs to $\mathrm{add}(D)$ is well. By our choice of $\{v_i, v_i'\}$, the module K is contained in the kernel of ψ. However, since $Q \in \mathrm{add}(D)$, all $\psi(w_j)$ and $\psi(w_j')$ are generators of some projective D-modules. Hence the restriction of ψ to the submodule M of P, generated by $\{w_j, w_j'\}$, is injective. Hence $K = \mathrm{Ker}(\psi) \in \mathrm{add}(D)$. Similarly one shows that the cokernel of ψ is isomorphic to the submodule of Q, generated by $\{z_i, z_i'\}$, which also belongs to $\mathrm{add}(D)$. The claim follows. □

Since the shift of grading is an autoequivalence, the proposition follows from Lemma 5.77. □

For a complex $\mathcal{X}^\bullet$ and $i \in \mathbb{Z}$ we denote by $[i]$ the functor, which shifts the complex i steps to the left. In particular, $\mathcal{X}[i]^j = \mathcal{X}^{j+i}$ for all $j \in \mathbb{Z}$. For example, we have:

$$\mathcal{X}^\bullet : \qquad \cdots \xrightarrow{d_{i-2}} \mathcal{X}^{i-1} \xrightarrow{d_{i-1}} \mathcal{X}^i \xrightarrow{d_i} \mathcal{X}^{i+1} \xrightarrow{d_{i+1}} \cdots$$

$$\mathcal{X}[1]^\bullet : \qquad \cdots \xrightarrow{d_{i-1}} \mathcal{X}^i \xrightarrow{d_i} \mathcal{X}^{i+1} \xrightarrow{d_{i+1}} \mathcal{X}^{i+2} \xrightarrow{d_{i+2}} \cdots$$

From the definition of $\mathfrak{LP}$ we have that for every $i \in \mathbb{Z}$ the functor $[-i]\langle i \rangle$ preserves the category $\mathfrak{LP}$. This defines a free action of $\mathbb{Z}$ on $\mathfrak{LP}$.

Exercise 5.78. Show that for all $i \in \mathbb{Z}$ the complexes $\mathcal{C}(1)[-i]\langle i \rangle^\bullet$ (see Figure 5.5) and $\mathcal{C}(2)[-i]\langle i \rangle^\bullet$ (see Figure 5.6) are simple objects in $\mathfrak{LP}$. Show further that every simple object in $\mathfrak{LP}$ is isomorphic to one of these complexes (and that these complexes are pairwise non-isomorphic).

Proposition 5.79.

(i) For all $i \in \mathbb{Z}$ the complexes $\mathcal{C}(4)[-i]\langle i \rangle^\bullet$ (see Figure 5.8) and $\mathcal{C}(5)[-i]\langle i \rangle^\bullet$ (see Figure 5.9) are injective objects in $\mathfrak{LP}$.

(ii) The category $\mathfrak{LP}$ has enough injective objects.

Proof. Let

$$0 \to \mathcal{C}(4)^\bullet \to \mathcal{X}^\bullet \to \mathcal{C}(1)[-i]\langle i \rangle^\bullet \to 0 \qquad (5.8)$$

be a short exact sequence in $\mathfrak{LP}$. If $i \neq -2$, we have

$$\mathrm{Hom}_{\mathrm{D\text{-}gmod}}(P(\mathrm{p})\langle i \rangle, \mathcal{C}(4)^{i+1}) = 0$$

and hence the sequence (5.8) splits. If $i = -2$, there is a unique (up to scalar) non-zero map ψ from $P(\mathrm{p})\langle -2 \rangle$ to $\mathcal{C}(4)^{-1}$. However, the composition of ψ with the differential in $\mathcal{C}(4)^\bullet$ is not zero, and hence does not produce a complex (an object in $\mathfrak{LP}$). Therefore the sequence (5.8) splits also for $i = -2$.

Consider now the short exact sequence

$$0 \to \mathcal{C}(4)^\bullet \to \mathcal{X}^\bullet \to \mathcal{C}(2)[-i]\langle i \rangle^\bullet \to 0 \qquad (5.9)$$

in $\mathfrak{LP}$. If $i \neq -1$, we have

$$\mathrm{Hom}_{\mathrm{D\text{-}gmod}}(P(\mathrm{q})\langle i \rangle, \mathcal{C}(4)^{i+1}) = 0$$

and hence the sequence (5.9) splits. If $i = -1$ we first observe the isomorphism $P(\mathrm{q})\langle -1 \rangle \cong \mathcal{C}(4)^{-1}$. There is a unique (up to scalar) injective map

ψ from $P(\mathsf{q})\langle -1\rangle$ to $\mathcal{C}(4)^0$. However, the image of ψ coincides with the image of $\mathcal{C}(4)^{-1}$ (under the differential in $\mathcal{C}(4)^\bullet$). Hence the sum of ψ and the differential in $\mathcal{C}(4)^\bullet$ has a nontrivial kernel, isomorphic to $P(\mathsf{q})\langle -1\rangle$. Thus the sequence (5.9) splits also for $i = -1$.

Taking Exercise 5.78 into account, the above shows that every extension from a simple object in $\mathcal{LP}$ to $\mathcal{C}(4)^\bullet$ vanishes. Thus $\mathcal{C}(4)^\bullet$ is an injective object. We can similarly show that $\mathcal{C}(5)^\bullet$ is injective. The claim (i) now follows from the observation that $[-i]\langle i\rangle$ is an autoequivalence for every i.

The natural embedding of $\mathcal{C}(1)^\bullet$ to $\mathcal{C}(4)^\bullet$ (which identifies the projective modules in position 0) is a monomorphism of complexes. Hence $\mathcal{C}(4)^\bullet$ is the injective envelope (see (i)) of the simple object $\mathcal{C}(1)^\bullet$. Similarly $\mathcal{C}(5)^\bullet$ is the injective envelope of the simple object $\mathcal{C}(2)^\bullet$. Using Exercise 5.78 and the autoequivalences $[-i]\langle i\rangle$ we get that all simple objects in $\mathcal{LP}$ have injective envelopes. Since every complex in $\mathcal{LP}$ is bounded by definition, such a complex has finite length in $\mathcal{LP}$. Now the claim (ii) follows by arguments dual to the those used in the proof of Theorem 5.16. $\qquad\square$

Now we are ready to describe the structure of $\mathcal{LP}$ completely.

Theorem 5.80. *The category $\mathcal{LP}$ is equivalent to the category* D-gmod.

Proof. By Propositions 5.76 and 5.79 the category $\mathcal{LP}$ is abelian with enough injectives. It also has a natural free action of $\mathbb{Z}$. Hence $\mathcal{LP}$ is equivalent to the category of graded modules over the opposite of the endomorphism algebra of some injective cogenerator of $\mathcal{LP}$ (up to graded shifts); see Chapter II, §1 of [7] (one could instead use arguments similar to the ones used in the proof of Theorem 5.31). From Exercise 5.78 and Proposition 5.79 we have that the complex $\mathcal{C}(4)^\bullet \oplus \mathcal{C}(5)^\bullet$ is an injective cogenerator of $\mathcal{LP}$ up to graded shifts. Let A denote the endomorphism algebra of this complex (taken up to graded shift in the category $\mathcal{LP}$).

From Figures 5.8 and 5.9 it is easy to see that the identification of two left terms of $\mathcal{C}(5)^\bullet$ with the complex $\mathcal{C}(4)[1]\langle -1\rangle^\bullet$ defines a surjection φ from $\mathcal{C}(5)^\bullet$ to $\mathcal{C}(4)[1]\langle -1\rangle^\bullet$. Similarly, the identification of the left term of $\mathcal{C}(4)[-1]\langle 1\rangle^\bullet$ with the right term in $\mathcal{C}(5)^\bullet$ defines a homomorphism ψ from $\mathcal{C}(4)[-1]\langle 1\rangle^\bullet$ to $\mathcal{C}(5)^\bullet$. From the construction we have $\varphi \circ \psi = 0$. Hence, sending a to φ and b to ψ defines a homomorphism from D to A. A straightforward computation (which is left to the reader) shows that this homomorphism is, in fact, an isomorphism of graded algebras. The claim of the theorem now follows from Lemma 5.33. $\qquad\square$

Corollary 5.81. *Each indecomposable object in* $\mathfrak{LP}$ *is isomorphic to one of the complexes* $\mathcal{C}(i)^\bullet$, $i = 1, \ldots, 5$, *up to a shift* $[i]\langle -i \rangle$, $i \in \mathbb{Z}$, *of grading.*

Proof.   This follows directly from Theorems 5.80 and 5.40.   □

One might observe that indecomposable injective objects in $\mathfrak{LP}$, as given by Proposition 5.79, are (up to shift of grading) exactly the projective resolutions of simple D-modules. As we mentioned above, the category $\mathfrak{LP}$ coincides with the corresponding homotopy category. This means that the homomorphisms between indecomposable injective objects in $\mathfrak{LP}$ correspond to extensions between simple D-modules. Furthermore, by Corollary 5.71 any extension between simple D-modules can be realized as a homomorphism between two indecomposable injective objects in $\mathfrak{LP}$. For a Koszul algebra A, the Yoneda extension algebra of the direct sum of all simple A-modules is called the *Koszul dual* of A (this is motivated by the result that for Koszul algebras the second Koszul dual of A is isomorphic to A). From the above we have the following result:

Corollary 5.82 (Koszul self-duality of D**).** *The algebra* D *is isomorphic to its Koszul dual.*

Note that the isomorphism given by Corollary 5.82 is not trivial, in particular, it swaps indices of primitive idempotents.

Exercise 5.83.

(a) Define the category $\mathfrak{LJ}$ of (bounded) linear complexes of graded injective D-modules.
(b) Show that $\mathfrak{LJ}$ is an abelian category with a free action of $\mathbb{Z}$ on it.
(c) Show that simple objects in $\mathfrak{LJ}$ are given (up to the action of $\mathbb{Z}$ from (b)) by indecomposable injective D-modules.
(d) Show that injective resolutions of simple D-modules are projective objects in $\mathfrak{LJ}$ and derive from this that $\mathfrak{LJ}$ has enough projectives.
(e) Show that the category $\mathfrak{LJ}$ is equivalent to the category D-gmod.

5.8   Projective functors on $\mathcal{O}_0$

As we saw in Section 5.2, the functor $V \otimes {}_-$, where V is a finite-dimensional $\mathfrak{g}$-module, restricts to an exact and self-adjoint functor on $\mathcal{O}$ (Lemma 5.20). Unfortunately, in all nontrivial cases the functor $V \otimes {}_-$ does not preserve the principal block $\mathcal{O}_0$ (or any other block of $\mathcal{O}_0$). Therefore to be able to study the action of $V \otimes {}_-$ on $\mathcal{O}_0$ we are forced to consider direct summands of the functor $V \otimes {}_-$. Functors that are isomorphic to direct summand of the endofunctors $V \otimes {}_-$ of $\mathcal{O}$, where V is a finite-dimensional $\mathfrak{g}$-module, are called *projective functors*. We use the convention that the zero functor is a projective functor. Partially this name is motivated by the property that projective functors send projective modules to projective modules (Corollary 5.21).

To proceed we will need to simplify notation. First we set $\mathbb{N}_{-1} = \{-1, 0, 1, 2, \ldots\}$ and for $i \in \mathbb{N}_{-1}$ denote by $\mathcal{O}_i$ the block $\mathcal{O}^{i+2\mathbb{Z},(i+1)^2}$. These blocks are called *integral blocks*, as they are exactly those blocks of $\mathcal{O}$ which contain modules whose support is a subset of $\mathbb{Z}$. The direct sum $\displaystyle\bigoplus_{i \in \mathbb{N}_{-1}} \mathcal{O}_i$ is called the *integral part* of $\mathcal{O}$ and is denoted by $\mathcal{O}_{\mathrm{int}}$.

For $i \in \mathbb{N}_{-1}$ we denote by $\mathfrak{i}_i : \mathcal{O}_i \to \mathcal{O}_{\mathrm{int}}$ the natural inclusion functor and by $\mathfrak{p}_i : \mathcal{O}_{\mathrm{int}} \to \mathcal{O}_i$ the natural projection functor. Due to the direct sum decomposition of $\mathcal{O}_{\mathrm{int}}$, the functors $\mathfrak{i}_i$ and $\mathfrak{p}_i$ are exact and both left and right adjoint to each other.

Proposition 5.84. *For every finite-dimensional $\mathfrak{g}$-module V the functor $V \otimes {}_-$ preserves the category $\mathcal{O}_{\mathrm{int}}$ and hence restricts to an exact and self-adjoint endofunctor on $\mathcal{O}_{\mathrm{int}}$.*

Proof. If V is finite-dimensional, then $\mathrm{supp}(V) \subset \mathbb{Z}$ by (1.9) and Theorems 1.22 and 1.29. If $M \in \mathcal{O}$ is such that $\mathrm{supp}(M) \subset \mathbb{Z}$, then we have $\mathrm{supp}(V \otimes M) \subset \mathbb{Z}$ by Exercise 3.10. This proves that $V \otimes {}_-$ preserves the category $\mathcal{O}_{\mathrm{int}}$. The rest of the claim follows from Lemma 5.20. $\square$

Some elementary properties of projective functors are collected in the following statement:

Proposition 5.85.

(i) For every $i \in \mathbb{N}_{-1}$ the functor $\mathrm{ID}_{\mathcal{O}_i}$ is a projective functor.
(ii) Any direct sum of projective functors is a projective functor.
(iii) Any composition of projective functors is a projective functor.

Proof. First we note that the functor $\mathrm{ID}_{\mathcal{O}_i}$ can be realized as the composition $\mathfrak{p}_i(\mathbf{V}^{(1)} \otimes \mathfrak{i}_i(\_))$ (see Exercise 3.75) and hence is a projective functor, proving (i). The claim (ii) follows from the definitions and the fact that a direct sum of two finite-dimensional $\mathfrak{g}$-modules is a finite-dimensional $\mathfrak{g}$-module. The claim (iii) follows from the definitions, Exercise 3.74 and the fact that the tensor product of two finite-dimensional $\mathfrak{g}$-modules is a finite-dimensional $\mathfrak{g}$-module. $\square$

Let $\mathfrak{PF}$ denote the category, whose set of objects is $\mathbb{N}_{-1}$, and for $i, j \in \mathbb{N}_{-1}$ the set of morphisms $\mathfrak{PF}(i, j)$ is the set of isomorphism classes of projective functors from $\mathcal{O}_i$ to $\mathcal{O}_j$. Define the composition of morphisms in $\mathfrak{PF}$ as the composition of functors. From Proposition 5.85 it follows that $\mathfrak{PF}$ is indeed a category. We call $\mathfrak{PF}$ the *category of projective functors* on $\mathcal{O}_{\text{int}}$. From Proposition 5.85(ii) it follows that each set $\mathfrak{PF}(i, j)$ is equipped with the natural structure of a commutative monoid with respect to taking the direct sum of functors.

Exercise 5.86. Check that the above definition does define on $\mathfrak{PF}$ the structure of a category.

For $i \in \mathbb{N}_{-1}$ define the following projective functors:

$$\theta_i^{i+1} = \mathfrak{p}_{i+1}(\mathbf{V}^{(2)} \otimes \mathfrak{i}_i(-)), \quad \theta_{i+1}^i = \mathfrak{p}_i(\mathbf{V}^{(2)} \otimes \mathfrak{i}_{i+1}(-)).$$

These functors will be called *elementary* projective functors. Our first goal is to show that these projective functors generate the category $\mathfrak{PF}$. This will require some preparation.

Lemma 5.87. *Let $i \in \mathbb{N}_{-1}$. Then the functor θ_i^{i+1} is both left and right adjoint to the functor θ_{i+1}^i. In particular, both θ_i^{i+1} and θ_{i+1}^i, are exact, send projective modules to projective modules, and send injective modules to injective modules.*

Proof. The endofunctor $\mathbf{V}^{(2)} \otimes -$ of $\mathcal{O}_{\text{int}}$ is self-adjoint by Proposition 5.84. For every $i \in \mathbb{N}_{-1}$ the functors $\mathfrak{p}_i$ and $\mathfrak{i}_i$ are both left and right adjoint to each other. This yields that the functor θ_i^{i+1} is both left and right adjoint to the functor θ_{i+1}^i. It follows that both θ_i^{i+1} and θ_{i+1}^i are exact. As both θ_i^{i+1} and θ_{i+1}^i are both left and right adjoint to exact functors, they send projective modules to projective modules, and injective modules to injective modules (see proof of Corollary 5.21 and Exercise 5.22). $\square$

Lemma 5.88.

(i) *If $i \in \mathbb{N}_0$ and $M \in \mathcal{O}_i$, then $\mathbf{V}^{(2)} \otimes M \in \mathcal{O}_{i-1} \oplus \mathcal{O}_{i+1}$.*
(ii) *If $M \in \mathcal{O}_{-1}$, then $\mathbf{V}^{(2)} \otimes M \in \mathcal{O}_0$.*

Proof. The functor $\mathbf{V}^{(2)} \otimes -$ is exact by Lemma 5.20 and hence it is enough to prove the claim in the situation when the module M is simple.

If $i \in \mathbb{N}_0$, then the category $\mathcal{O}_i$ has two simple modules, namely $L(i)$ and $L(-i-2)$ (Exercise 5.15(c)). For the module $L(i)$ the claim follows from Theorem 1.39. For the module $L(-i-2)$ the claim follows from Exercise 3.110.

The category $\mathcal{O}_{-1}$ has one simple module, namely $L(-1) = M(-1)$ (Exercise 5.15(b)). For this module the claim of the lemma follows from Exercise 3.111. This completes the proof. $\qquad\square$

Proposition 5.89. *Every projective functor is a direct summand of a direct sum of compositions of elementary projective functors.*

Proof. If $i \in \mathbb{N}_0$, then from Lemma 5.88(i) it follows that the functor

$$\mathbf{V}^{(2)} \otimes \mathfrak{i}_i(\_) : \mathcal{O}_i \to \mathcal{O}_{\mathrm{int}}$$

is isomorphic to the direct sum $\theta_i^{i+1} \oplus \theta_i^{i-1}$ (where we disregard the additional zero summands). Furthermore, from Lemma 5.88(ii) it follows that the functor

$$\mathbf{V}^{(2)} \otimes \mathfrak{i}_{-1}(\_) : \mathcal{O}_{-1} \to \mathcal{O}_{\mathrm{int}}$$

is isomorphic to the functor θ_{-1}^0. Thus the endofunctor $\mathbf{V}^{(2)} \otimes \_$ of $\mathcal{O}_{\mathrm{int}}$ is a direct sum of the functor θ_i^{i+1} and θ_{i+1}^i, $i \in \mathbb{N}_{-1}$. The claim now follows from Exercise 3.74 and the fact that every simple finite-dimensional $\mathfrak{g}$-module is a direct summand of some tensor power of $\mathbf{V}^{(2)}$ (see Exercises 1.62 and 1.63). $\qquad\square$

Proposition 5.90. *For $i \in \mathbb{N}_0$ there exist the following isomorphisms of functors:*

$$\theta_0^{-1} \circ \theta_{-1}^0 \cong \mathrm{ID}_{\mathcal{O}_{-1}} \oplus \mathrm{ID}_{\mathcal{O}_{-1}}; \tag{5.10}$$

$$\theta_i^{i+1} \circ \theta_{i+1}^i \cong \mathrm{ID}_{\mathcal{O}_{i+1}}; \tag{5.11}$$

$$\theta_{i+1}^i \circ \theta_i^{i+1} \cong \mathrm{ID}_{\mathcal{O}_i}. \tag{5.12}$$

Proof. From Theorem 1.39 we get

$$\theta_i^{i+1} L(i) \cong L(i+1) \quad \text{and} \quad \theta_{i+1}^i L(i+1) \cong L(i).$$

From Exercise 3.110 we have

$$\theta_i^{i+1} L(-i-2) \cong L(-i-3) \quad \text{and} \quad \theta_{i+1}^i L(-i-3) \cong L(-i-2).$$

As θ_i^{i+1} is left adjoint to θ_{i+1}^i, we have the adjunction morphism

$$\mathrm{adj} : \theta_i^{i+1} \circ \theta_{i+1}^i \to \mathrm{ID}_{\mathcal{O}_{i+1}}.$$

This morphism is non-zero on any $X \in \mathcal{O}_{i+1}$ such that $\theta_{i+1}^i X \neq 0$. From the previous paragraph it follows that this morphism is non-zero and, moreover, is an isomorphism for all simple modules. Since the involved functors are exact, from the Five Lemma it follows that adj is an isomorphism of functors (see proof of Theorem 3.58). This gives us the isomorphism (5.11) and the isomorphism (5.12) is proved similarly. To prove the isomorphism (5.10) we will need the following lemma:

Lemma 5.91.

(i) *For* $n \in \mathbb{N}$ *set* $\vartheta_n = \mathfrak{p}_{-1}(\mathbf{V}^{(n)} \otimes \mathfrak{i}_{-1}(\_))$. *Then we have*

$$\vartheta_n \cong \begin{cases} 0, & n \text{ is even;} \\ \mathrm{ID}_{\mathcal{O}_{-1}}, & n \text{ is odd.} \end{cases}$$

(ii) *Every projective endofunctor of* $\mathcal{O}_{-1}$ *is a direct sum of some copies of the identity functor.*

Proof. The claim (ii) follows from the claim (i) and Theorems 1.22 and 1.29. Let us prove the claim (i).

The support of the unique simple module $M(-1)$ in $\mathcal{O}_{-1}$ consists of odd numbers. If n is even, the support of $\mathbf{V}^{(n)}$ consists of odd numbers as well (see (1.9)). By Exercise 3.10, the support of $\mathbf{V}^{(n)} \otimes M(-1)$ therefore consists of even numbers and hence $\vartheta_n = 0$.

Assume now that n is odd. Similar to the proof of Lemma 5.87 one shows that ϑ_n is self-adjoint. By Exercise 3.114, the module $\mathbf{V}^{(n)} \otimes M(-1)$ has a filtration with subquotients $M(n-2-2i)$, $i = 0, 1, \ldots, n-1$, each occurring with multiplicity one. Since $M(n-2-2i)$ does not belong to $\mathcal{O}_{-1}$ unless $n - 2 - 2i = -1$, it follows that $\vartheta_n M(-1) \cong M(-1)$. Similar to the proof of (5.11) one obtains that $\vartheta_n \circ \vartheta_n \cong \mathrm{ID}_{\mathcal{O}_{-1}}$. In particular, ϑ_n is an autoequivalence of $\mathcal{O}_{-1}$ and $\vartheta_n M(-1) \cong M(-1)$. However, the category $\mathcal{O}_{-1}$ is just the category of finite-dimensional complex vector spaces by Theorem 5.31(ii). Hence, any isomorphism $M(-1) \cong \vartheta_n M(-1)$ gives an isomorphism from $\mathrm{ID}_{\mathcal{O}_{-1}}$ to ϑ_n. This completes the proof. $\square$

By Exercise 3.114 we have a short exact sequence

$$0 \to M(0) \to \theta_{-1}^0 M(-1) \to M(-2) \to 0. \tag{5.13}$$

From Exercise 3.110 it follows that both $\theta_0^{-1} M(0)$ and $\theta_0^{-1} M(-2)$ are isomorphic to $M(-1)$. Hence, applying θ_0^{-1} to (5.13) and using the fact that the category $\mathcal{O}_{-1}$ is semi-simple (Theorem 5.31(ii)), we conclude that $\theta_0^{-1} \circ \theta_{-1}^0 M(-1) \cong M(-1) \oplus M(-1)$. Now the isomorphism (5.10) follows from Lemma 5.91(ii). $\square$

Exercise 5.92. Show that $\theta^0_{-1} M(-1) \cong P(-2)$.

Lemma 5.93. *The functors θ^0_{-1} and $\theta^0_{-1} \circ \theta^{-1}_0$ are indecomposable.*

Proof. Assume that we have a decomposition $\theta^0_{-1} \circ \theta^{-1}_0 \cong A \oplus B$. First we observe that $\mathbf{V}^{(2)} \otimes L(0) \cong \mathbf{V}^{(2)}$ is indecomposable and belongs to $\mathcal{O}_1$. Hence $\theta^{-1}_0 L(0) = 0$ and thus both $A L(0) = 0$ and $B L(0) = 0$.

From Exercise 3.110 we have that $\theta^{-1}_0 L(-2) \cong M(-1)$ is projective in $\mathcal{O}$. By Exercise 5.92 we thus have

$$N = \theta^0_{-1} \circ \theta^{-1}_0 L(-2) = \theta^0_{-1} M(-1) = P(-2),$$

which is indecomposable. Hence either $A L(-2) = 0$ or $B L(-2) = 0$. This yields that either $A = 0$ or $B = 0$ and thus the functor $\theta^0_{-1} \circ \theta^{-1}_0$ is indecomposable. As the latter functor factors through θ^0_{-1} and the functor θ^0_{-1} does not annihilate any module from $\mathcal{O}_{-1}$, the functor θ^0_{-1} is also indecomposable. This completes the proof. $\qquad \square$

Theorem 5.94 (Structure of projective functors).

(i) *Every projective functor is a direct sum of indecomposable projective functors and this decomposition is unique up to permutation and isomorphism of summands.*

(ii) *Every projective functor from $\mathfrak{PF}(-1,-1)$ is a direct sum of some copies of the identity functor.*

(iii) *Every projective functor from $\mathfrak{PF}(-1,i)$, $i \in \mathbb{N}$, is a direct sum of some copies of the indecomposable functor $\theta^i_{-1} = \theta^i_{i-1} \theta^{i-1}_{i-2} \cdots \theta^1_0 \theta^0_{-1}$.*

(iv) *Every projective functor from $\mathfrak{PF}(i,-1)$, $i \in \mathbb{N}$, is a direct sum of some copies of the indecomposable functor $\theta^{-1}_i = \theta^{-1}_0 \theta^0_1 \cdots \theta^{i-2}_{i-1} \theta^{i-1}_i$.*

(v) *Every projective functor from $\mathfrak{PF}(i,i)$, $i \in \mathbb{N}$, is a direct sum of some copies of the identity functor and the indecomposable functor $\vartheta^i_i = \theta^i_{-1} \theta^{-1}_i$ (not isomorphic to the identity).*

(vi) *Every projective functor from $\mathfrak{PF}(i,j)$, $i, j \in \mathbb{N}$, $i \neq j$, is a direct sum of some copies of the the following non-isomorphic indecomposable functors:*

$$\vartheta^j_i = \theta^j_{-1} \theta^{-1}_i, \qquad \theta^j_i = \begin{cases} \theta^j_{j+1} \theta^{j+1}_{j+2} \cdots \theta^{i-2}_{i-1} \theta^{i-1}_i, & i > j; \\ \theta^j_{j-1} \theta^{j-1}_{j-2} \cdots \theta^{i+2}_{i+1} \theta^{i+1}_i, & i < j. \end{cases}$$

(vii) *If $i, j \in \mathbb{N}$, $i \neq j$, then the functor θ^j_i is an equivalence of categories with inverse θ^i_j.*

Proof. The category $\mathfrak{PF}$ is additive by definition and all idempotents in $\mathfrak{PF}$ split. Hence the claim (i) follows from the abstract Krull–Schmidt Theorem (Theorem 3.6 in [7]). The claim (ii) follows from Lemma 5.91.

Let us prove that the functor θ^i_{-1} is indecomposable by induction on $i \in \mathbb{N}$. For $i = 0$ the claim follows from Lemma 5.93. If $i > 0$, by (5.11) and (5.12) we find that θ^{i-1}_i and θ^i_{i-1} are mutually inverse equivalence between $\mathcal{O}_{i-1}$ and $\mathcal{O}_i$. Hence the functors θ^i_{-1} and θ^{i-1}_{-1} are either both decomposable or indecomposable. Thus we can apply the inductive assumption and conclude that θ^i_{-1} is indecomposable for all $i \in \mathbb{N}$.

Consider the following graph (of the category $\mathfrak{PF}$):

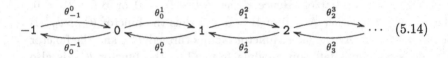

$$\cdots \quad (5.14)$$

By Proposition 5.89, every projective functor is a direct summand of a direct sum of compositions of elementray projective functors. Compositions of elementary projective functors are paths in the graph (5.14). Compositions of elementary projective functors which belong to $\mathfrak{PF}(-1, i)$ are paths which start at -1 and terminate at i. Let p be such a path. Assume that we can write $p = p_1 p_2$, where the path p_2 terminates at -1 and is the shortest path with this property. Then, using the isomorphisms (5.10)–(5.12), we conclude that the path p_2 corresponds to the functor $\mathrm{ID}_{\mathcal{O}_{-1}} \oplus \mathrm{ID}_{\mathcal{O}_{-1}}$, and hence the projective functor p is a direct sum of two copies of the projective functor p_1. If the path p never returns to -1, then, using the isomorphisms (5.11) and (5.12), we obtain that the projective functor p is isomorphic to θ^i_{-1}. From (i) we thus get that θ^i_{-1} is the only indecomposable projective functor in $\mathfrak{PF}(-1, i)$, which completes the proof of the claim (iii).

The proofs of the claims (iv)–(vii) similarly reduce to the analysis of the paths in the graph (5.14) using the isomorphisms (5.10)–(5.12) and Proposition 5.89. We leave the details to the reader. $\qquad \square$

The block $\mathcal{O}_{-1}$ is usually called the block *on the wall* (meaning that the highest weight of the unique simple module in this block belongs to the stabilizer of the shifted action of the Weyl group). The functors θ^{-1}_i, $i \in \mathbb{N}$, are called *translations onto the wall*. The functors θ^i_{-1}, $i \in \mathbb{N}$, are called *translations out of the wall*. The functors ϑ^i_i, $i \in \mathbb{N}$, are called *translations through the wall*. The functors θ^j_i, $i, j \in \mathbb{N}$, $i \neq j$, are called *translation* functors.

Corollary 5.95.

(i) *Let* $i, j \in \mathbb{N}_{-1}$, $i \neq -1$. *Then the correspondence*
$$\mathfrak{PF}(i,j) \longrightarrow \left\{ \begin{matrix} \textit{isomorphism classes of} \\ \textit{projective modules in } \mathcal{O}_j \end{matrix} \right\}$$
$$\mathrm{X} \mapsto \mathrm{X} M(i)$$
is an isomorphism of additive monoids. In particular, indecomposable projective functors correspond to indecomposable projective modules.

(ii) *Let* $j \in \mathbb{N}_{-1}$. *Then the correspondence*
$$\mathfrak{PF}(-1,j) \longrightarrow \left\{ \begin{matrix} \textit{isomorphism classes of} \\ \textit{projective-injective modules in } \mathcal{O}_j \end{matrix} \right\}$$
$$\mathrm{X} \mapsto \mathrm{X} M(-1)$$
is an isomorphism of additive monoids. In particular, indecomposable projective functors correspond to indecomposable projective-injective modules.

Proof. It is enough to verify that indecomposable projective functors send $M(i)$ (respectively $M(-1)$) to indecomposable projective (respectively projective-injective) modules in $\mathcal{O}_j$.

Let $j \in \mathbb{N}_{-1}$. There is a unique indecomposable projective functor in $\mathfrak{PF}(-1,j)$, given by Theorem 5.94(ii) and (iii). From the computation in Exercise 5.92 and Theorem 5.94(vii) it follows that this functor sends $M(-1)$ to the indecomposable projective-injective module in $\mathcal{O}_j$.

If $i \in \mathbb{N}_0$, then there is a unique indecomposable projective functor in $\mathfrak{PF}(i,-1)$, given by Theorem 5.94(iv). It sends $M(i)$ to $M(-1)$ by Exercise 3.110.

If $i, j \in \mathbb{N}_0$, then there are two indecomposable projective functors in $\mathfrak{PF}(i,j)$, given by Theorem 5.94(v) and (vi). One of them is an equivalence (Theorem 5.94(vii)) and hence sends $M(i)$ to $M(j)$. Another factors as $\theta^j_{-1} \circ \theta^{-1}_i$, sending $M(i)$ to the indecomposable projective-injective module in $\mathcal{O}_j$ by the previous paragraphs. This completes the proof. $\square$

Exercise 5.96. Show that for every $i \in \mathbb{N}_0$ we have $\vartheta^i_i \circ \vartheta^i_i \cong \vartheta^i_i \oplus \vartheta^i_i$.

Exercise 5.97. Let $i \in \mathbb{N}_0$.

(a) Show that $\mathrm{Ext}_{\mathcal{O}}(M(-i-2), M(i)) \cong \mathbb{C}$.

(b) Show that $\vartheta^i_i M(i) \cong \vartheta^i_i M(-i-2)$ and that there exist the following non-split short exact sequences:
$$0 \to M(i) \to \vartheta^i_i M(i) \to M(-i-2) \to 0$$
$$0 \to M(i) \to \vartheta^i_i M(-i-2) \to M(-i-2) \to 0.$$

Exercise 5.98. Let $i \in \mathbb{N}_0$. Show that the following is the table of values of the functor ϑ_i^i on indecomposable modules from $\mathcal{O}_i$:

M	$L(-i-2)$	$L(i)$	$P(-i-2)$	$P(i)$	$I(i)$
$\vartheta_i^i M$	$P(-i-2)$	0	$P(-i-2) \oplus P(-i-2)$	$P(-i-2)$	$P(-i-2)$.

5.9 Addenda and comments

5.9.1

Category $\mathcal{O}$ was introduced (for arbitrary simple finite-dimensional complex Lie algebra) by I. Bernstein, I. Gelfand and S. Gelfand in [18]. One of the main motivations for their study was the BGG-reciprocity (Exercise 5.44), which is true in full generality and even naturally generalizes to arbitrary quasi-hereditary algebras. Several monographs are dedicated to the study of this category, from different perspectives (see for example [63, 64, 66, 96]). A good introduction to the category $\mathcal{O}$ is the recent textbook [58].

The material presented in Section 5.1 is taken from [18], where it is presented in full generality, but mostly without proofs. The fact that $\mathcal{O}$ has enough projectives is also proved in [18]. However, in Section 5.2 we follow an alternative approach using projective functors, which is largely due to Jantzen. There is no straightforward generalization of the description of the blocks of $\mathcal{O}$ via quiver with relations to the general case. The quiver for each block of the category $\mathcal{O}$ can be computed using the Kazhdan–Lusztig combinatorics ([70, 12, 26]). Using the combinatorial description of blocks of $\mathcal{O}$ obtained by Soergel in [109], one can formulate an algorithm to compute the relations (see [113, 117]). For example, in [113] one finds quivers and relations for all integral blocks of $\mathcal{O}$ for Lie algebras of rank two and even for the regular block of $\mathcal{O}$ for the Lie algebra $\mathfrak{sl}_4$.

The structure of a highest weight category on $\mathcal{O}$ was established in [18], but the name and the abstract set up appeared much later in [31]. The algebraic counterpart of this theory (quasi-hereditary algebras) was defined in [36]. Tilting modules for quasi-hereditary algebras were discovered in [101], but for the category $\mathcal{O}$ they had already been constructed in [32].

The grading on $\mathcal{O}$ is defined following the combinatorial description of [109]. However, it is not obvious that this grading is positive. This was proved only in [13]. The graded version of the category $\mathcal{O}$ was first studied in [114]. Generally, not all modules from $\mathcal{O}$ are gradeable (see [113]).

Basic homological properties of the category $\mathcal{O}$ (for example, an upper bound for the global dimension) were already obtained in [18]. Some further results in this direction appeared in [85]. Koszulity and Koszul self-duality for $\mathcal{O}$ was established in [109]. For proof that standard and costandard modules over arbitrary quasi-hereditary algebras form homologically dual families, see [101]. The category of linear complexes of projective modules appears in [79, 87] and was subsequently studied, especially for the category $\mathcal{O}$, in [86, 88].

Projective functors were formally introduced and indecomposable projective functors were classified by I. Bernstein and S. Gelfand in [16]. However, they had already appeared in the original paper [17] during the study of Verma modules and in [62] during the study of some generalized Verma modules.

5.9.2

Category $\mathcal{O}$ was also studied for other generalizations of $U(\mathfrak{g})$; for example, for algebras similar to $U(\mathfrak{g})$ in [107], for down-up algebras in [14], and for some other generalizations in [81].

5.9.3

The condition (II) from the definition of the category $\mathcal{O}$ is fairly controversial. It is natural to weaken it as follows:

(II') M is generalized weight.

The full subcategory of the category of all $\mathfrak{g}$-modules in which all modules satisfy conditions (I), (II') and (III) is called the *thick* category $\mathcal{O}$ and is denoted by $\tilde{\mathcal{O}}$ (see [108, 110]).

The category $\mathcal{O}$ is obviously a subcategory of $\tilde{\mathcal{O}}$. By Exercise 3.90, simple objects in the categories $\mathcal{O}$ and $\tilde{\mathcal{O}}$ coincide. Furthermore, the action of $Z(\mathfrak{g})$ on all objects from $\tilde{\mathcal{O}}$ is locally finite and hence the decomposition (3.40) induces the decomposition

$$\tilde{\mathcal{O}} = \bigoplus_{\substack{\xi \in \mathbb{C}/2\mathbb{Z} \\ \tau \in \mathbb{C}}} \tilde{\mathcal{O}}^{\xi,\tau},$$

where $\tilde{\mathcal{O}}^{\xi,\tau} = \tilde{\mathcal{O}} \cap \overline{\mathfrak{G}\mathfrak{W}}^{\xi,\tau}$.

The advantage of the category $\tilde{\mathcal{O}}$ is that it is extension closed in $\mathfrak{g}$-mod in the sense that, given a short exact sequence

$$0 \to X \to Y \to Z \to 0$$

in $\mathfrak{g}$-mod such that $X, Z \in \tilde{\mathcal{O}}$, one automatically gets $Y \in \tilde{\mathcal{O}}$ (the category $\mathcal{O}$ does not have this property, see Exercise 5.6). The disadvantage of the category $\tilde{\mathcal{O}}$ is that, unlike $\mathcal{O}$, it does not have enough projective modules (actually it does not contain any projective modules at all).

The blocks $\tilde{\mathcal{O}}^{\xi,\tau}$ can be described using quivers and relations just like the blocks $\mathcal{O}^{\xi,\tau}$. This can be deduced from the results mentioned in 3.12.3. If the block $\tilde{\mathcal{O}}^{\xi,\tau}$ contains one simple module, it is equivalent to the category $\mathbb{C}[[x]]$-mod. If the block $\tilde{\mathcal{O}}^{\xi,\tau}$ contains two simple modules, it is equivalent to the category of finite-dimensional representations of the following quiver with relations (see [51]):

$$c\,\bigcirc\,p\;\underset{b}{\overset{a}{\rightleftharpoons}}\;q \qquad ac = cb = 0; \quad c, ba \text{ nilpotent}.$$

Notably, the associative algebras describing the blocks $\tilde{\mathcal{O}}^{\xi,\tau}$ are positively graded. One can show that they are also Koszul. These algebras are no longer quasi-hereditary, but rather properly stratified in the sense of [35] (see [84, 74]). Naturally, the algebra describing $\mathcal{O}^{\xi,\tau}$ is the quotient of the algebra describing $\tilde{\mathcal{O}}^{\xi,\tau}$.

Projective functors restrict to $\tilde{\mathcal{O}}$. Theorem 5.94 is true in the more general set up of the category $\tilde{\mathcal{O}}$ as well.

5.9.4

Consider the integral group algebra $\mathbb{Z}[\mathbf{S}_2]$ of the symmetric group $\mathbf{S}_2$ (recall that $\mathbf{S}_2$ is the Weyl group of $\mathfrak{g}$). This algebra has a natural basis consisting of the identity element 1 and the simple reflection $s = (1,2)$. Choose another basis in $\mathbb{Z}[\mathbf{S}_2]$ as follows: $a = 1$, $b = 1 + s$. In the basis $\{a, b\}$ the multiplication table of $\mathbb{Z}[\mathbf{S}_2]$ has the following form:

$$
\begin{array}{c||c|c}
\cdot & a & b \\
\hline\hline
a & a & b \\
\hline
b & b & 2b
\end{array}
\tag{5.15}
$$

From Exercise 5.96 we have that for every $i \in \mathbb{N}$, the multiplication table in $\mathfrak{PF}(0,0)$ with respect to the generating system $\mathrm{ID}_{\mathcal{O}_0}$, ϑ_0^0 has the following form:

$$
\begin{array}{c|c|c}
\cdot & \mathrm{ID} & \vartheta_0^0 \\
\hline
\mathrm{ID} & \mathrm{ID} & \vartheta_0^0 \\
\hline
\vartheta_0^0 & \vartheta_0^0 & \vartheta_0^0 \oplus \vartheta_0^0
\end{array}
\tag{5.16}
$$

Comparing (5.15) and (5.16) we conclude that the assignment

$$
a \mapsto \mathrm{ID}_{\mathcal{O}_0}, \qquad b \mapsto \vartheta_0^0
\tag{5.17}
$$

defines a *weak action* of the algebra $\mathbb{Z}[\mathbf{S}_2]$ on the category $\mathcal{O}_0$ via projective functors (here by *weak* we mean that the relations for functors are given by some isomorphisms of functors). For arbitrary simple finite-dimensional complex Lie algebra, the existence of such action was established in [16].

Let $\mathrm{Gr}[\mathcal{O}_0]$ denote the *Grothendieck group* of $\mathcal{O}_0$, which is the quotient of the free abelian group generated by $[M]$, $M \in \mathcal{O}_0$, modulo the subgroup, generated by all possible elements of the form $[X] - [Y] + [Z]$ such that there is a short exact sequence

$$
0 \to X \to Y \to Z \to 0
$$

in $\mathcal{O}_0$. The group $\mathrm{Gr}[\mathcal{O}_0]$ is a free abelian group with basis $\{[L(-2)], [L(0)]\}$.

Exercise 5.99. Check that both $\{[P(-2)], [P(0)]\}$ and $\{[M(-2)], [M(0)]\}$ also form bases of $\mathrm{Gr}[\mathcal{O}_0]$.

From Exercise 5.99 we have that the assignment

$$
[M(0)] \mapsto 1, \qquad [M(-2)] \mapsto s
\tag{5.18}
$$

uniquely extends to an isomorphism $\psi : \mathrm{Gr}[\mathcal{O}_0] \cong \mathbb{Z}[\mathbf{S}_2]$. Since both functors $\mathrm{ID}_{\mathcal{O}_0}$ and ϑ_0^0 are exact, they send short exact sequences to short exact sequences and hence induce well-defined homomorphisms $[\mathrm{ID}]$ and $[\vartheta_0^0]$ of $\mathrm{Gr}[\mathcal{O}_0]$, respectively. From Exercise 5.97(b) it follows that the matrices of the operators $[\mathrm{ID}]$ and $[\vartheta_0^0]$ in the basis $\{[M(-2)], [M(0)]\}$ are as follows:

$$
[\mathrm{ID}] = \begin{pmatrix} 1 & 0 \\ 0 & 1 \end{pmatrix}, \qquad [\vartheta_0^0] = \begin{pmatrix} 1 & 1 \\ 1 & 1 \end{pmatrix}.
$$

These matrices coincide with the matrices of the action of the elements $a, b \in \mathbb{Z}[\mathbf{S}_2]$ on the right regular module $\mathbb{Z}[\mathbf{S}_2]_{\mathbb{Z}[\mathbf{S}_2]}$ in the basis $\{1, s\}$. This means that the assignments (5.17) and (5.18) define a *categorification* (see Chapter 7) of the right regular module $\mathbb{Z}[\mathbf{S}_2]_{\mathbb{Z}[\mathbf{S}_2]}$, in the sense that the following diagrams commute:

$$
\begin{array}{ccc}
\mathrm{Gr}[\mathcal{O}_0] & \xrightarrow{\ [\mathrm{ID}]\ } & \mathrm{Gr}[\mathcal{O}_0] \\
\Big\downarrow{\psi} & & \Big\downarrow{\psi} \\
\mathbb{Z}[\mathbf{S}_2] & \xrightarrow{\ a\ } & \mathbb{Z}[\mathbf{S}_2]
\end{array}
\qquad
\begin{array}{ccc}
\mathrm{Gr}[\mathcal{O}_0] & \xrightarrow{\ [\vartheta_0^0]\ } & \mathrm{Gr}[\mathcal{O}_0] \\
\Big\downarrow{\psi} & & \Big\downarrow{\psi} \\
\mathbb{Z}[\mathbf{S}_2] & \xrightarrow{\ b\ } & \mathbb{Z}[\mathbf{S}_2]
\end{array}
$$

5.9.5

The endofunctor ϑ_0^0 of $\mathcal{O}_0$ is defined as the composition $\theta_{-1}^0 \circ \theta_0^{-1}$ of two functors, which are both left and right adjoint to each other (see Lemma 5.87). Hence we have two adjunction morphisms

$$\text{adj} : \text{ID}_{\mathcal{O}_0} \to \vartheta_0^0, \qquad \overline{\text{adj}} : \vartheta_0^0 \to \text{ID}_{\mathcal{O}_0}.$$

Define the endofunctor C of $\mathcal{O}_0$ as the cokernel of the adjunction morphism adj, and the endofunctor K of $\mathcal{O}_0$ as the kernel of the adjunction morphism $\overline{\text{adj}}$. The functor C is called *shuffling* and the functor K is called *coshuffling* (see [29, 59]). These functors form a pair (C, K) of adjoint functors. In particular, the functor C is right exact, while the functor K is left exact. One can also show that $K \cong \circledast \circ C \circ \circledast$ (see [29, 73, 89]).

Taking into account Exercise 5.98 and the fact that both adj and $\overline{\text{adj}}$ are non-zero on $L(-2)$ (as $\theta_0^{-1} L(-2) \cong M(-1) \neq 0$, see Exercise 3.110), one can produce the following table, which gives the values of the functors C and K on indecomposable objects from $\mathcal{O}$:

M	$L(-2)$	$L(0)$	$P(-2)$	$P(0)$	$I(0)$
$\vartheta_0^0 M$	$P(-2)$	0	$P(-2) \oplus P(-2)$	$P(-2)$	$P(-2)$
$C M$	$I(0)$	0	$P(-2)$	$L(-2)$	$I(0)$
$K M$	$P(0)$	0	$P(-2)$	$P(0)$	$L(-2)$.

(5.19)

It is interesting to note that the functor C maps indecomposable projective modules $P(-2)$ and $P(0)$ to the indecomposable tilting modules $T(0) \cong P(-2)$ and $T(-2) \cong L(-2)$. Moreover, it maps indecomposable tilting modules $T(0)$ and $T(-2)$ to the indecomposable injective modules $P(-2)$ and $I(0)$, respectively. The functor K does the opposite.

We refer the reader to [29, 59, 73, 89, 90] for further properties of these functors, the most exciting of which are related to the corresponding derived functors. We will mention some of them later on in 5.9.9, where we will discuss the derived category of $\mathcal{O}$.

5.9.6

The correspondence

$$\mathbf{e}^{\square} = \mathbf{f}, \quad \mathbf{f}^{\square} = \mathbf{e}, \quad \mathbf{h}^{\square} = -\mathbf{h}$$

extends uniquely to an involution on both $\mathfrak{g}$ and $U(\mathfrak{g})$, which we will denote by $\square$.

Consider the associative algebra $U^{(f)}$ defined in Section 3.5. By Corollary 3.43, we have an inclusion of $U(\mathfrak{g})$-bimodules via $\iota : U(\mathfrak{g}) \hookrightarrow U^{(f)}$.

Consider the vector space $\overline{U}^{(f)} = U^{(f)}/U(\mathfrak{g})$ and define on $\overline{U}^{(f)}$ the structure of a $U(\mathfrak{g})$-bimodule via:

$$x \cdot (u + U(\mathfrak{g})) \cdot y = x^{\square}uy + U(\mathfrak{g}),$$

where $x, y \in U(\mathfrak{g})$ and $u \in U^{(f)}$.

Denote by T the endofunctor of $\mathfrak{g}$-mod, defined as the tensor product with $\overline{U}^{(f)}$. The functor T is called (Arkhipov's) *twisting functor* (see [3, 1, 111, 73, 90]). From the construction it follows that the functor T is right exact and is left adjoint to the functor $\mathrm{Hom}_{\mathfrak{g}}(U^{(f)}, \_)$. Main properties of the functor T are collected in the following statement (see [1, 73]):

Theorem 5.100.

(i) *The functor* T *restricts to a right exact endofunctor of the category* $\mathcal{O}$ *and even of the category* $\mathcal{O}_0$.

(ii) *The right adjoint of the functor* T, *viewed as an endofunctor of* $\mathcal{O}_0$, *is isomorphic to the functor* $\circledast \circ \mathrm{T} \circ \circledast$.

(iii) *The functor* T *commutes with projective functors.*

The adjoint functor $\mathrm{G} = \circledast \circ \mathrm{T} \circ \circledast$ is called (Joseph's) *completion functor* and has been considered, in a completely different guise, in [44, 68] already (see [73] for details).

Using Theorem 3.52(iii) and Theorem 5.100(ii) one can compute the following table of values of the functors T and G on indecomposable modules of the block $\mathcal{O}_0$:

M	$L(-2)$	$L(0)$	$P(-2)$	$P(0)$	$I(0)$
$\mathrm{T}\,M$	$I(0)$	0	$P(-2)$	$L(-2)$	$I(0)$
$\mathrm{G}\,M$	$P(0)$	0	$P(-2)$	$P(0)$	$L(-2)$.

(5.20)

In particular, comparing (5.19) and (5.20) one concludes that the functors T and C (as well as the functors K and G) have isomorphic values on all modules from $\mathcal{O}_0$. This is an accident of a small example; the functors themselves are *not* isomorphic (see [90]). Both functors admit an alternative description in terms of partial (co)approximation functors with respect to certain projective (injective) modules, see [73]. We refer the reader to [1, 73, 90] for further properties of these functors. Again, the most exciting properties are related to the corresponding derived functors, some of which we will mention in 5.9.9.

5.9.7

The condition (III) from the definition of the category $\mathcal{O}$ looks rather non-symmetric. It involves only the element e and does not involve the element f at all. If one substitutes (III) by the following condition:

(III') $\mathbb{C}[e]v, \mathbb{C}[f]v < \infty$ for all $v \in M$,

one obtains the category $\mathcal{FD}$ of all finite-dimensional $\mathfrak{g}$-modules. This is a special case of the *parabolic* category $\mathcal{O}$, defined and studied in [102].

The category $\mathcal{FD}$ is abelian and has enough projectives. It also inherits from $\mathcal{O}$ a block decomposition. In our very small example, all non-zero blocks of $\mathcal{FD}$ are equivalent to the category of $\mathbb{C}$-modules by Weyl's theorem. In fact, on the level of the algebra D, the category $\mathcal{FD}$ has a natural description as the category of D/Did_pD-modules. Generally, blocks of the parabolic category $\mathcal{O}$ are also described by quasi-hereditary algebras (however, "smaller" ones than those which describe blocks of $\mathcal{O}$). The embedding $i : \mathcal{FD} \hookrightarrow \mathcal{O}$ is exact and has the left adjoint Z and the right adjoint $\hat{Z}$. The functors Z and $\hat{Z}$ are given by taking, for a module $M \in \mathcal{O}$, the maximal finite-dimensional quotient or submodule of M, respectively. The functor Z is called *Zuckerman's* functor and the functor $\hat{Z}$ is called *co-Zuckerman's* functor, as it was first considered in [118].

The functor Z is right exact and the functor $\hat{Z}$ is left exact. It is easy to see that $\hat{Z} \cong \circledast \circ Z \circ \circledast$. Using the definition one can easily compute the following table of values of the functors Z and $\hat{Z}$ on indecomposable modules of the block $\mathcal{O}_0$:

M	$L(-2)$	$L(0)$	$P(-2)$	$P(0)$	$I(0)$
$Z\,M$	0	$L(0)$	0	$L(0)$	0
$\hat{Z}\,M$	0	$L(0)$	0	0	$L(0)$.

(5.21)

We refer the reader to [90] for further properties of these functors. Again, the most exciting properties are related to the corresponding derived functors and we will discuss some of them later in 5.9.9.

5.9.8

All projective functors, as well as the functors C, K, T, G, Z and $\hat{Z}$, admit a lift to the category D-gmod.

For the functors Z and $\hat{Z}$ this is evident. Since id_p is a homogeneous element of D, the algebra D/Did_pD inherits from D a natural positive grading.

Hence the category $\mathcal{FD}$ admits a graded lift such that the inclusion functor i into D-gmod becomes homogeneous of degree zero (hence graded). Taking the adjoints of the graded inclusion, we get graded lifts of both Z and $\hat{Z}$.

For all other functors this requires substantial work. For projective functors this was carried out in [114], based on the combinatorial description of projective functors from [109, 110]. It turns out that all projective functors admit graded lifts. Moreover, the adjunction morphisms between projective functors, used to define the functors C and K, become homogeneous maps. Taking graded cokernels and kernels, one obtains graded lifts of C and K, respectively. Grading on twisting and completion functors can be defined using the fact that they commute with (gradable) projective functors, see Appendix of [87]. For indecomposable functors, their graded lifts are unique up to isomorphism and shift of gradings. In the table below we collected the values of the *standard* graded lifts of all the above functors on standard graded lifts of indecomposable objects from D-gmod. Abusing notation, we will denote standard graded lifts of the functors in the same way as the functors themselves.

M	$L(\mathsf{p})$	$L(\mathsf{q})$	$P(\mathsf{p})$	$P(\mathsf{q})$	$I(\mathsf{q})$
$\vartheta_0^0 M$	$P(\mathsf{p})\langle 1\rangle$	0	$P(\mathsf{p})\langle 1\rangle \oplus P(\mathsf{p})\langle -1\rangle$	$P(\mathsf{p})$	$P(\mathsf{p})\langle 2\rangle$
$C M$	$I(\mathsf{q})$	0	$P(\mathsf{p})\langle 1\rangle$	$L(\mathsf{p})$	$I(\mathsf{q})\langle 1\rangle$
$K M$	$P(\mathsf{q})$	0	$P(\mathsf{p})\langle -1\rangle$	$P(\mathsf{q})\langle -1\rangle$	$L(\mathsf{p})$
$T M$	$I(\mathsf{q})$	0	$P(\mathsf{p})\langle 1\rangle$	$L(\mathsf{p})$	$I(\mathsf{q})\langle 1\rangle$
$G M$	$P(\mathsf{q})$	0	$P(\mathsf{p})\langle -1\rangle$	$P(\mathsf{q})\langle -1\rangle$	$L(\mathsf{p})$
$Z M$	0	$L(\mathsf{q})$	0	$L(\mathsf{q})$	0
$\hat{Z} M$	0	$L(\mathsf{q})$	0	0	$L(\mathsf{q}).$

$$(5.22)$$

We also have the following graded lifts of the short exact sequences from Exercise 5.97(b):

$$0 \to M(\mathsf{q})\langle -1\rangle \to \vartheta_0^0 M(\mathsf{q}) \to M(\mathsf{p}) \to 0$$
$$0 \to M(\mathsf{q}) \to \vartheta_0^0 M(\mathsf{p}) \to M(\mathsf{p})\langle 1\rangle \to 0.$$

We will return to graded lifts of our functors in 5.9.9 and 5.9.10.

5.9.9

Consider the bounded derived categories $\mathcal{D}^b(\mathcal{O}_0)$ and $\mathcal{D}^b(\text{D-gmod})$ of $\mathcal{O}_0$ and D-gmod, respectively. The (graded) projective functors ID and ϑ_0^0 are exact and thus induce well-defined functors on both $\mathcal{D}^b(\mathcal{O}_0)$ and $\mathcal{D}^b(\text{D-gmod})$. Consider the left derived functors $\mathcal{L}C$, $\mathcal{L}T$ and $\mathcal{L}Z$ of the right exact functors

C, T and Z, respectively. Consider also the right derived functors $\mathcal{R}$K, $\mathcal{R}$G and $\mathcal{R}\hat{Z}$ of the left exact functors K, G and $\hat{Z}$, respectively. All these derived functors (and their graded lifts) are functors on $\mathcal{D}^b(\mathcal{O}_0)$ and $\mathcal{D}^b$(D-gmod), respectively. Taking (5.22) into account, the following statement is not at all surprising:

Theorem 5.101 ([1, 89]).

(i) *The functor $\mathcal{L}$T is a self-equivalence of $\mathcal{D}^b(\mathcal{O}_0)$ with inverse $\mathcal{R}$G.*
(ii) *The functor $\mathcal{L}$T is a self-equivalence of $\mathcal{D}^b$(D-gmod) with inverse $\mathcal{R}$G.*
(iii) *The functor $\mathcal{L}$C is a self-equivalence of $\mathcal{D}^b(\mathcal{O}_0)$ with inverse $\mathcal{R}$K.*
(iv) *The functor $\mathcal{L}$C is a self-equivalence of $\mathcal{D}^b$(D-gmod) with inverse $\mathcal{R}$K.*

Another nice property of both $\mathcal{L}$T and $\mathcal{L}$C is that we have

$$\mathcal{L}\text{T} \circ \mathcal{L}\text{T} \cong \mathcal{L}\text{C} \circ \mathcal{L}\text{C}.$$

This composition is the *Serre functor* on $\mathcal{D}^b(\mathcal{O}_0)$ in the sense that it is an autoequivalence of $\mathcal{D}^b(\mathcal{O}_0)$ and for all $X, Y \in \mathcal{D}^b(\mathcal{O}_0)$ there exists an isomorphism

$$\text{Hom}_{\mathcal{D}^b(\mathcal{O}_0)}(X, \mathcal{L}\text{T} \circ \mathcal{L}\text{T}(Y)) \cong \text{Hom}_{\mathcal{D}^b(\mathcal{O}_0)}(Y, X)^*,$$

natural in both X and Y (see [92]).

In the case of an arbitrary simple finite-dimensional complex Lie algebras $\mathfrak{a}$, one defines twisting, shuffling, coshuffling and completion functors associated to all simple positive roots of the algebra $\mathfrak{a}$. These functors satisfy braid relations ([68, 73, 89]). Theorem 5.101 extends to the general case and hence derived twisting, shuffling, coshuffling and completeion functors define a weak action of the braid group on the categories $\mathcal{D}^b(\mathcal{O}_0)$ and $\mathcal{D}^b$(D-gmod).

For $i \in \mathbb{N}_0$ we denote by $\mathcal{L}_i$ and $\mathcal{R}^i$ the corresponding cohomology part of the derived functors. A connection between Zuckerman's functors on the one side and twisting and shuffling functors on the other side is given by the following:

Theorem 5.102 ([1, 90]).

(i) *We have*

$$\mathcal{L}_i\text{T} = \begin{cases} \text{T}, & i = 0; \\ \hat{\text{Z}}, & i = 1; \\ 0, & otherwise; \end{cases} \qquad \mathcal{R}^i\text{G} = \begin{cases} \text{G}, & i = 0; \\ \text{Z}, & i = 1; \\ 0, & otherwise. \end{cases}$$

(ii) We have

$$\mathcal{L}_i C = \begin{cases} T, & i = 0; \\ \hat{Z}, & i = 1; \\ 0, & otherwise; \end{cases} \qquad \mathcal{R}^i K = \begin{cases} G, & i = 0; \\ Z, & i = 1; \\ 0, & otherwise. \end{cases}$$

Description of the cohomology of Zuckerman's functors is more complex:

Theorem 5.103 ([46, 90]). *The functors* $\mathcal{L}Z[-1]$ *and* $\mathcal{R}\hat{Z}[1]$ *are isomorphic. Moreover, we have*

$$\mathcal{L}_i Z = \begin{cases} Z, & i = 0; \\ Q, & i = 1; \\ \hat{Z}, & i = 1; \\ 0, & otherwise; \end{cases} \qquad \mathcal{R}^i \hat{Z} = \begin{cases} \hat{Z}, & i = 0; \\ Q, & i = 1; \\ Z, & i = 2; \\ 0, & otherwise; \end{cases}$$

where the functor Q *satisfies* $Q \cong \circledast \circ Q \circ \circledast$ *and has the following table of values on indecomposable modules in* $\mathcal{O}_0$:

M	$L(-2)$	$L(0)$	$P(-2)$	$P(0)$	$I(0)$
QM	$L(0)$	0	0	0	$0.$

The functor Q admits also a description as the cokernel of the natural transformation from the identity functor to G, see [90].

Another interplay between these functors is given by the Koszul duality. As we saw in Corollary 5.82, the algebra D is isomorphic to its Koszul dual. This yields the existence of the so-called Koszul duality functor, which is a nontrivial self-equivalence of the category $\mathcal{D}^b(\text{D-gmod})$, see [13, 88]. It turns out that this equivalence swaps the following pairs of functors (see [105, 88]):

$$\mathcal{L}Z[-1] \overset{\text{Koszul dual}}{\longleftrightarrow} \vartheta_0^0; \qquad \mathcal{L}T \overset{\text{Koszul dual}}{\longleftrightarrow} \mathcal{L}C; \qquad \mathcal{R}K \overset{\text{Koszul dual}}{\longleftrightarrow} \mathcal{R}G.$$

Consider the Grothendieck group $\text{Gr}[\mathcal{D}^b(\mathcal{O}_0)]$ of $\mathcal{D}^b(\mathcal{O}_0)$. This is again a free abelian group, which has several basis, for example $\{[L(-2)], [L(0)]\}$, $\{[M(-2)], [M(0)]\}$ or $\{[P(-2)], [P(0)]\}$. The functor $\mathcal{L}T$ induces an endomorphism $[\mathcal{L}T]$ of $\text{Gr}[\mathcal{D}^b(\mathcal{O}_0)]$. Using Theorem 5.102(i), (5.20) and (5.21), one computes the the action of $[\mathcal{L}T]$ on $\text{Gr}[\mathcal{D}^b(\mathcal{O}_0)]$ is given in the basis $\{[M(-2)], [M(0)]\}$ by the matrix

$$\begin{pmatrix} 0 & 1 \\ 1 & 0 \end{pmatrix}.$$

Consider the isomorphism $\psi : \mathrm{Gr}[\mathcal{D}^b(\mathcal{O}_0)] \to \mathbb{Z}[\mathbf{S}_2]$ given by (5.18). Then, assigning to 1 the identity functor, and to the involution $s = (1,2) \in \mathbf{S}_2$ the functor $\mathcal{L}T$, we obtain a categorification of the left regular representation $_{\mathbb{Z}[\mathbf{S}_2]}\mathbb{Z}[\mathbf{S}_2]$ in the sense that the following diagrams commute:

$$
\begin{array}{ccc}
\mathrm{Gr}[\mathcal{D}^b(\mathcal{O}_0)] & \xrightarrow{\;[\mathrm{ID}]\;} & \mathrm{Gr}[\mathcal{D}^b(\mathcal{O}_0)] \\
\downarrow{\scriptstyle \psi} & & \downarrow{\scriptstyle \psi} \\
\mathbb{Z}[\mathbf{S}_2] & \xrightarrow{\;\;1\;\;} & \mathbb{Z}[\mathbf{S}_2]
\end{array}
\qquad
\begin{array}{ccc}
\mathrm{Gr}[\mathcal{D}^b(\mathcal{O}_0)] & \xrightarrow{\;[\mathcal{L}T]\;} & \mathrm{Gr}[\mathcal{D}^b(\mathcal{O}_0)] \\
\downarrow{\scriptstyle \psi} & & \downarrow{\scriptstyle \psi} \\
\mathbb{Z}[\mathbf{S}_2] & \xrightarrow{\;\;s\;\;} & \mathbb{Z}[\mathbf{S}_2]
\end{array}
$$

Note the following difference to the categorification from 5.9.4: in the present picture, the functor $\mathcal{L}T$ corresponds to an element from $\mathbf{S}_2$, while in 5.9.4 the functor ϑ_0^0 was associated with the linear combination $1 + s$.

Since the functor ϑ_0^0 is exact, in 5.9.4 we can substitute $\mathcal{O}_0$ by $\mathcal{D}^b(\mathcal{O}_0)$, keeping the same result. If we combine this with the above and note that twisting and projective functors commute (Theorem 5.100(iii)), we obtain a categorification of the $\mathbb{Z}[\mathbf{S}_2]$-bimodule $_{\mathbb{Z}[\mathbf{S}_2]}\mathbb{Z}[\mathbf{S}_2]_{\mathbb{Z}[\mathbf{S}_2]}$.

5.9.10

The categorification of the right regular representation of the group algebra $\mathbb{Z}[\mathbf{S}_2]$, which is described in 5.9.4, also admits a "graded lift". This connects the action of projective functors on D-gmod with the Hecke algebra of the group $\mathbf{S}_2$.

Consider the ring $\mathbb{Z}[v, v^{-1}]$ of Laurent polynomials in v with integer coefficients. The *Hecke algebra* $\mathcal{H} = \mathcal{H}_2$ of $\mathbf{S}_2$ is a free $\mathbb{Z}[v, v^{-1}]$-module with basis $\{H_1, H_s\}$. The module $\mathcal{H}$ is equipped with an associative multiplication, for which H_1 is the unit element and

$$H_s^2 = H_1 + (v^{-1} - v)H_s.$$

The element H_s is invertible and $H_s^{-1} = H_s + (v - v^{-1})H_1$. There is a unique involution $\blacklozenge$ on $\mathcal{H}$ such that

$$v^{\blacklozenge} = v^{-1}, \qquad H_s^{\blacklozenge} = H_s^{-1}.$$

The elements $\underline{H}_1 = H_1$ and $\underline{H}_s = H_s + vH_1$ form a basis of $\mathcal{H}$ and satisfy $\underline{H}_1^{\blacklozenge} = \underline{H}_1$ and $\underline{H}_s^{\blacklozenge} = \underline{H}_s$. This basis is called the *Kazhdan–Lusztig* basis of $\mathcal{H}$ (see [70]).

Let $\tau : \mathcal{H} \to \mathbb{Z}[v, v^{-1}]$ be the $\mathbb{Z}[v, v^{-1}]$-linear map satisfying $\tau(H_1) = 1$ and $\tau(H_s) = 0$. The map τ is called the *trace map*. Consider the elements

$\underline{\hat{H}}_1 = H_1 - vH_s$ and $\underline{\hat{H}}_s = H_s$. These elements satisfy $\tau(\underline{H}_x\underline{\hat{H}}_y) = \delta_{x,y^{-1}}$ and form the *dual* Kazhdan–Lusztig basis (with respect to τ).

The map

$$\Psi : \mathrm{Gr}[\mathcal{D}^b(\text{D-gmod})] \to \mathcal{H},$$
$$[M(\mathsf{p})\langle i\rangle] \mapsto v^{-i}H_s,$$
$$[M(\mathsf{q})\langle i\rangle] \mapsto v^{-i}H_1,$$

uniquely extends to an isomorphism of abelian groups by linearity. We have:

$$\Psi([P(\mathsf{p})]) = \underline{H}_s, \quad \Psi([P(\mathsf{q})]) = \underline{H}_1, \quad \Psi([L(\mathsf{p})]) = \underline{\hat{H}}_s, \quad \Psi([L(\mathsf{q})]) = \underline{\hat{H}}_1.$$

All these equalities have categorical interpretations. The first two say that projective modules have filtrations whose quotients are isomorphic to (correspondingly shifted) Verma modules. The last two say that both simple modules have resolutions by Verma modules.

The exact functor ϑ_0^0 induces an endomorphism of $\mathrm{Gr}[\mathcal{D}^b(\text{D-gmod})]$. From the exactness of ϑ_0^0 and the second line of (5.22) it follows that

$$\Psi([\mathrm{ID}\,M]) = \Psi([M]) \cdot \underline{H}_1, \quad \Psi([\vartheta_0^0\,M]) = \Psi([M]) \cdot \underline{H}_s$$

for all $M \in$ D-gmod. Thus the action of graded projective functors on D-gmod gives us a categorification of the right regular $\mathcal{H}$-module $\mathcal{H}_{\mathcal{H}}$. Note that $\underline{H}_s^2 = v\underline{H}_s + v^{-1}\underline{H}_s$, which can be interpreted as the following isomorphism of functors:

$$\vartheta_0^0 \circ \vartheta_0^0 \cong \vartheta_0^0\langle 1\rangle \oplus \vartheta_0^0\langle -1\rangle.$$

The above correspondence between the Grothendieck group of the category $\mathcal{O}_0$ and the Hecke algebra of the Weyl group extends to the case of an arbitrary simple finite-dimensional complex Lie algebra (see [91]). In particular, multiplicities of simple modules in Verma modules can be computed using the coefficients of elements from the Kazhdan–Lusztig basis (this is the Kazhdan–Lusztig conjecture from [70], proved in [12, 26]).

5.10   Additional exercises

Exercise 5.104. Denote by $\mathcal{O}'$ the full subcategory of the category $\mathfrak{g}$-mod, which consists of all modules M, satisfying the conditions (I), (II) and the condition $\dim \mathbb{C}[f]v < \infty$ for all $v \in M$.

(a) Show that the categories $\mathcal{O}'$ and $\mathcal{O}$ are equivalent.

(b) Show that $\mathcal{O} \cap \mathcal{O}' = \mathcal{FD}$, the category of all finite-dimensional $\mathfrak{g}$-modules.
(c) Show that the duality $\circledcirc$ restricts to a contravariant equivalence from $\mathcal{O}$ to $\mathcal{O}'$ and vice versa.

Exercise 5.105 ([18]). Let $i \in \mathbb{N}_0$. Denote by I be the left ideal of $U(\mathfrak{g})$, generated by $h + i + 2$ and e^{i+2}, and set $M = U(\mathfrak{g})/I$.

(a) Show that $M \in \mathcal{O}$.
(b) Show that for any $N \in \mathcal{O}_i$ we have

$$\mathrm{Hom}_\mathfrak{g}(M, N) = N_{-i-2}.$$

(c) Show that $\mathfrak{p}_i(M)$ is projective in $\mathcal{O}$.

Exercise 5.106. Let $\xi \in \mathbb{C}/2\mathbb{Z}$. Show that every subcategory $\mathcal{O}^\xi = \mathcal{O} \cap \overline{\mathfrak{W}}^\xi$ contains a simple projective module.

Exercise 5.107. Show that $(c - 1)^2 M = 0$ for any $M \in \mathcal{O}_0$.

Exercise 5.108. Let $\hat{\mathcal{O}}_0 = \mathcal{O}_0 \cap \overline{\mathfrak{C}}^{2\mathbb{Z},1}$ (see Exercise 3.107).

(a) Show that $\hat{\mathcal{O}}_0$ is an abelian category with enough projective objects and two simple objects $L(-2)$ and $L(0)$.
(b) Show that the only indecomposable objects of $\hat{\mathcal{O}}_0$ are $L(-2)$, $L(0)$, $P(0)$ and $I(0)$.
(c) Show that the category $\hat{\mathcal{O}}_0$ is equivalent to the category of modules over the following $\mathbb{C}$-category:

$$\hat{\mathfrak{D}} : \qquad \mathtt{p} \underset{\mathtt{b}}{\overset{\mathtt{a}}{\rightleftharpoons}} \mathtt{q}, \qquad \mathtt{ab} = \mathtt{ba} = 0.$$

(d) Show that the category $\hat{\mathcal{O}}_0$ is not closed with respect to the action of ϑ^0_0.

Exercise 5.109 ([101]). Show that

$$\mathcal{F}(\Delta) = \{M \in \mathcal{O}_0 : \mathrm{Ext}^1_\mathcal{O}(M, \nabla(\lambda)) = 0, \lambda \in \mathtt{I}\},$$
$$\mathcal{F}(\Delta) = \{M \in \mathcal{O}_0 : \mathrm{Ext}^1_\mathcal{O}(M, T(\lambda)) = 0, \lambda \in \mathtt{I}\},$$
$$\mathcal{F}(\nabla) = \{M \in \mathcal{O}_0 : \mathrm{Ext}^1_\mathcal{O}(\Delta(\lambda), M) = 0, \lambda \in \mathtt{I}\},$$
$$\mathcal{F}(\nabla) = \{M \in \mathcal{O}_0 : \mathrm{Ext}^1_\mathcal{O}(T(\lambda), M) = 0, \lambda \in \mathtt{I}\}.$$

Exercise 5.110.

(a) Show that assigning a degree one and b degree two defines a positive grading on the algebra D.

(b) Show that, with respect to this grading, all standard graded lifts of standard modules have linear minimal projective resolution.

(c) Show that, with respect to this grading, not all standard graded lifts of costandard and simple modules have linear minimal projective resolution. In particular, deduce that the algebra D is not Koszul with respect to this grading.

Exercise 5.111. Let $M \in \mathcal{O}_0$. Show that there exists a complex of tilting modules from $\mathcal{O}_0$, whose only non-zero homology is concentrated in degree zero and is isomorphic to M.

Exercise 5.112.

(a) Define the category $\mathfrak{LT}$ of bounded linear complexes of graded tilting modules from $\mathcal{O}_0$ using standard lifts of indecomposable tilting modules.

(b) Show that $\mathfrak{LT}$ is an abelian category with enough projectives.

(c) Show that simple objects in $\mathfrak{LT}$ correspond to indecomposable tilting modules from $\mathcal{O}_0$.

(d) Show that (up to a shift of grading) for every indecomposable graded module $M \in$ D-gmod there exists an object in $\mathfrak{LT}$, whose only non-zero homology is concentrated in degree zero and is isomorphic to M.

(e) Show that $\mathfrak{LT}$ is equivalent to D-gmod, in particular, that $\mathfrak{LT}$ is a highest weight category.

(f) Show that tilting objects in $\mathfrak{LT}$ correspond to simple modules from $\mathcal{O}_0$.

(g) Show that standard and costandard objects in $\mathfrak{LT}$ correspond to standard and costandard modules from $\mathcal{O}_0$, respectively.

(h) Use the functor T to establish equivalence between $\mathfrak{LP}$ and $\mathfrak{LT}$.

(i) Use the functor T to establish equivalence between $\mathfrak{LT}$ and $\mathfrak{LJ}$.

Exercise 5.113. Determine the center of the algebra D and show that it is isomorphic to $\mathbb{C}[x]/(x^2)$.

Exercise 5.114. Recall that the *center* of a $\mathbb{C}$-category $\mathcal{C}$ is the endomorphism algebra of the identity functor on $\mathcal{C}$. Let $Z_{\mathcal{O}_0}$ denote the center of $\mathcal{O}_0$. Show that the evaluation at the indecomposable projective object $P(-2)$

defines an isomorphism

$$Z_{\mathcal{O}_0} \longrightarrow \mathrm{End}_{\mathcal{O}}(P(-2))$$

$$\psi \mapsto \psi_{P(-2)}$$

from $Z_{\mathcal{O}_0}$ to the algebra $\mathrm{End}_{\mathcal{O}}(P(-2))$. In particular, show that the algebra $Z_{\mathcal{O}_0}$ is isomorphic to the algebra $\mathbb{C}[x]/(x^2)$.

Exercise 5.115. Show that $\{[T(-2)], [T(0)]\}$ forms a basis of both $\mathrm{Gr}[\mathcal{O}_0]$ and $\mathrm{Gr}[\mathcal{D}^b(\mathcal{O}_0)]$.

Exercise 5.116. Consider the Yoneda extension algebra

$$A = \bigoplus_{i=0}^{\infty} \mathrm{Ext}_{\mathcal{O}}^i(M(-2) \oplus M(0), M(-2) \oplus M(0))$$

of standard modules in $\mathcal{O}_0$.

(a) Show that A is isomorphic to the path algebra of the following quiver:

$$p \underset{\longrightarrow}{\overset{\longrightarrow}{\rule{0pt}{0pt}}} q.$$

Show that A is positively graded in the natural way (by defining the degree of each arrow to be one).

(b) Show that the algebra A is Koszul and even Koszul self-dual.

(c) Show that the algebra A is isomorphic to the opposite algebra A^{op}.

(d) Show that, using $\circledast$, the algebra A^{op} can be naturally identified with the Yoneda extension algebra of costandard modules in $\mathcal{O}_0$.

Exercise 5.117. Let $\lambda, \mu \in \mathbb{C}$. Show that the $\mathfrak{g}$-module $M(\lambda) \otimes M(\mu)$ does not belong to $\mathcal{O}$, but can be written as an infinite direct sum of modules from $\mathcal{O}$, and thus that it has well-defined projections on blocks of $\mathcal{O}$.

Exercise 5.118. Show that for any projective module $P \in \mathcal{O}_0$ there exists an exact sequence

$$0 \to P \to X \to Y,$$

where $X, Y \in \mathrm{add}(P(-2))$.

Exercise 5.119. Show that for every $i, j \in \mathbb{N}$, $i \neq j$, we have

$$\underbrace{\mathcal{L}\mathrm{T} \circ \mathcal{L}\mathrm{T} \circ \cdots \circ \mathcal{L}\mathrm{T}}_{i \text{ times}} \not\cong \underbrace{\mathcal{L}\mathrm{T} \circ \mathcal{L}\mathrm{T} \circ \cdots \circ \mathcal{L}\mathrm{T}}_{j \text{ times}}$$

as functors on $\mathcal{D}^b(\mathcal{O}_0)$. Derive from this that twisting and completion functors define a faithful action of $\mathbb{Z}$ on $\mathcal{D}^b(\mathcal{O}_0)$.

Exercise 5.120. Show that for every $i, j \in \mathbb{N}$, $i \neq j$, we have

$$\underbrace{\mathcal{L}C \circ \mathcal{L}C \circ \cdots \circ \mathcal{L}C}_{i \text{ times}} \not\cong \underbrace{\mathcal{L}C \circ \mathcal{L}C \circ \cdots \circ \mathcal{L}C}_{j \text{ times}}$$

as functors on $\mathcal{D}^b(\mathcal{O}_0)$. Derive from this that shuffling and coshuffling functors define a faithful action of $\mathbb{Z}$ on $\mathcal{D}^b(\mathcal{O}_0)$.

Exercise 5.121.

(a) Show that for every finite-dimensional $\mathfrak{g}$-module V the endofunctors $(V \otimes (\_)^\circledast)^\circledast$ and $V \otimes \_$ of $\mathcal{O}$ are isomorphic.

(b) Use (a) to show that projective functors commute with the duality $\circledast$ in the sense that, for every projective functor θ, the functors $\theta \circ \circledast$ and $\circledast \circ \theta$ are isomorphic.

Exercise 5.122. Show that Z commutes with all projective functors.

Exercise 5.123. Show that $\hat{Z}$ commutes with all projective functors.

Exercise 5.124. Show that Z commutes with both C and K.

Exercise 5.125. Show that $\hat{Z}$ commutes with both C and K.

Exercise 5.126. Show that T commutes with C.

Exercise 5.127 ([72]). Show that Q commutes with all projective functors.

Exercise 5.128 ([92]).

(a) Show that the restriction of T^2 to the category of projective-injective modules in $\mathcal{O}_0$ is isomorphic to the identity functor.

(b) Show that T^2 defines an equivalence between the additive categories of projective and injective modules in $\mathcal{O}_0$.

Exercise 5.129 ([92]).

(a) Show that the restriction of C^2 to the category of projective-injective modules in $\mathcal{O}_0$ is isomorphic to the identity functor.

(b) Show that C^2 defines an equivalence between the additive categories of projective and injective modules in $\mathcal{O}_0$.

Exercise 5.130. Show that the endomorphism algebra of $P(-2)$ is isomorphic to the *coinvariant* algebra $\mathbb{C}[x, y]/(x + y, xy)$ of $\mathbf{S}_2$.

Exercise 5.131. For $\lambda \in \mathbb{C}$ and $n \in \mathbb{N}$ define the *thick* Verma module $\tilde{M}_n(\lambda)$ as follows:

$$\tilde{M}_n(\lambda) = U(\mathfrak{g})/(e, (h - \lambda)^n).$$

(a) Show that $\tilde{M}_n(\lambda)$ has a (unique) filtration of length n, whose subquotients are isomorphic to $M(\lambda)$.

(b) Determine all composition subquotients of $\tilde{M}_n(\lambda)$ and their multiplicities.

(c) Show that $\tilde{M}_n(\lambda)$ has a unique composition series.

Chapter 6

Description of all simple modules

6.1 Weight and nonweight modules

In this chapter we will describe all simple $\mathfrak{g}$-modules, although less explicitly than all simple weight modules as described in Section 3. The results of Section 3 are, of course, a part of this description, and our first aim is to determine some characteristic properties of other simple modules that we want to describe. Our description is obtained in terms of the algebra $U(\mathfrak{g})$, so from the very beginning we will work with this algebra.

Let M be a $U(\mathfrak{g})$-module. The module M is called $\mathbb{C}[h]$-*torsion*, provided that for any $v \in M$ there exists $g(h) \in \mathbb{C}[h]$ such that $g(h) \cdot v = 0$. The module M is called $\mathbb{C}[h]$-*torsion-free* or simply *torsion-free* provided that $M \neq 0$ and $g(h) \cdot v \neq 0$ for all non-zero $v \in M$ and $g(h) \in \mathbb{C}[h]$.

Example 6.1. The left regular module $_{U(\mathfrak{g})}U(\mathfrak{g})$ is torsion free by Corollary 2.26.

Example 6.2. Any weight module is $\mathbb{C}[h]$-torsion.

The importance of this notion is clarified by the following result:

Theorem 6.3 (Dichotomy for simple modules). *Let L be a simple $\mathfrak{g}$-module. Then the module L is either a weight module or a torsion-free module.*

Proof. Assume that L is not torsion-free. Then there exist $v \in L$, $v \neq 0$, and $g(h) \in \mathbb{C}[h]$, $g(h) \neq 0$, such that $g(h) \cdot v = 0$. As $\mathbb{C}$ is algebraically closed, we may write

$$g(h) = \alpha(h - \lambda_1)(h - \lambda_2) \cdots (h - \lambda_n)$$

for some $\alpha, \lambda_1, \ldots, \lambda_n \in \mathbb{C}$ such that $\alpha \neq 0$. Set $v_{n+1} = v$ and for $i = 1, \ldots, n$ put $v_i = (h - \lambda_i)(h - \lambda_{i+1}) \cdots (h - \lambda_n) \cdot v$. This gives us $v_1 = 0$ and $v_{n+1} \neq 0$. Hence there exists $i \in \{1, \ldots, n\}$ such that $v_i = 0$ and $v_{i+1} \neq 0$. In particular, $(h - \lambda_i)v_{i+1} = 0$, which means that v_{i+1} is a weight vector.

As L is simple, it is generated by every non-zero vector, including v_{i+1}. Since v_{i+1} is a weight vector, the module L is a weight module by Proposition 3.12. This completes the proof. $\qquad\square$

Some further properties of torsion-free modules are collected in the following statement:

Proposition 6.4.

(i) Any non-zero submodule of a torsion-free module is torsion-free.

(ii) Let $0 \to X \to Y \to Z \to 0$ be a short exact sequence of $\mathfrak{g}$-modules. If the modules X and Z are torsion-free, then the module Y is torsion-free as well. In particular, a direct sum of torsion-free modules is torsion-free.

(iii) Let M be a non-zero $\mathfrak{g}$-module. Then M is torsion-free if, and only if, $\mathrm{soc}(M)$ is torsion-free.

Proof. The claim (i) follows directly from the definitions. To prove the claim (ii) we assume that both X and Z are torsion-free and fix some non-zero $v \in Y$ and $g(h) \in \mathbb{C}[h]$. If $v \in X$, then $g(h) \cdot v \neq 0$ follows from the fact that X is torsion-free. If $v \notin X$, the image $v + X$ in $Z/X \cong Y$ is non-zero. As Y is torsion-free, we have $g(h) \cdot (v + X) \neq 0$ in Y and hence $g(h) \cdot v \notin X$. Therefore $g(h) \cdot v \neq 0$ and the claim (i) follows.

The necessity of the claim (iii) follows from the claim (i). To prove the sufficiency, assume that M is a $\mathfrak{g}$-module, which is not torsion-free. As with the proof of Theorem 6.3 we get that M contains a non-zero weight element, say v. The submodule N of M, generated by v, is a weight module by Proposition 3.12. This submodule must intersect with $\mathrm{soc}(M)$, which yields that $\mathrm{soc}(M)$ is not torsion-free. This completes the proof. $\qquad\square$

Exercise 6.5. Let M be a non-zero $U(\mathfrak{g})$-module. Show that M is torsion-free if, and only if, $\mathrm{Hom}_{U(\mathfrak{g})}(N, M) = 0$ for any weight $U(\mathfrak{g})$-module N.

Exercise 6.6. Let M be a non-zero $U(\mathfrak{g})$-module. Consider the vector space $M^{\square} = M$. For $u \in U(\mathfrak{g})$ and $v \in M$ set $u \cdot v = u^{\square}(v)$. Show that this defines on $M^{\square}$ the structure of a $U(\mathfrak{g})$-module and that the module M is torsion-free if, and only if, $M^{\square}$ is torsion-free.

Theorem 6.3 reduces the description of simple $U(\mathfrak{g})$-modules to that of simple weight modules (which was covered in Chapter 3) and of simple torsion-free modules. The latter is the main goal of this chapter. However, before we proceed, we can do another small simplification.

Let L be a simple $U(\mathfrak{g})$-module. By Theorem 4.7(ii) we have $\mathcal{I}_\lambda L = 0$ for some $\lambda \in \mathbb{C}$. In particular, the module L is a simple module over the algebra $U(\mathcal{I}_\lambda) = U(\mathfrak{g})/\mathcal{I}_\lambda$. Therefore, to describe all simple torsion-free $U(\mathfrak{g})$-modules, it is enough to describe all simple torsion-free $U(\mathcal{I}_\lambda)$-modules for all $\lambda \in \mathbb{C}$.

6.2 Embedding into a Euclidean algebra

Consider the field $\mathbb{F} = \mathbb{C}(h)$ of rational functions in h and the set

$$\mathbb{A} = \mathbb{F}[X, X^{-1}] = \left\{ \sum_{i \in \mathbb{Z}} a_i(h) X^i \; : \; a_i(h) \in \mathbb{F}, a_i(h) = 0 \text{ if } |i| \gg 0 \right\}.$$

Set

$$\left(\sum_{i \in \mathbb{Z}} a_i(h) X^i \right) + \left(\sum_{i \in \mathbb{Z}} b_i(h) X^i \right) = \sum_{i \in \mathbb{Z}} (a_i(h) + b_i(h)) X^i, \qquad (6.1)$$

$$\left(\sum_{i \in \mathbb{Z}} a_i(h) X^i \right) \cdot \left(\sum_{j \in \mathbb{Z}} b_j(h) X^j \right) = \sum_{i,j \in \mathbb{Z}} a_i(h) b_j(h - 2i) X^{i+j}, \qquad (6.2)$$

and for $\lambda \in \mathbb{C}$ set

$$\lambda \cdot \left(\sum_{i \in \mathbb{Z}} a_i(h) X^i \right) = \sum_{i \in \mathbb{Z}} \lambda a_i(h) X^i. \qquad (6.3)$$

Exercise 6.7. Check that the formulae (6.1), (6.2) and (6.3) define on the set $\mathbb{A}$ the structure of an associative and unital $\mathbb{C}$-algebra.

The elements of $\mathbb{A}$ are called *skew Laurent* polynomials over $\mathbb{F}$ and the algebra $\mathbb{A}$ is called the algebra of *skew Laurent* polynomials over $\mathbb{F}$.

Exercise 6.8. Check that the algebra $\mathbb{A}$ is a domain.

Define the *norm* function $\mathfrak{n} : \mathbb{A} \setminus \{0\} \to \mathbb{N}$ as follows:

$$\mathfrak{n}(a_m(h) X^m + a_{m+1}(h) X^{m+1} + \cdots + a_{n-1}(h) X^{n-1} + a_n(h) X^n) = n - m$$

provided that $m, n \in \mathbb{Z}$, $m \leq n$, and $a_i(h) \in \mathbb{F}$, $i = m, m+1, \ldots, n$, are such that $a_m(h), a_n(h) \neq 0$.

Proposition 6.9.

(i) *For any $\alpha, \beta \in \mathbb{A}$ such that $\beta \neq 0$, there exist $\gamma, \rho \in \mathbb{A}$ such that $\alpha = \gamma\beta + \rho$ and either $\rho = 0$ or $n(\rho) < n(\beta)$.*

(ii) *For any $\alpha, \beta \in \mathbb{A}$ such that $\beta \neq 0$, there exist elements $\gamma, \rho \in \mathbb{A}$ such that $\alpha = \beta\gamma + \rho$ and either $\rho = 0$ or $n(\rho) < n(\beta)$.*

Proof. We prove the claim (i) and the claim (ii) is proved similarly. We proceed by induction on $n(\alpha)$. If $n(\alpha) < n(\beta)$, then we can take $\gamma = 0$ and $\rho = \alpha$. Assume now that $n(\alpha) \geq n(\beta)$ and that

$$\alpha = \sum_{i=m}^{n} a_i(h)X^i, \quad \beta = \sum_{j=m'}^{n'} b_j(h)X^j$$

such that $a_m(h)$, $a_n(h)$, $b_{m'}(h)$, $b_{n'}(h)$ are non-zero. Then for the element

$$\gamma_1 = \frac{a_n(h)}{b_{n'}(h - 2(n - n'))}X^{n-n'}$$

we have $n(\alpha - \gamma_1\beta) < n(\alpha)$. By induction, we can write $\alpha - \gamma_1\beta = \gamma_2\beta + \rho$, where either $\rho = 0$ or $n(\rho) < n(\beta)$. This gives $\alpha = (\gamma_1 + \gamma_2)\beta + \rho$ and the claim of the proposition follows. $\square$

Exercise 6.10. Show that the elements γ and ρ in both Proposition 6.9(i) and Proposition 6.9(ii) are uniquely defined.

Proposition 6.9 says that the algebra $\mathbb{A}$ is both left and right *Euclidean* with respect to the norm n. A non-zero element α of $\mathbb{A}$ is called *irreducible* provided that α is not invertible, but for every factorization $\alpha = \beta\gamma$ we have that either β or γ is invertible in $\mathbb{A}$.

Corollary 6.11. *The algebra $\mathbb{A}$ is both left and right principal ideal domain.*

Proof. We prove the claim for left ideals. For right ideals the proof is similar. Let $I \subset \mathbb{A}$ be a non-zero left ideal. Let $\alpha \in I$ be a non-zero element such that $n(\alpha) \leq n(\beta)$ for any non-zero $\beta \in I$. We claim that $I = \mathbb{A}\alpha$. Indeed, if $\beta \in I$ is non-zero, by Proposition 6.9(i) we have $\beta = \gamma\alpha + \rho$, where either $\rho = 0$ or $n(\rho) < n(\alpha)$. On the other hand $\rho = \beta - \gamma\alpha \in I$ and hence $n(\rho) < n(\alpha)$ is not possible by our choice of α. This yields $\rho = 0$ and the claim follows. $\square$

Exercise 6.12. Show that the element $1 + X \in \mathbb{A}$ is not invertible.

Lemma 6.13. *Let $\alpha, \beta \in \mathbb{A}$ be non-zero. Then $\mathbb{A}\alpha = \mathbb{A}\beta$ if, and only if, there exists an invertible $\gamma \in \mathbb{A}$ such that $\alpha = \gamma\beta$.*

Proof. The sufficiency of the condition is obvious. To prove the necessity we assume that $\mathbb{A}\alpha = \mathbb{A}\beta$. Then $\alpha = \gamma\beta$ and $\beta = \gamma'\alpha$ for some elements $\gamma, \gamma' \in \mathbb{A}$. Hence $\alpha = \gamma\gamma'\alpha$ and $\beta = \gamma'\gamma\beta$, which can be written in the form $(1 - \gamma\gamma')\alpha = 0$, $(1 - \gamma'\gamma)\beta = 0$. As $\mathbb{A}$ is a domain (Exercise 6.8) and both α and β are non-zero, we get $1 = \gamma\gamma'$ and $1 = \gamma'\gamma$. This implies that γ is invertible and completes the proof. $\square$

Proposition 6.14. *Every simple $\mathbb{A}$-module has the form $\mathbb{A}/(\mathbb{A}\alpha)$ for some irreducible element $\alpha \in \mathbb{A}$.*

Proof. Let L be a simple $\mathbb{A}$-module and $v \in L$ be a non-zero element. The map $\varphi : \alpha \mapsto \alpha \cdot v$ is a homomorphism from the free $\mathbb{A}$-module $_{\mathbb{A}}\mathbb{A}$ to L. The map φ is surjective as L is simple (and hence generated by the element v). The kernel of φ is the left ideal

$$I = \{\alpha \in \mathbb{A} : \alpha \cdot v = 0\}$$

of A. Hence $L \cong \mathbb{A}/I$.

Note that the left regular module $\mathbb{A}$ is not simple as $\mathbb{A}$ contains non-invertible elements (Exercise 6.12) and any submodule of $\mathbb{A}$, generated by a non-invertible element, must be proper. Therefore I is non-zero and hence has the form $I = \mathbb{A}\alpha$ for some non-zero $\alpha \in \mathbb{A}$ by Corollary 6.11. Assume that α is not irreducible and let $\alpha = \beta\gamma$ be a factorization of α into a product of non-invertible elements. Then, by Lemma 6.13, we have $I \subsetneq \mathbb{A}\gamma \neq \mathbb{A}$ and hence $\mathbb{A}\gamma/I$ is a proper submodule of $\mathbb{A}/I \cong L$. This contradicts the assumption that the module L is simple. Thus α must be irreducible.

On the other hand, let $\alpha \in \mathbb{A}$ be irreducible. Set $I = \mathbb{A}\alpha$ and consider the $\mathbb{A}$-module $M = \mathbb{A}/I$. Let $\beta \in \mathbb{A} \setminus I$ and J denote the left ideal of $\mathbb{A}$, generated by α and β. Then $J = \mathbb{A}\gamma$ for some $\gamma \in \mathbb{A}$ and thus we have $\alpha = \delta\gamma$ for some $\delta \in \mathbb{A}$. As α is irreducible, either δ or γ must be invertible. However, δ invertible would imply $J = I$ by Lemma 6.13, which is not possible as J contains the element $\beta \notin I$. Therefore γ is invertible, which yields $J = \mathbb{A}$. This means that M is generated by $\beta + I$ and, since $\beta + I$ was an arbitrary element of M, it follows that M is generated by any non-zero element, hence simple. This completes the proof. $\square$

For an irreducible element $\alpha \in \mathbb{A}$ we denote by L_α the simple $\mathbb{A}$-module $\mathbb{A}/(\mathbb{A}\alpha)$.

Proposition 6.15. *Let α and β be two irreducible elements of $\mathbb{A}$. Then $L_\alpha \cong L_\beta$ if, and only if, there exist $\gamma \in \mathbb{A} \setminus \mathbb{A}\beta$ and $\delta \in \mathbb{A}$ such that $\alpha\gamma = \delta\beta$.*

Proof. Let $\varphi : L_\alpha \to L_\beta$ be an isomorphism. Then $\varphi(1 + \mathbb{A}\alpha) = \gamma + \mathbb{A}\beta$ for some $\gamma \in \mathbb{A} \setminus \mathbb{A}\beta$. We have

$$
\begin{aligned}
0 &= \varphi(0) \\
&= \varphi(\alpha \cdot (1 + \mathbb{A}\alpha)) \\
&= \alpha \cdot \varphi(1 + \mathbb{A}\alpha) \\
&= \alpha \cdot (\gamma + \mathbb{A}\beta) \\
&= \alpha\gamma + \mathbb{A}\beta,
\end{aligned}
$$

which yields $\alpha\gamma \in \mathbb{A}\beta$, that is $\alpha\gamma = \delta\beta$ for some $\delta \in \mathbb{A}$.

On the other hand, assume that there exist $\gamma \in \mathbb{A} \setminus \mathbb{A}\beta$ and $\delta \in \mathbb{A}$ such that $\alpha\gamma = \delta\beta$. Consider the non-zero homomorphism $\psi : \mathbb{A} \to L_\beta$, given by $\psi(x) = x \cdot (\gamma + \mathbb{A}\beta)$. Then $\psi(\alpha) = \alpha\gamma + \mathbb{A}\beta = 0$ and hence ψ factors through $\mathbb{A}/(\mathbb{A}\alpha) = L_\alpha$. As L_α is simple, the induced homomorphism from L_α to L_β is an isomorphism by Schur's lemma. This completes the proof of the proposition. $\qquad\square$

Two irreducible elements α and β of $\mathbb{A}$ are called *similar* provided that there exist $\gamma \in \mathbb{A} \setminus \mathbb{A}\beta$ and $\delta \in \mathbb{A}$ such that $\alpha\gamma = \delta\beta$. Thus Proposition 6.15 can be reformulated as follows: Two irreducible elements α and β from $\mathbb{A}$ the modules L_α and L_β are isomorphic if and only if α and β are similar.

Exercise 6.16. Show that two irreducible elements α and β of $\mathbb{A}$ are similar if, and only if, there exist $\gamma \in \mathbb{A} \setminus \mathbb{A}\alpha$ and $\delta \in \mathbb{A}$ such that $\beta\gamma = \delta\alpha$.

Exercise 6.17. Let $\alpha, \beta \in \mathbb{A}$ be two similar irreducible elements. Let further $\gamma \in \mathbb{A} \setminus \mathbb{A}\beta$ and $\delta \in \mathbb{A}$ be such that $\alpha\gamma = \delta\beta$. Show that $\mathbb{A}\beta \cap \mathbb{A}\gamma = \mathbb{A}\delta\beta = \mathbb{A}\alpha\gamma$. Show also that there exist $x, y \in \mathbb{A}$ such that $x\gamma + y\beta = 1$.

Theorem 6.18.

(i) *Let $\lambda \in \mathbb{C}$. There then exists a unique homomorphism $\Phi_\lambda : U(\mathfrak{g}) \to \mathbb{A}$ of associative algebras such that*

$$
\Phi_\lambda(h) = h, \quad \Phi_\lambda(e) = X, \quad \Phi_\lambda(f) = \frac{(\lambda+1)^2 - (h+1)^2}{4} X^{-1}. \quad (6.4)
$$

(ii) *The kernel of Φ_λ coincides with $\mathcal{I}_\lambda$.*

(iii) *The homomorphism Φ_λ induces a monomorphism $\overline{\Phi}_\lambda : U(\mathcal{I}_\lambda) \to \mathbb{A}$.*

Proof. Due to the definition of $U(\mathfrak{g})$, we have to check the relations (2.1) for the elements h, X and $Y = \frac{(\lambda+1)^2-(h+1)^2}{4}X^{-1}$. By (6.2) we have the relation $Xh = (h-2)X$, which implies the relation $hX - Xh = 2X$. Multiplying $Xh = (h-2)X$ with X^{-1} from both sides, we get $hX^{-1} = X^{-1}(h-2)$. This implies the relation $hY - Yh = -2Y$. To check the relation $XY - YX = h$ we compute:

$$
\begin{aligned}
XY - YX &= X\frac{(\lambda+1)^2-(h+1)^2}{4}X^{-1} - \frac{(\lambda+1)^2-(h+1)^2}{4}X^{-1}X \\
\text{(by (6.2))} &= \frac{(\lambda+1)^2-(h-1)^2}{4}XX^{-1} - \frac{(\lambda+1)^2-(h+1)^2}{4} \\
&= \frac{2h}{4} + \frac{2h}{4} \\
&= h.
\end{aligned}
$$

The claim (i) follows.

To prove the claim (ii) we compute:

$$
\begin{aligned}
\Phi_\lambda(c) &= \Phi_\lambda((h+1)^2 + 4fe) \\
&= (h+1)^2 + 4\frac{(\lambda+1)^2-(h+1)^2}{4}X^{-1}X \\
&= (\lambda+1)^2.
\end{aligned}
$$

Hence $\Phi_\lambda(c-(\lambda+1)^2) = 0$ and $\mathcal{I}_\lambda$ belongs to the kernel of Φ_λ. On the other hand, for $\lambda \notin \mathbb{Z}$, the algebra $U(\mathcal{I}_\lambda)$ is simple by Theorem 4.15(iv). Hence for such λ the kernel of the non-zero homomorphism Φ_λ must coincide with $U(\mathcal{I}_\lambda)$.

Finally, assume that $\lambda \in \mathbb{Z}$. By Theorem 4.15(v), the algebra $U(\mathcal{I}_\lambda)$ has a unique two-sided ideal which, moreover, has finite codimension. However, the image of Φ_λ contains the infinite-dimensional space $\mathbb{C}[h]$. This yields that the kernel of Φ_λ coincides with $\mathcal{I}_\lambda$, also for $\lambda \in \mathbb{Z}$, and completes the proof of the claim (ii).

The claim (iii) follows from the claims (i) and (ii). □

We will identify the algebra $U(\mathcal{I}_\lambda)$ with its image in $\mathbb{A}$ under the homomorphism $\overline{\Phi}_\lambda$.

6.3 Description of simple nonweight modules

For this section, we fix $\lambda \in \mathbb{C}$. The monomorphism $\overline{\Phi}_\lambda : U(\mathcal{I}_\lambda) \to \mathbb{A}$, constructed in Theorem 6.18(iii), defines on the algebra $\mathbb{A}$ the structure of an A–$U(\mathcal{I}_\lambda)$-bimodule ${}_A A_{U(\mathcal{I}_\lambda)}$ via

$$a \cdot x \cdot b = a x \overline{\Phi}_\lambda(b)$$

for all $a, x \in A$ and $b \in U(\mathcal{I}_\lambda)$. The bimodule ${}_A A_{U(\mathcal{I}_\lambda)}$ allows us to consider the functor

$$F = {}_A A_{U(\mathcal{I}_\lambda)} \bigotimes_{U(\mathcal{I}_\lambda)} - : \quad U(\mathcal{I}_\lambda)\text{-Mod} \to A\text{-Mod}.$$

The right adjoint functor

$$G = \operatorname{Hom}_A({}_A A_{U(\mathcal{I}_\lambda)}, -) : \quad A\text{-Mod} \to U(\mathcal{I}_\lambda)\text{-Mod}$$

is just the functor, which restricts the action from A to $U(\mathcal{I}_\lambda)$.

Exercise 6.19. Show that for any $\alpha \in A$ there exists $g(h) \in \mathbb{C}[h]$ such that $g(h) \neq 0$ and $g(h)\alpha \in U(\mathcal{I}_\lambda)$.

Exercise 6.20. Consider A as a $\mathbb{C}(h)$–$U(\mathcal{I}_\lambda)$-bimodule by restriction. Show that the multiplication map mult defines an isomorphism of the following $\mathbb{C}(h)$–$U(\mathcal{I}_\lambda)$-bimodules:

$$\operatorname{mult} : \mathbb{C}(h) \bigotimes_{\mathbb{C}[h]} U(\mathcal{I}_\lambda) \to A.$$

Exercise 6.21. Consider the endofunctor T of $\mathbb{C}[h]$-Mod given by tensoring with the $\mathbb{C}[h]$–$\mathbb{C}[h]$-bimodule $\mathbb{C}(h)$. Show that the embedding $\mathbb{C}[h] \hookrightarrow \mathbb{C}(h)$ defines a natural transformation from the identity functor to T, whose kernel coincides with the functor of taking the maximal $\mathbb{C}[h]$-torsion submodule.

Consider the endofunctor $\overline{F}$ of $U(\mathcal{I}_\lambda)$-Mod given by the tensoring with the $U(\mathcal{I}_\lambda)$–$U(\mathcal{I}_\lambda)$-bimodule A.

Exercise 6.22. Show that $\overline{F} = G \circ F$.

Lemma 6.23. *The embedding $U(\mathcal{I}_\lambda) \hookrightarrow A$ defines a natural transformation from the identity functor to $\overline{F}$, whose kernel coincides with the functor of taking the maximal generalized weight submodule.*

Proof. First we remark that $\mathbb{C}[h]$-torsion $U(\mathcal{I}_\lambda)$-modules are exactly generalized weight modules. Let $M \in U(\mathcal{I}_\lambda)$-Mod. If we consider FM as a $\mathbb{C}(h)$-module, then, applying Exercise 6.20, we obtain that

$$FL \cong \mathbb{C}(h) \bigotimes_{\mathbb{C}[h]} L.$$

The claim follows from Exercise 6.21. $\square$

Now we are ready to describe simple nonweight $U(\mathcal{I}_\lambda)$-modules.

Theorem 6.24.

(i) *The functor* F *induces a bijection,* $\hat{\text{F}}$, *from the set of isomorphism classes of simple nonweight* $U(\mathcal{I}_\lambda)$-modules *to the set of isomorphism classes of simple* A-modules.

(ii) *The inverse of the bijection from* (i) *is the map, which sends a simple* A-module N *to its* $U(\mathcal{I}_\lambda)$-socle $\text{soc}_{U(\mathcal{I}_\lambda)}(N)$.

Proof. Let L be a simple torsion-free $U(\mathcal{I}_\lambda)$-module. From Lemma 6.23 we obtain that $L \subset \text{F} L$ as a $U(\mathcal{I}_\lambda)$-module. In particular, $\text{F} L \neq 0$.

Now we claim that the A-module $\text{F} L$ is simple. Since $L = U(\mathcal{I}_\lambda) \cdot L$, by Exercise 6.20 we have $\text{F} L = \mathbb{C}(h)L$. If $w \in \mathbb{C}(h)L$ is non-zero, then, by Exercise 6.19, there exists $g(h) \in \mathbb{C}[h]$ such that $g(h) \cdot w \in L$. The element $g(h) \cdot w$ is non-zero as the action of $g(h)$ on $\text{F} L$ is invertible. This means that $U(\mathcal{I}_\lambda) \cdot (g(h) \cdot w) = L$ and hence $\mathbb{C}(h)U(\mathcal{I}_\lambda) \cdot (g(h) \cdot w) = \text{F} L$. This shows that the A-module $\text{F} L$ is generated by any non-zero element and is, therefore, simple.

It follows that the functor F induces a well-defined map $\hat{\text{F}}$ from the set of isomorphism classes of simple nonweight $U(\mathcal{I}_\lambda)$-modules to the set of isomorphism classes of simple A-modules.

Lemma 6.25. *The module* $\text{F} L$ *has a simple* $U(\mathcal{I}_\lambda)$-socle, *isomorphic to* L.

Proof. The module $\text{F} L$ is non-zero by the previous paragraph. Let $w \in \text{F} L$, $w \neq 0$. Then w is a finite linear combination of the elements of the form $u \otimes x$, where $u \in \text{A}$ and $x \in L$. For every non-zero $g(h) \in \mathbb{C}[h]$ the action of $g(h)$ on $\text{F} L$ is bijective (as $g(h)$ is invertible in A). By Exercise 6.19, there exists $g(h) \in \mathbb{C}[h]$ such that $g(h) \neq 0$ and $g(h)u \in U(\mathcal{I}_\lambda)$ for all u which occur in the decomposition of w. This means that the element $g(h) \cdot w$ is non-zero (as the action of $g(h)$ is bijective) and belongs to L (as all $g(h)u \in U(\mathcal{I}_\lambda)$). Therefore any $U(\mathcal{I}_\lambda)$-submodule of $\text{F} L$ intersects L and hence L is a simple $U(\mathcal{I}_\lambda)$-socle of $\text{F} L$. □

Lemma 6.26. *Any simple* A-module *has a simple torsion-free* $U(\mathcal{I}_\lambda)$-submodule.

Proof. Let $\alpha \in \text{A}$ be an irreducible element. By Exercise 6.19, there exists $g(h) \in \mathbb{C}[h]$ such that $g(h) \neq 0$ and $g(h)\alpha \in U(\mathcal{I}_\lambda)$. This means that the element $1 + \mathbb{A}\alpha$ of the simple $\mathbb{A}$-module $\mathbb{A}/(\mathbb{A}\alpha)$ is annihilated by the element $g(h)\alpha \in U(\mathcal{I}_\lambda)$. In particular, the $U(\mathcal{I}_\lambda)$-submodule of $\mathbb{A}/(\mathbb{A}\alpha)$, generated by $1 + \mathbb{A}\alpha$, is a quotient of the $U(\mathcal{I}_\lambda)$-module $U(\mathcal{I}_\lambda)/(U(\mathcal{I}_\lambda)g(h)\alpha)$. By Theorem 4.26, the module $U(\mathcal{I}_\lambda)/(U(\mathcal{I}_\lambda)g(h)\alpha)$ has finite length and hence a simple $U(\mathcal{I}_\lambda)$-submodule, say N. As the action of every $p(h) \in \mathbb{C}[h]$, $p(h) \neq 0$, is invertible on $\mathbb{A}/(\mathbb{A}\alpha)$, the kernel of $p(h)$ on N is zero and thus N is torsion-free. This completes the proof. $\square$

Now let N be a simple $\mathbb{A}$-module and L be a simple $U(\mathcal{I}_\lambda)$-submodule of N, which exists by Lemma 6.26. Using adjunction, we have

$$0 \neq \operatorname{Hom}_{U(\mathcal{I}_\lambda)}(L, \mathrm{G}\, N) = \operatorname{Hom}_{\mathbb{A}}(\mathrm{F}\, L, N).$$

The module $\mathrm{F}\, L$ is simple by the above and the module N is simple by our assumption. Hence $\mathrm{F}\, L \cong N$. It follows that the map $\hat{\mathrm{F}}$ is surjective.

Assume $\mathrm{F}\, L' \cong N$ for some simple torsion-free $U(\mathcal{I}_\lambda)$-module L'. Then, by Lemma 6.25, both L and L' are isomorphic to the simple $U(\mathcal{I}_\lambda)$-socle of N and hence $L \cong L'$. Therefore the map $\hat{\mathrm{F}}$ is injective, which completes the proof. $\square$

Corollary 6.27. *For every* $\lambda \in \mathbb{C}$ *there is a natural bijection between the isomorphism classes of simple* $U(\mathcal{I}_\lambda)$*-modules and the similarity classes of irreducible elements in* $\mathbb{A}$.

Proof.   This statement follows from Theorem 6.24 and Propositions 6.14 and 6.15. $\square$

Exercise 6.28. Show that for any simple $U(\mathcal{I}_\lambda)$-module L there exists an irreducible (as element of $\mathbb{A}$) element $\alpha \in U(\mathcal{I}_\lambda) \subset \mathbb{A}$ such that $L \cong U(\mathcal{I}_\lambda)/(U(\mathcal{I}_\lambda) \cap \mathbb{A}\alpha)$.

6.4   Finite-dimensionality of kernels and cokernels

Throughout this section we fix $\lambda \in \mathbb{C}$. Let $\mathfrak{FL}_\lambda$ denote the full subcategory of the category of all $U(\mathcal{I}_\lambda)$-modules, which consists of all finite length modules. For a $U(\mathcal{I}_\lambda)$-module M and $u \in U(\mathcal{I}_\lambda)$ we denote by u_M the linear operator, representing the action of u on M.

Exercise 6.29. Show that $\mathfrak{FL}_\lambda$ is an abelian Krull–Schmidt category in which simple objects are simple $U(\mathcal{I}_\lambda)$-modules and every object has finite length.

Our goal for the present section is to prove the following result:

Theorem 6.30. *Let $M \in \mathfrak{FL}_\lambda$ and $u \in U(\mathcal{I}_\lambda)$, $u \neq 0$. Then both the kernel and the cokernel of the linear operator u_M are finite-dimensional.*

First we prove the claim of Theorem 6.30 for kernels.

Proof. We will need the following lemmas:

Lemma 6.31. *For any $M, N \in \mathfrak{FL}_\lambda$ the vector space $\mathrm{Hom}_{\mathfrak{FL}_\lambda}(M, N)$ is finite-dimensional.*

Proof. Let $0 \to V \to Y \to Z \to 0$ be a short exact sequence in $\mathfrak{FL}_\lambda$. Then the following two sequences are exact:

$$\mathrm{Hom}_{\mathfrak{FL}_\lambda}(Z, N) \to \mathrm{Hom}_{\mathfrak{FL}_\lambda}(Y, N) \to \mathrm{Hom}_{\mathfrak{FL}_\lambda}(V, N),$$
$$\mathrm{Hom}_{\mathfrak{FL}_\lambda}(M, V) \to \mathrm{Hom}_{\mathfrak{FL}_\lambda}(M, Y) \to \mathrm{Hom}_{\mathfrak{FL}_\lambda}(M, Z).$$

In particular, the middle term in both sequences is finite-dimensional provided that both the left and the rights terms are. Since every module in $\mathfrak{FL}_\lambda$ has finite length, it is enough to prove the assertion for simple $U(\mathcal{I}_\lambda)$-modules. For such modules the assertion follows from Schur's lemma and Exercise 4.10. $\square$

Denote by M_u the left $U(\mathcal{I}_\lambda)$-module $U(\mathcal{I}_\lambda)/(U(\mathcal{I}_\lambda)u)$.

Lemma 6.32. *Let M be a $U(\mathcal{I}_\lambda)$-module. Then there is an isomorphism*

$$\mathrm{Ker}(u_M) \cong \mathrm{Hom}_{U(\mathcal{I}_\lambda)}(M_u, M).$$

Proof. Let $\varphi : M_u \to M$ be a homomorphism. Then

$$u_M \varphi(1 + U(\mathcal{I}_\lambda)u) = \varphi(u + U(\mathcal{I}_\lambda)u) = 0$$

and hence $\varphi(1 + U(\mathcal{I}_\lambda)u) \in \mathrm{Ker}(u_M)$. This gives us a map

$$g : \mathrm{Hom}_{U(\mathcal{I}_\lambda)}(M_u, N) \to \mathrm{Ker}(u_M).$$

On the other hand, if $v \in \mathrm{Ker}(u_M)$, we have a unique homomorphism ψ from the free $U(\mathcal{I}_\lambda)$-module $U(\mathcal{I}_\lambda)$ to M, given by $\psi(1) = v$. As $u(v) = 0$, the homomorphism ψ factors through M_u. This gives us a map

$$g' : \mathrm{Ker}(u_M) \to \mathrm{Hom}_{U(\mathcal{I}_\lambda)}(M_u, N).$$

From the definitions it follows that g and g' are mutually inverse linear maps. This completes the proof. $\square$

Let now $M \in \mathfrak{FL}_\lambda$. By Lemma 6.32 we have

$$\mathrm{Ker}(u_M) \cong \mathrm{Hom}_{U(\mathcal{I}_\lambda)}(M_u, M).$$

Since $u \neq 0$, by Theorem 4.26 the module M_u has finite length. This means that

$$\mathrm{Hom}_{U(\mathcal{I}_\lambda)}(M_u, M) = \mathrm{Hom}_{\mathfrak{FL}_\lambda}(M_u, M),$$

which is finite-dimensional by Lemma 6.31. Hence $\mathrm{Ker}(u_M)$ is finite-dimensional. $\qquad\square$

To prove the finite-dimensionality of the cokernels we will have to work much harder. Our first reduction is the following:

Exercise 6.33. Let $u \in U(\mathcal{I}_\lambda)$, $u \neq 0$. Then the cokernel of u_M is finite-dimensional for any $M \in \mathfrak{FL}_\lambda$ if and only if the cokernel of u_L is finite-dimensional for any simple $U(\mathcal{I}_\lambda)$-module L.

Our next reduction is given by the following observation:

Lemma 6.34. *Assume that for any* $\alpha, \beta \in U(\mathcal{I}_\lambda)$, $\alpha, \beta \neq 0$, *the vector space*

$$U(\mathcal{I}_\lambda)/(U(\mathcal{I}_\lambda)\alpha + \beta U(\mathcal{I}_\lambda))$$

is finite-dimensional. Then for any $u \in U(\mathcal{I}_\lambda)$, $u \neq 0$, *the cokernel of* u_L *is finite-dimensional for any simple module* L.

Proof. Let L be a simple $U(\mathcal{I}_\lambda)$-module and $v \in L$ be non-zero. Then the map $1 \mapsto v$ extends uniquely to an epimorphism φ from the free module $U(\mathcal{I}_\lambda)$ to L. As the free module $U(\mathcal{I}_\lambda)$ is not simple (the algebra $U(\mathcal{I}_\lambda)$ contains non-invertible elements, for instance h by Theorem 4.15(ii)), the morphism φ is not injective. Let α be any element in the kernel. The map φ factors through the module $U(\mathcal{I}_\lambda)/(U(\mathcal{I}_\lambda)\alpha)$. Let $\overline{\varphi}$ be the induced epimorphism. This gives us the following short exact sequence:

$$0 \to \mathrm{Ker}(\overline{\varphi}) \to U(\mathcal{I}_\lambda)/(U(\mathcal{I}_\lambda)\alpha) \to L \to 0. \qquad (6.5)$$

Consider now the module $N = U(\mathcal{I}_\lambda)/(U(\mathcal{I}_\lambda)\alpha)$. The cokernel of u_N is isomorphic to the vector space $U(\mathcal{I}_\lambda)/(U(\mathcal{I}_\lambda)\alpha + uU(\mathcal{I}_\lambda))$, which is finite-dimensional by assumption. Factoring out the subspace $\mathrm{Ker}(\overline{\varphi})$ from (6.5), we obtain that the cokernel of u_L is finite-dimensional as well. This completes the proof. $\qquad\square$

Lemma 6.34 reduces our problem to the study of $U(\mathcal{I}_\lambda)$-modules of the form $U(\mathcal{I}_\lambda)/(U(\mathcal{I}_\lambda)\alpha)$, where $\alpha \in U(\mathcal{I}_\lambda)$, $\alpha \neq 0$.

Lemma 6.35. *Let* $\alpha = a_0(h) + a_1(h)e + \cdots + a_k(h)e^k$, *where* $k \geq 0$, $a_i(h) \in \mathbb{C}[h]$ *and* $a_0(h), a_k(h) \neq 0$. *Let* V_α *denote the subspace of* $U(\mathcal{I}_\lambda)$, *spanned by the following monomials:*

$$h^i f^j, \ i \in \mathbb{N}_0, \ j \in \mathbb{N}, \ i < \deg(a_0(h)); \tag{6.6}$$

$$h^i e^j, \ i \in \mathbb{N}_0, \ j \in \mathbb{N}_0, \ i < \deg(a_k(h)), \ j \geq k; \tag{6.7}$$

$$h^i e^j, \ i \in \mathbb{N}_0, \ j \in \mathbb{N}_0, \ j < k. \tag{6.8}$$

Then $U(\mathcal{I}_\lambda) = V_\alpha \oplus U(\mathcal{I}_\lambda)\alpha$.

Proof. Let us show that, modulo $U(\mathcal{I}_\lambda)\alpha$, every element from $U(\mathcal{I}_\lambda)$ can be written as a linear combination of the elements from the formulation of the lemma. By Theorem 4.15(ii), it is enough to prove the claim for the monomials of the form $h^i f^j$ and $h^i e^j$, where $i, j \in \mathbb{N}_0$. We will prove the claim for the monomials of the form $h^i e^j$. For the monomials of the form $h^i f^j$ the arguments are similar.

We proceed by double induction on i and j. If we have $j < k$ or $i < \deg(a_k(h))$, the monomial $h^i e^j$ appears in the above list and we have nothing to prove. Assume that $j \geq k$ and $i \geq \deg(a_k(h))$. If we have $a_k(h) = c_0 h^{\deg(a_k(h))} + c_1 h^{\deg(a_k(h))-1} + \ldots$ (here $c_i \in \mathbb{C}$ for all i and $c_0 \neq 0$), then the element

$$h^i e^j - \frac{1}{c_0} h^{i-\deg(a_k(h))} e^{j-k}\alpha$$

is equal to $h^i e^j$ modulo $U(\mathcal{I}_\lambda)\alpha$ and is a linear combination of monomials of the form $h^{i'} e^{j'}$ such that either $i' < i$ or $j' < j$. This completes our induction.

On the other hand, for any $\beta \in U(\mathcal{I}_\lambda)$, $\beta \neq 0$, we either have the equality $\beta = g_s(h)f^s + g_{s-1}(h)f^{s-1} + \ldots$ for some $s \in \mathbb{N}$ and $g_s(h) \neq 0$, or the equality $\beta = g_t(h)e^t + g_{t-1}(h)e^{t-1} + \ldots$ for some $t \in \mathbb{N}_0$ and some $g_t(h) \neq 0$. In the first case, the product $\beta\alpha$ contains in its decomposition the monomial $h^{\deg(g_s)+\deg(a_0)} f^s$. In the second case, the product $\beta\alpha$ contains in its decomposition the monomial $h^{\deg(g_t)+\deg(a_k)} e^{k+t}$. These monomials are not listed in the formulation. This yields $V_\alpha \cap U(\mathcal{I}_\lambda)\alpha = 0$ and the claim follows. $\square$

Lemma 6.36. *Let*

$$\alpha = a_0(h) + a_1(h)e + \cdots + a_k(h)e^k, \beta = b_0(h) + b_1(h)e + \cdots + b_m(h)e^m, \tag{6.9}$$

where $k, m \geq 0$, $a_i(h), b_i(h) \in \mathbb{C}[h]$ and $a_0(h), a_k(h), b_0(h), b_m(h) \neq 0$. Set $W = U(\mathcal{I}_\lambda)\alpha + \beta U(\mathcal{I}_\lambda)$. The space $U(\mathcal{I}_\lambda)/W$ is finite-dimensional.

Proof.   As $U(\mathcal{I}_\lambda)\alpha \subset W$, by Lemma 6.35 the space $U(\mathcal{I}_\lambda)/W$ is spanned by the images of the monomials from the lists (6.6), (6.7) and (6.8).

For all $j \in \mathbb{N}$ that is sufficiently large, the polynomials $a_k(h - 2(j - k))$ and $b_m(h)$ are coprime and hence there exist $x_j(h)$ and $y_j(h)$ such that

$$x_j(h)a_k(h - 2(j - k)) + b_m(h)y_j(h - 2m) = 1$$

Using $g(h)e = eg(h+2)$ for all $g(h) \in \mathbb{C}[h]$ (see (2.1)), by a direct calculation we obtain for all $i \in \mathbb{N}_0$ the following:

$$h^i x_j(h)e^{j-k}\alpha + \beta(h + 2m)^i y_j(h)e^{j-m} = h^i e^j + \text{ terms of smaller degree.}$$

By induction we get that there exists $n \in \mathbb{N}$ such that for all $j > n$, all monomials from the list (6.7) can be reduced modulo W to monomials with smaller j. Similarly treating the list (6.6), we can assume that n is a common bound for both lists and that $n > k + m$. Let Y denote the finite-dimensional subspace of $U(\mathcal{I}_\lambda)$, spanned by the monomials from the lists (6.6) and (6.7) such that $j \leq n$.

For $s \in \mathbb{N}_0$, denote by V_s the linear span of $h^i e^j$, $0 \leq i \leq s$, $0 \leq j < k$. Then V_s is a subspace in $U(\mathcal{I}_\lambda)$ of dimension $k(s + 1)$. Furthermore, we have the flag $V_0 \subset V_1 \subset V_2 \subset \ldots$ and can thus define $V = \bigcup_{i \in \mathbb{N}_0} V_i$. From Lemma 6.35 and the previous paragraph we obtain

$$U(\mathcal{I}_\lambda) = W + Y + V. \tag{6.10}$$

Let a be the maximal degree among the degrees of the polynomials $a_i(h)$ and b be the maximal degree among the degrees of the polynomials $b_i(h)$.

Lemma 6.37. *For every $l \geq 0$ we have*

$$\beta V_l \subset V_{l+ma+b} + Y + U(\mathcal{I}_\lambda)\alpha. \tag{6.11}$$

Proof.   Let $h^i e^j$ be some monomial from V_l. Then $i \leq l$ and $j \leq k$ and we have

$$\beta h^i e^j = b_0(h)h^i e^j + b_1(h)(h - 2)^i e^{j+1} + \cdots + b_m(h)(h - 2m)^i e^{m+j}.$$

All coefficients from $\mathbb{C}[h]$ in the above element have degrees at most $b + l$. In particular, $\beta h^i e^j \in V_{b+l}$ if $m + j < k$.

Assume now that $m + j \geq k$. As $m + j < m + k$, the space Y contains, by construction, all monomials $h^s e^{m+j}$ where $s < \deg(a_k(h))$. Hence, there exists some $y \in Y$ and $g(h) \in \mathbb{C}[h]$ of degree at most $b + l$, such that

$$b_m(h)(h - 2m)^i e^{m+j} = y + g(h)e^{m+j-k}a_k(h)e^k$$

(as writing y and $g(h)$ with unknown coefficients reduces the above equality to a triangular system of linear equations with non-zero elements on the diagonal). Therefore, modulo Y and $U(\mathcal{I}_\lambda)\alpha$, the element $\beta h^i e^j$ is equal to the element

$$\beta_1 = \beta h^i e^j - y - g(h)e^{m+j-k}\alpha.$$

The element β_1 has the form $q(h) + r(h)e + \cdots + p(h)e^{m+j-1}$ where all coefficients from $\mathbb{C}[h]$ have degrees, at most, $b + l + a$. Now we can proceed inductively at most m times. We end up with an element from V, in which all coefficients from $\mathbb{C}[h]$ have degrees at most $b + l + ma$. The claim of Lemma 6.37 follows. $\qquad\square$

Now let d denote the dimension of the kernel of the linear operator β on the module M_α. We have $d < \infty$ by the first part of Theorem 6.30, proved above. By (6.11), the rank of $\beta(V_l + U(\mathcal{I}_\lambda)\alpha)$ in $V_{l+ma+b} + Y + U(\mathcal{I}_\lambda)\alpha$ is at least $\dim(V_l) - d$. Hence, using (6.10) and (6.11), we get

$$\dim(V_{l+ma+b} + Y/((V_{l+ma+b} + Y) \cap W))$$
$$\leq (l + ma + b + 1)k + \dim(Y) - (l + 1)k + d$$
$$= (ma + b)k + \dim(Y) + d.$$

Observe that the bound we get does not depend on l. As $U(\mathcal{I}_\lambda)\alpha \subset W$, taking the limit we thus obtain

$$\dim(V + Y/((V + Y) \cap W)) \leq (ma + b)k + \dim(X) + d < \infty.$$

From (6.10) it follows that $\dim(U(\mathcal{I}_\lambda)/W) < \infty$, which completes the proof of Lemma 6.36. $\qquad\square$

Exercise 6.38. Show that for any non-zero $\alpha, \beta \in U(\mathcal{I}_\lambda)$ there exists non-zero $x, y \in U(\mathcal{I}_\lambda)$ such that $x\alpha$ and βy have the form (6.9).

Now we can prove Theorem 6.30 for cokernels.

Proof. Let $\alpha, \beta \in U(\mathcal{I}_\lambda)$, $\alpha, \beta \neq 0$. Then for any non-zero $x, y \in U(\mathcal{I}_\lambda)$ we have the inclusions $U(\mathcal{I}_\lambda)x\alpha \subset U(\mathcal{I}_\lambda)\alpha$ and $\beta y U(\mathcal{I}_\lambda) \subset \beta U(\mathcal{I}_\lambda)$. In particular,

$$U(\mathcal{I}_\lambda)/(U(\mathcal{I}_\lambda)x\alpha + \beta y U(\mathcal{I}_\lambda)) \twoheadrightarrow U(\mathcal{I}_\lambda)/(U(\mathcal{I}_\lambda)\alpha + \beta U(\mathcal{I}_\lambda)). \qquad (6.12)$$

By Exercise 6.38 we can choose x and y such that $x\alpha$ and βy have the form (6.9). By Lemma 6.36 we get that $U(\mathcal{I}_\lambda)/(U(\mathcal{I}_\lambda)x\alpha + \beta y U(\mathcal{I}_\lambda))$ is finite-dimensional. From (6.12) it follows that $U(\mathcal{I}_\lambda)/(U(\mathcal{I}_\lambda)\alpha + \beta U(\mathcal{I}_\lambda))$ is finite-dimensional. Theorem 6.30 now follows from Lemma 6.34 and Exercise 6.33. $\qquad\square$

6.5 Finite-dimensionality of extensions

Let $\mathfrak{FL}$ denote the full subcategory of the category of all $\mathfrak{g}$-modules, which consists of all $\mathfrak{g}$-modules of finite length.

Exercise 6.39. Show that $\mathfrak{FL}$ is an abelian Krull–Schmidt category in which simple objects are simple $U(\mathfrak{g})$-modules and every object has finite length.

Our aim in the present section is to prove the following result:

Theorem 6.40. *For every* $M, N \in \mathfrak{FL}$ *the vector space*

$$\bigoplus_{i \in \mathbb{N}_0} \mathrm{Ext}^i_{U(\mathfrak{g})}(M, N)$$

is finite-dimensional.

To prove this theorem we need some preparation. We start with the following standard reduction:

Exercise 6.41. Assume that the statement of Theorem 6.40 is true for all simple modules M and N. Show, by induction on the length of M and N, that the statement is true for all $M, N \in \mathfrak{FL}$.

Let $\lambda \in \mathbb{C}$. Set $c_\lambda = c - (\lambda + 1)^2$. As the element c_λ is central in $U(\mathfrak{g})$, the left ideal generated by c_λ coincides with the two-sided ideal generated by c_λ. Therefore, from the definition we have the following free resolution of the left $U(\mathfrak{g})$-module $U(\mathcal{I}_\lambda)$:

$$0 \to U(\mathfrak{g}) \xrightarrow{\ -\cdot c_\lambda\ } U(\mathfrak{g}) \xrightarrow{\mathrm{proj}} U(\mathcal{I}_\lambda) \to 0, \tag{6.13}$$

where proj denotes the canonical projection. Let now $\alpha \in U(\mathfrak{g}) \setminus \mathcal{I}_\lambda$ be arbitrary. Then the element $\overline{\alpha} = \alpha + \mathcal{I}_\lambda$ is non-zero in $U(\mathcal{I}_\lambda)$. From the definition of the $U(\mathcal{I}_\lambda)$-module $M_\alpha = U(\mathcal{I}_\lambda)/U(\mathcal{I}_\lambda)\overline{\alpha}$ we have the following free resolution of M_α over $U(\mathcal{I}_\lambda)$:

$$0 \to U(\mathcal{I}_\lambda) \xrightarrow{\ -\cdot\overline{\alpha}\ } U(\mathcal{I}_\lambda) \xrightarrow{\mathrm{proj}} M_\alpha \to 0, \tag{6.14}$$

again where proj denotes the canonical projection. Resolving each copy of $U(\mathcal{I}_\lambda)$ in (6.14), using the resolution (6.13), we obtain the following free resolution of M_α over $U(\mathfrak{g})$:

$$0 \to U(\mathfrak{g}) \xrightarrow{\ \varphi\ } U(\mathfrak{g}) \oplus U(\mathfrak{g}) \xrightarrow{\ \psi\ } U(\mathfrak{g}) \xrightarrow{\mathrm{proj}} M_\alpha \to 0, \tag{6.15}$$

where for $x, y \in U(\mathfrak{g})$ the maps φ and ψ are given by the following:

$$\varphi(x) = \begin{pmatrix} -x\overline{\alpha} \\ xc_\lambda \end{pmatrix}, \quad \psi\begin{pmatrix} x \\ y \end{pmatrix} = xc_\lambda + y\overline{\alpha}.$$

Lemma 6.42. *Let L be a simple $U(\mathfrak{g})$-module, $\lambda \in \mathbb{C}$ and $\alpha \in U(\mathfrak{g}) \setminus \mathcal{I}_\lambda$. Then*

$$\mathrm{Ext}^i_{U(\mathfrak{g})}(M_\alpha, L) = \begin{cases} \mathrm{Ker}(\alpha_L), & c_\lambda L = 0, i = 0; \\ \mathrm{Ker}(\alpha_L) \oplus \mathrm{Coker}(\alpha_L), & c_\lambda L = 0, i = 1; \\ \mathrm{Coker}(\alpha_L), & c_\lambda L = 0, i = 2; \\ 0, & otherwise. \end{cases} \tag{6.16}$$

Proof. As $\mathrm{Hom}_{U(\mathfrak{g})}(U(\mathfrak{g}), L) \cong L$, applying the functor $\mathrm{Hom}_{U(\mathfrak{g})}(\_, L)$ to the resolution part of the sequence (6.15) we obtain the following complex:

$$0 \to L \xrightarrow{\overline{\psi}} L \oplus L \xrightarrow{\overline{\varphi}} L \to 0, \tag{6.17}$$

where the maps $\overline{\psi}$ and $\overline{\varphi}$ are given by

$$\overline{\psi}(x) = \begin{pmatrix} c_\lambda x \\ \overline{\alpha} x \end{pmatrix}, \quad \overline{\varphi}\begin{pmatrix} x \\ y \end{pmatrix} = -\overline{\alpha} x + c_\lambda y.$$

The element c_λ is central and hence defines an endomorphism of the simple module L. Notably, if $c_\lambda L \neq 0$, then this endomorphism is invertible and the sequence (6.17) is exact. In this case all extensions vanish. In the case $c_\lambda L = 0$, the homology of (6.17) is obviously given by (6.16). $\square$

Corollary 6.43. *The claim of Theorem 6.40 is true in the case $M = M_\alpha$ and $N = L$ is a simple module.*

Proof. This follows from Lemma 6.42 and Theorem 6.30. $\square$

Corollary 6.44. *The claim of Theorem 6.40 is true in the case M is a simple dense weight module and $N = L$ is a simple module.*

Proof. Let $\lambda \in \mathbb{C}$, and assume that $M \cong \mathbf{V}(\xi, (\lambda+1)^2)$ for some $\xi \in \mathbb{C}/2\mathbb{Z}$ such that $\lambda, -\lambda - 2 \notin \xi$ (see Theorem 3.32). Fix $\mu \in \xi$ and let $v \in \mathbf{V}(\xi, (\lambda+1)^2)_\mu$ be non-zero. Consider $\alpha = h - \mu \in U(\mathfrak{g})$. Then $(h-\mu)v = 0$ and hence we obtain a surjection $M_\alpha \twoheadrightarrow \mathbf{V}(\xi, (\lambda+1)^2)$. Since $\mathbf{V}(\xi, (\lambda+1)^2)$ is dense and has one-dimensional weight spaces (see (3.8)), the elements $\{b \cdot v : b \in \mathbf{B}_1\}$ form a basis in $\mathbf{V}(\xi, (\lambda + 1)^2)$. Applying Lemma 6.35 we find that the surjection $M_\alpha \twoheadrightarrow \mathbf{V}(\xi, (\lambda + 1)^2)$ is an isomorphism. Now the necessary claim follows from Corollary 6.43. $\square$

Exercise 6.45. Let $\lambda \in \mathbb{C}$. Show that for $\alpha = e$ the module M_α is isomorphic to the direct sum of the Verma modules $M(\lambda)$ and $M(-\lambda - 2)$.

Corollary 6.46. *The claim of Theorem 6.40 is true in the case where M is a Verma module and $N = L$ is a simple module.*

Proof. Using Exercise 6.45 and the additivity of $\text{Ext}^i_{U(\mathfrak{g})}(\_, L)$ we can prove this corollary using the same arguments as in the proof of Corollary 6.44. $\qquad\square$

Exercise 6.47. Show that the claim of Theorem 6.40 is true in the case when M is the universal lowest weight module $\overline{M}(\lambda)$ for some $\lambda \in \mathbb{C}$ and $N = L$ is a simple module.

Corollary 6.48. *The claim of Theorem 6.40 is true in the case when M is the simple finite-dimensional module $\mathbf{V}^{(n)}$ for some $n \in \mathbb{N}$ and $N = L$ is a simple module.*

Proof. This follows by applying $\text{Hom}_{U(\mathfrak{g})}(\_, L)$ to the Verma resolution

$$0 \to M(-n-1) \to M(n-1) \to \mathbf{V}^{(n)} \to 0$$

of $\mathbf{V}^{(n)}$ (see Theorem 3.16) and using Corollary 6.46 and the long exact sequence in homology. $\qquad\square$

Exercise 6.49. Show, by induction on the length of M, that the claim of Theorem 6.40 is true in the case when M is a generalized weight module of finite length and $N = L$ is a simple module.

Now we are ready to prove Theorem 6.40.

Proof. By Exercise 6.41, it is enough to prove the claim in the case when both M and N are simple. If M is a simple weight module, the claim follows from the classification of simple weight modules (Theorem 3.32), Exercise 6.47 and Corollaries 6.44, 6.46 and 6.48.

Assume now that M is a simple nonweight module. We then have that $M \cong \text{soc}_{U(\mathcal{I}_\alpha)}(\mathbb{A}/(\mathbb{A}\alpha))$ for some non-zero element $\alpha \in U(\mathcal{I}_\alpha)$, which is irreducible in $\mathbb{A}$ (Theorem 6.24). We may assume that the canonical generator $v = 1 + \mathbb{A}\alpha$ of $\mathbb{A}/(\mathbb{A}\alpha)$ belongs to M. Then v is annihilated by α. Consider the canonical map from $U(\mathfrak{g})$ to $\mathbb{A}/(\mathbb{A}\alpha)$, which sends 1 to v. This map factors through M_α, giving us the following short exact sequence:

$$0 \to K \to M_\alpha \twoheadrightarrow M \to 0, \tag{6.18}$$

where K denotes the kernel of the projection $M_\alpha \twoheadrightarrow M$. From the definition of M_α we have that

$$K \cong (U(\mathcal{I}_\alpha) \cap \mathbb{A}\alpha)/(U(\mathcal{I}_\alpha)\alpha).$$

Every element from $\mathbb{A}$ is a linear combination of elements of the form $g(h)^{-1}u$ for some $g(h) \in \mathbb{C}[h]$ and $u \in U(\mathcal{I}_\alpha)$ (Exercise 6.20). For any $g(h)^{-1}u\alpha$ we have $g(h) \cdot g(h)^{-1}u\alpha \in U(\mathcal{I}_\alpha)\alpha$. Hence the module K is $\mathbb{C}[h]$-torsion, that is a generalized weight module. It has finite length by Theorem 4.26.

By Exercise 6.49, the claim of Theorem 6.40 is true for the modules K and N. By Corollary 6.43, the claim of Theorem 6.40 is true for the modules M_α and N. From the long exact sequence in homology, which we obtain by applying $\mathrm{Hom}_{U(\mathfrak{g})}(-, N)$ to the short exact sequence (6.18), we thus conclude that the claim of Theorem 6.40 is true for the modules M and N. This completes the proof. □

6.6 Addenda and comments

6.6.1

Simple $\mathfrak{sl}_2$-modules were described by R. Block in [21, 22]. This description was extended by V. Bavula in [10] to generalized Weyl algebras. Finite-dimensionality of kernels, cokernels and extensions for finite length modules over the first Weyl algebra was proved by J. McConnell and J. Robson in [94]. For finite length $\mathfrak{sl}_2$-modules this was proved by V. Bavula in [8] and then extended to generalized Weyl algebras in [9, 10]. We mostly followed [10] during our exposition in this chapter.

6.6.2

The wording "description of simple modules" instead of "classification of simple modules" is used intentionally in order to avoid any controversy over the fact that the presented description is given only up to a classification of similarity classes of irreducible elements in the Euclidean algebra $\mathbb{A}$. Many years after [21, 22, 10] there still appear papers where some new classes of irreducible elements in $\mathbb{A}$ and hence new simple $\mathfrak{sl}_2$-modules are constructed (see for example [99]).

6.6.3

Note that our description of torsion-free simple $U(\mathcal{I}_\lambda)$-modules does not depend on λ.

6.6.4

One of the main problems in the description of simple $U(\mathfrak{g})$-modules is that they are easily described on the level of the algebra $\mathbb{A}$, but not on the level of the algebra $U(\mathcal{I}_\lambda)$. Since $\mathbb{A}$ is Euclidean, every simple $\mathbb{A}$-module has the form $\mathbb{A}/(\mathbb{A}\alpha)$ for some irreducible element $\alpha \in \mathbb{A}$. As any element in $\mathbb{A}$ is a linear combination of elements of the form $g(h)^{-1}u$ for some $g(h) \in \mathbb{C}[h]$ and $u \in U(\mathcal{I}_\lambda)$ (Exercise 6.20), and all elements $g(h)^{-1}$ are invertible in $\mathbb{A}$, we may always assume that $\alpha \in U(\mathcal{I}_\lambda)$. The main problem is that this does not guarantee that the $U(\mathcal{I}_\lambda)$-module $U(\mathcal{I}_\lambda)/(U(\mathcal{I}_\lambda)\alpha)$ is simple (see, for example, (6.18)). To determine when the module $U(\mathcal{I}_\lambda)/(U(\mathcal{I}_\lambda)\alpha)$ is simple is a subtle task. In [8, 10] one finds several very explicit conditions on α, which guarantee that the module $U(\mathcal{I}_\lambda)/(U(\mathcal{I}_\lambda)\alpha)$ is simple. Here is one example:

Theorem 6.50 ([10]). *Let* $g(h), p(h) \in \mathbb{C}[h]$ *be non-zero polynomials such that the following condition is satisfied: for any* $z \in \mathbb{C}$ *such that* $p(z) = 0$ *and for any* $i \in \mathbb{Z}$ *we have* $g(z + i) \neq 0$ *and* $(\lambda + 1)^2 \neq (z + i + 1)^2$. *Then the* $U(\mathcal{I}_\lambda)$-*modules*

$$U(\mathcal{I}_\lambda)/(U(\mathcal{I}_\lambda)(g(h)e + p(h))) \quad and \quad U(\mathcal{I}_\lambda)/(U(\mathcal{I}_\lambda)(g(h)f + p(h)))$$

are simple.

6.6.5

The first series of non-weight simple $U(\mathfrak{g})$-modules was constructed by D. Arnal and G. Pinczon in [4, 5], then put into a more general context by B. Kostant in [78]. These modules are called *Whittaker* modules because of their connection to the Whittaker equation in number theory. Whittaker modules have many important properties and are studied for several classes of algebras.

Whittaker modules are the modules $N_\alpha = U(\mathcal{I}_\lambda)/(U(\mathcal{I}_\lambda)(1 - \alpha e))$, where $\alpha \in \mathbb{C} \setminus \{0\}$. The module N_α is simple by Lemma 6.50. By Lemma 6.35, as a vector space the module N_α can be identified with $\mathbb{C}[h]$. The action of h on N_α is then given by the left multiplication. Taking into account that

$(1 - \alpha e) \cdot 1 = 0$, the action of e on $h^j \in N_\alpha$ can be computed as follows:

$$
\begin{aligned}
e \cdot h^j &= eh^j \cdot 1 \\
&= (h - 2)^j e \cdot 1 \\
&= \frac{1}{\alpha}(h - 2)^j.
\end{aligned}
$$

Similarly, we have $f(1 - \alpha e) \cdot 1 = 0$ and hence

$$
\begin{aligned}
f \cdot 1 &= \alpha f e \\
&= \frac{\alpha}{4}((\lambda + 1)^2 - (h + 1)^2).
\end{aligned}
$$

This yields that

$$
\begin{aligned}
f \cdot h^j &= fh^j \cdot 1 \\
&= (h + 2)^j f \cdot 1 \\
&= \frac{\alpha(h + 2)^j}{4}((\lambda + 1)^2 - (h + 1)^2).
\end{aligned}
$$

In [5] D. Arnal and G. Pinczon also considered more general modules which correspond to the case when $\alpha = h^n \in \mathbb{C}[h]$, $n \in \mathbb{N}$.

6.6.6

The Lie algebra $\mathfrak{sl}_2$ is the only simple complex Lie algebra, for which some description of all simple modules, analogous to the one presented in this chapter, exists. For all other simple and semi-simple finite-dimensional complex Lie algebras this question is wide open.

6.7 Additional exercises

Exercise 6.51. Construct an example of a torsion-free $\mathfrak{g}$-module with a non-zero quotient, which is not torsion-free.

Exercise 6.52. Let M be a torsion-free $\mathfrak{g}$-module and N be a non-zero finite-dimensional $\mathfrak{g}$-module. Prove that the module $M \otimes N$ is torsion-free.

Exercise 6.53. Show that for any non-zero element $g(h) \in \mathbb{C}(h)$ the element $1 + g(h)X \in \mathbb{A}$ is non-invertible and irreducible.

Exercise 6.54. Show that for every $\lambda \in \mathbb{C}$ there exists a unique monomorphism $\Phi'_\lambda : U(\mathcal{I}_\lambda) \to \mathbb{A}$ of associative algebras such that

$$\Phi'_\lambda(h) = h, \quad \Phi'_\lambda(e) = \frac{(\lambda+1)^2 - (h-1)^2}{4}X, \quad \Phi'_\lambda(f) = X^{-1}.$$

Exercise 6.55. Show that for any $\alpha \in \mathbb{A}$ there exists $g(h) \in \mathbb{C}[h]$ such that $g(h) \neq 0$ and $\alpha g(h) \in U(\mathcal{I}_\lambda)$.

Exercise 6.56. Consider $\mathbb{A}$ as a $U(\mathcal{I}_\lambda)$–$\mathbb{C}(h)$-bimodule by restriction. Show that the multiplication map mult defines an isomorphism of the following $U(\mathcal{I}_\lambda)$–$\mathbb{C}(h)$-bimodules:

$$\text{mult} : U(\mathcal{I}_\lambda) \bigotimes_{\mathbb{C}[h]} \mathbb{C}(h) \to \mathbb{A}.$$

Exercise 6.57. Show that every simple $\mathbb{A}$-module has infinite length, when considered as a $U(\mathcal{I}_\lambda)$-module.

Exercise 6.58. Show that every simple $\mathbb{A}$-module is not finitely generated, when considered as a $U(\mathcal{I}_\lambda)$-module.

Exercise 6.59. Prove the assertion of Theorem 6.30 for simple weight $U(\mathfrak{g})$-modules using a direct calculation and the classification of simple weight modules (Theorem 3.32).

Exercise 6.60. Consider the left regular $U(\mathfrak{g})$-module $_{U(\mathfrak{g})}U(\mathfrak{g})$. Show that the cokernel of the element e on this module is infinite-dimensional.

Exercise 6.61. Construct an example of a $U(\mathfrak{g})$-module M and a non-zero element $u \in U(\mathfrak{g})$ such that the kernel of u_M is infinite-dimensional.

Exercise 6.62. Let $\lambda \in \mathbb{C}$ and L be a simple $U(\mathcal{I}_\lambda)$-module. Show that for any $\alpha \in U(\mathcal{I}_\lambda)$, $\alpha \neq 0$, we have

$$\text{Ext}^i_{U(\mathcal{I}_\lambda)}(M_\alpha, L) = \begin{cases} \text{Ker}(\alpha_L), & i = 0; \\ \text{Coker}(\alpha_L), & i = 1; \\ 0, & \text{otherwise.} \end{cases}$$

Exercise 6.63. Let $\lambda \in \mathbb{C}$. Prove that for every $M, N \in \mathfrak{FL}_\lambda$ the vector space

$$\bigoplus_{i \in \mathbb{N}_0} \mathrm{Ext}^i_{U(\mathcal{I}_\lambda)}(M, N)$$

is finite-dimensional.

Exercise 6.64 ([10]). Let $\lambda \in \mathbb{C}$ and $\mathbf{m}$ be a left maximal ideal of $\mathbb{A}$. Show that the $U(\mathcal{I}_\lambda)$-module $N_{\mathbf{m}} = U(\mathcal{I}_\lambda)/(U(\mathcal{I}_\lambda) \cap \mathbf{m})$ is simple if and only if

$$\mathrm{Hom}_{U(\mathcal{I}_\lambda)}(N_{\mathbf{m}}, M) = 0$$

for any simple weight $U(\mathcal{I}_\lambda)$-module M.

Exercise 6.65 ([10]). Let $\lambda \in \mathbb{C}$ and $\mathbf{m}$ be a left maximal ideal of $\mathbb{A}$. Assume that $U(\mathcal{I}_\lambda) \cap \mathbf{m}$ contains an element that acts injectively on any simple weight $U(\mathcal{I}_\lambda)$-module. Show that the $U(\mathcal{I}_\lambda)$-module $N_{\mathbf{m}}$ from Exercise 6.64 is simple.

Exercise 6.66 ([10]). For two polynomials $g(h), p(h) \in \mathbb{C}[h]$ write $g(h) < p(h)$ provided that there do not exist $a \in \mathbb{C}$ and $i \in \mathbb{N}_0$ such that $g(a) = p(a - i) = 0$. Show that every element of the form

$$f^m g_m(h) + f^{m-1} g_{m-1}(h) + \cdots f g_1(h) + g_0(h),$$

$g_i(h) \in \mathbb{C}[h]$, $g_0(h), g_m(h) \neq 0$, $g_0(h) < g_m(h)$, $g_0(h) < (\lambda + 1)^2 - (h + 1)^2$, acts injectively on any simple weight $U(\mathcal{I}_\lambda)$-module.

Exercise 6.67. Use Exercises 6.64–6.66 to construct simple $U(\mathcal{I}_\lambda)$-modules.

Chapter 7

Categorification of simple finite-dimensional modules

7.1 Decategorification and categorification

Let $\mathcal{C}$ be an abelian category, $\Bbbk$ a unital commutative ring and $\mathrm{Gr}[\mathcal{C}]$ the Grothendieck group of $\mathcal{C}$ (see 5.9.4). Then the $\Bbbk$-module $[\mathcal{C}]^{\Bbbk} = \Bbbk \otimes_{\mathbb{Z}} \mathrm{Gr}[\mathcal{C}]$ is called the $\Bbbk$-*decategorification* of $\mathcal{C}$. If $\Bbbk = \mathbb{Z}$, we will call $[\mathcal{C}] = [\mathcal{C}]^{\mathbb{Z}}$ simply the *decategorification* of $\mathcal{C}$. For $M \in \mathcal{C}$ we denote by $[M]$ the class of M in $\mathrm{Gr}[\mathcal{C}]$ as well as the element $1 \otimes [M]$ of any $\Bbbk$-decategorification of $\mathcal{C}$.

Let V be a $\Bbbk$-module. A *categorification* of V is a pair $(\mathcal{C}, \varphi)$, where $\mathcal{C}$ is an abelian category and $\varphi : V \to [\mathcal{C}]^{\Bbbk}$ is a fixed isomorphism. Given a $\Bbbk$-module V, a categorification $(\mathcal{C}, \varphi)$ of V, and a $\Bbbk$-linear endomorphism $f \in \mathrm{End}_{\Bbbk}(V)$, a *categorification* of f is an exact functor F on $\mathcal{C}$, which induces the $\Bbbk$-linear endomorphism $[\mathrm{F}]$ of $[\mathcal{C}]^{\Bbbk}$, such that the following diagram commutes:

$$
\begin{array}{ccc}
V & \xrightarrow{\ \ f\ \ } & V \\
{\scriptstyle \varphi}\big\downarrow & & \big\downarrow{\scriptstyle \varphi} \\
[\mathcal{C}]^{\Bbbk} & \xrightarrow{\ [\mathrm{F}]\ } & [\mathcal{C}]^{\Bbbk}
\end{array}
\qquad (7.1)
$$

Example 7.1. Let $\mathcal{C} = \mathbb{C}\text{-mod}$ and $\varphi : \Bbbk \to [\mathcal{C}]^{\Bbbk}$ be given by $\varphi(1) = [\mathbb{C}]$. Then $(\mathcal{C}, \varphi)$ is a categorification of the free $\Bbbk$-module $\Bbbk$ of rank one.

Let now A be some associative $\Bbbk$-algebra with a fixed generating system $S = \{a_i : i \in I\}$ in A. Given some A-module M, every a_i defines a $\Bbbk$-linear endomorphism of M, which we will denote by a_i^M. A *naïve categorification* of the A-module M is a tuple $(\mathcal{C}, \varphi, \{\mathrm{F}_i : i \in I\})$ such that $(\mathcal{C}, \varphi)$ is a categorification of the $\Bbbk$-module M and F_i is a categorification of a_i^M for every $i \in I$. If the opposite is not explicitly stated, we always assume

219

that the unit element of A (if such an element exists) is categorified by the identity functor on $\mathcal{C}$.

Example 7.2. Let $A = \mathbb{C}[a]/(a^2 - 2a)$ and $M = \mathbb{C}$ be the A-module in which the action of the generator a is given by $a \cdot 1 = 2$. Let $(\mathcal{C}, \varphi)$ be the categorification of the $\mathbb{C}$-module $\mathbb{C}$ from Example 7.1 (here $\Bbbk = \mathbb{C}$). Let $\mathrm{ID}_{\mathcal{C}}$ denote the identity functor on $\mathcal{C}$. Then $(\mathcal{C}, \varphi, \mathrm{ID}_{\mathcal{C}} \oplus \mathrm{ID}_{\mathcal{C}})$ is a naïve categorification of the A-module M.

Exercise 7.3. Show that the algebra A from Example 7.2 is isomorphic to the group algebra $\mathbb{C}\mathbf{S}_2$ of the symmetric group $\mathbf{S}_2$.

Let A and M be as above. Assume that we are given two naïve categorifications $(\mathcal{C}, \varphi, \{\mathrm{F}_i : i \in I\})$ and $(\mathcal{D}, \psi, \{\mathrm{G}_i : i \in I\})$ of M. An exact functor $\Phi : \mathcal{C} \to \mathcal{D}$ is called a *naïve homomorphism* of naïve categorifications, provided that the following diagram commutes for every $i \in I$:

$$
\begin{array}{ccc}
[\mathcal{C}]^{\Bbbk} & \xrightarrow{\;[\mathrm{F}_i]\;} & [\mathcal{C}]^{\Bbbk} \\
{\scriptstyle[\Phi]}\downarrow & & \downarrow{\scriptstyle[\Phi]} \\
[\mathcal{D}]^{\Bbbk} & \xrightarrow{\;[\mathrm{G}_i]\;} & [\mathcal{D}]^{\Bbbk}
\end{array}
$$

The functor $\Phi : \mathcal{C} \to \mathcal{D}$ is called a *homomorphism* of naïve categorifications, provided that the following diagram commutes (up to isomorphism of functors) for every $i \in I$:

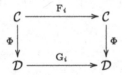

$$
\begin{array}{ccc}
\mathcal{C} & \xrightarrow{\;\mathrm{F}_i\;} & \mathcal{C} \\
{\scriptstyle\Phi}\downarrow & & \downarrow{\scriptstyle\Phi} \\
\mathcal{D} & \xrightarrow{\;\mathrm{G}_i\;} & \mathcal{D}
\end{array}
$$

The homomorphism Φ is said to be an *isomorphism* provided that it is an equivalence of categories.

Exercise 7.4. Show that every homomorphism of naïve categorifications is in fact a naïve homomorphism of naïve categorifications.

Example 7.5. Let $(\mathcal{C}, \varphi, \{\mathrm{F}_i : i \in I\})$ be a naïve categorification of an A-module M. Then the functor $\mathrm{ID}_{\mathcal{C}}$ is a homomorphism of naïve categorifications from $(\mathcal{C}, \varphi, \{\mathrm{F}_i : i \in I\})$ to $(\mathcal{C}, \varphi, \{\mathrm{F}_i : i \in I\})$.

Exercise 7.6. Let $(\mathcal{C}, \varphi, \{\mathrm{F}_i : i \in I\})$ and $(\mathcal{D}, \psi, \{\mathrm{G}_i : i \in I\})$ be two naïve categorifications of an A-module M. Show that the zero functor from $\mathcal{C}$ to $\mathcal{D}$ is a homomorphism of naïve categorifications.

Exercise 7.7. Let $(\mathcal{C}, \varphi, \{F_i : i \in I\})$ and $(\mathcal{D}, \psi, \{G_i : i \in I\})$ be two naïve categorifications of an A-module M and $\Phi : \mathcal{C} \to \mathcal{D}$ be a (naïve) homomorphism of naïve categorifications. Show that $\Phi \oplus \Phi$ is also a (naïve) homomorphism of naïve categorifications.

7.2 Naïve categorification of $\mathbf{V}^{(n)}$

Our goal in this chapter is to construct a nontrivial categorification of every simple finite-dimensional $U(\mathfrak{g})$-module with respect to the fixed generating system $\{e, f\}$ of $U(\mathfrak{g})$. The actions of e and f will be categorified via some exact functors E and F, respectively. To describe all possible categorifications of a given module and to understand the category of all such categorifications seems at the moment absolutely hopeless. However, we will attempt some small steps towards this and, at the very least, produce some examples.

To reasonably reduce the generality we will work, not with all abelian categories, but only with categories B-mod of finite-dimensional modules over a finite-dimensional associative and unital $\mathbb{C}$-algebra B. In this case $[B\text{-mod}]$ is a free abelian group of rank n, where n is the number of isomorphism classes of simple B-modules. If we fix a representative in each isomorphism class of simple B-modules, and consider their images in $[B\text{-mod}]$, we obtain a basis in $[B\text{-mod}]$. As we are only interested in the category B-mod, we may consider the algebra B up to Morita equivalence. In particular, if necessary, we may always assume that the algebra B is basic.

Let V be a simple finite-dimensional $U(\mathfrak{g})$-module. The natural problem to solve is the classification of all possible naïve categorifications of V up to isomorphism. Let $(B\text{-mod}, \varphi, \mathrm{E}, \mathrm{F})$ and $(B'\text{-mod}, \varphi', \mathrm{E}', \mathrm{F}')$ be two isomorphic naïve categorifications of V, and $\Phi : B\text{-mod} \to B'\text{-mod}$ be the corresponding isomorphism of naïve categorifications. Φ sends simple B-modules to simple B'-modules, and so it follows that the two bases

$$\{\varphi^{-1}([L]) : L \text{ is a simple } B\text{-module }\},$$
$$\{(\varphi')^{-1}([L]) : L \text{ is a simple } B'\text{-module }\}$$

of V coincide. This means that it is more natural to study naïve categorifications of a pair $(V, \{x_i\})$, where $\{x_i\}$ is a fixed basis of V. We will make the corresponding adjustment after the following observation, which sets out the types of base that actually appear:

Proposition 7.8. *Let* $n \in \mathbb{N}$. *Assume that* B *is a finite-dimensional associative* $\mathbb{C}$-*algebra and* $(B\text{-mod}, \varphi, \mathrm{E}, \mathrm{F})$ *is a naïve categorification of* $\mathbf{V}^{(n)}$. *Then, for every simple* B-*module* L, *the element* $\varphi^{-1}([L])$ *is a weight vector of* V.

Proof. Let $L_0, L_1, \ldots, L_{n-1}$ be a complete list of simple B-modules. Then $\mathbf{L} = \{[L_0], [L_1], \ldots, [L_{n-1}]\}$ is a basis of both $[B\text{-mod}]$ and $[B\text{-mod}]^{\mathbb{C}}$. We start with the following observations:

Exercise 7.9. Let X be a square matrix with non-negative integer coefficients. Assume that every column of X contains a positive entry. Show that for every $k \in \mathbb{N}$ every column of the matrix X^k contains a positive entry.

Lemma 7.10. *There exists* $i \in \{0, 1, \ldots, n-1\}$ *such that* $\mathrm{E}\, L_i = 0$.

Proof. Let X be the matrix of the linear operator $[\mathrm{E}]$ on $[B\text{-mod}]$ in the basis $\mathbf{L}$. The entries of X are non-negative integers. If $\mathrm{E}\, L_i \neq 0$ for all $i \in \{0, 1, \ldots, n-1\}$, then every column of X contains a positive entry. By Exercise 7.9, for every $k \in \mathbb{N}$ every column of the matrix X^k also contains a positive entry. In particular, $X^k \neq 0$ for all $k \in \mathbb{N}$. On the other hand, we have $E^n \cdot \mathbf{V}^{(n)} = 0$ by (1.9). Hence, by (7.1), we have $[\mathrm{E}]^n = \varphi\, E^n\, \varphi^{-1} = 0$ as well; a contradiction. The claim follows. $\qquad\square$

By Lemma 7.10, the functor E annihilates some simple B-module. Without loss of generality, we may assume that $\mathrm{E}\, L_0 = 0$. This yields $[\mathrm{E}][L_0] = 0$ and thus $E\varphi^{-1}([L_0]) = 0$ by (7.1). By (1.9), the kernel of E on $\mathbf{V}^{(n)}$ coincides with the linear span of v_0. Hence $\varphi^{-1}([L_0]) = \alpha v_0$ for some $\alpha \in \mathbb{C}$, which is a weight vector of weight $n-1$ by (1.9).

Lemma 7.11. *For every* $j \in \{0, 1, \ldots, n-1\}$ *there is a unique element* $i \in \{0, 1, \ldots, n-1\}$ *such that* $[\mathrm{F}^j\, L_0] = \alpha_j[L_i]$ *for some* $\alpha_j \in \mathbb{N}$. *Moreover,* $\varphi^{-1}([L_i])$ *is a weight vector of weight* $n-1-2j$.

Proof. We proceed by induction on j, the case $j = 0$ being trivial. Assume that $k \in \{0, 1, \ldots, n-1\}$ and that the statement is proved for all $j < k$. Then, without loss of generality, we may assume that for all $j < k$ we have $[\mathrm{F}^j\, L_0] = \alpha_j[L_j]$ for some $\alpha_j \in \mathbb{N}$. By (1.9), the element $\varphi^{-1}([L_j])$ forms a basis of $\mathbf{V}^{(n)}_{n-1-2j}$ for all $j < k$.

Let $[\mathrm{F}\, L_{k-1}] = \displaystyle\sum_{s=0}^{n-1} \beta_s[L_s]$ for some $\beta_s \in \mathbb{N}_0$. From (1.9) we have

$[EF\,L_{k-1}] = \beta[L_{k-1}]$ for some $\beta \in \mathbb{N}_0$. This means that for every s such that $\beta_s \neq 0$ we either have $E\,L_s = 0$ or $[E\,L_s] = \gamma_s[L_{k-1}]$ for some $\gamma_s > 0$.

However, from the above we know that the kernel of $[E]$ is one-dimensional and is generated by $[L_0]$. Hence $E\,L_s = 0$ is possible only for $s = 0$. At the same time, $\beta_0 > 0$ implies that the $[L_0]$-coefficient of $[F^{kt}\,L_0]$ is positive for any $t \in \mathbb{N}$. This contradicts the fact that the linear operator $[F]$ is nilpotent (see (1.9)). Therefore $\beta_0 = 0$.

If there existed two different $s, t > 0$ such that $[E\,L_s] = \gamma_s[L_{k-1}]$ and $[E\,L_t] = \gamma_t[L_{k-1}]$ with $\gamma_s, \gamma_t \neq 0$, then

$$[E](\gamma_t[L_s] - \gamma_s[L_t]) = 0,$$

which contradicts the fact that the kernel of $[E]$ is one-dimensional and is generated by $[L_0]$. This yields that there exists exactly one element $s \in \{0, 1, \ldots, n-1\}$ such that $\beta_s > 0$. For this s we also have $[E\,L_s] = \gamma_s[L_{k-1}]$ and $\gamma_s > 0$. From the inductive assumption we have $s > k - 1$. That $\varphi^{-1}([L_s])$ is a weight vector of weight $n - 1 - 2k$ follows from (1.9). This completes the proof. $\qquad\square$

The claim of Proposition 7.8 follows directly from Lemma 7.11. $\qquad\square$

Let now $n \in \mathbb{N}$. For every $i \in \{0, 1, \ldots, n-1\}$ fix some non-zero element $x_i \in \mathbf{V}^{(n)}_{n-1-2i}$. Then $\mathbf{x} = \{x_i : i \in \{0, 1, \ldots, n-1\}\}$ is a basis of $\mathbf{V}^{(n)}$ by Exercise 1.56. By (1.9), for every $i \in \{1, 2, \ldots, n-1\}$ we have $Ex_i = a_i x_{i-1}$ and $Fx_{i-1} = b_i x_i$ for some non-zero $a_i, b_i \in \mathbb{C}$. The tuple $(\{a_i\}, \{b_i\})$ is called the tuple of *structure constants* of $\mathbf{x}$. Two bases $\mathbf{x}$ and $\mathbf{y}$ as above are called *equivalent* provided that there exists $\alpha \in \mathbb{C} \setminus \{0\}$ such that $x_i = \alpha y_i$ for all $i \in \{0, 1, \ldots, n-1\}$.

By a *naïve categorification* of the pair $(\mathbf{V}^{(n)}, \mathbf{x})$, where $\mathbf{x}$ is as above, we mean a tuple $(B\text{-mod}, E, F)$, where B is a (basic) finite-dimensional associative $\mathbb{C}$-algebra with a fixed complete set $L_0, \ldots, L_{n-1}$ of pairwise non-isomorphic simple B-modules; and E and F are exact endofunctors of B-mod, such that $(B\text{-mod}, \varphi, E, F)$ is a naïve categorification of $\mathbf{V}^{(n)}$, where the isomorphism $\varphi : \mathbf{V}^{(n)} \to [B\text{-mod}]^{\mathbb{C}}$ is given by $\varphi(x_i) = [L_i]$, $i \in \{0, 1, \ldots, n-1\}$. Obviously, for a naïve categorification of the pair $(\mathbf{V}^{(n)}, \mathbf{x})$ to exist, we must have $a_i, b_i \in \mathbb{N}$ for all $i \in \{1, 2, \ldots, n-1\}$. We will call such $\mathbf{x}$ *admissible*. In what follows, we always assume that $\mathbf{x}$ is admissible. Three examples of admissible bases are the basis $\{v_i\}$ from (1.9), the basis $\{w_i\}$ from (1.10) and the basis $\{\hat{w}_i\}$ from (1.11).

Exercise 7.12. Show that for every $n \in \mathbb{N}$ there exist only finitely many equivalence classes of admissible bases $\mathbf{x}$ in $\mathbf{V}^{(n)}$.

Exercise 7.13. Let $v \in \mathbf{V}_0^{(1)}$ be non-zero. Show that every naïve categorification of $(\mathbf{V}^{(1)}, \{v\})$ has the form $(B\text{-mod}, \mathrm{E}, \mathrm{F})$, where B is a local algebra and both functors E and F are zero. Show further that two such categorifications $(B\text{-mod}, \mathrm{E}, \mathrm{F})$ and $(B'\text{-mod}, \mathrm{E}', \mathrm{F}')$ are isomorphic if, and only if, B and B' are Morita equivalent.

It might seem natural to extend the claim of Exercise 7.13 to all $\mathbf{V}^{(n)}$. However, this problem is still open and has proved very difficult. We would like to finish this section with an existence result showing that arbitrary algebras can be used to construct a naïve categorification of (some) $\mathbf{V}^{(n)}$ with respect to any admissible basis $\mathbf{x}$.

Let $\mathbf{x}$ be a fixed admissible basis of $\mathbf{V}^{(n)}$ and $(\{a_i\}, \{b_i\})$ be the tuple of structure constants of the basis $\mathbf{x}$. Let B be a basic finite-dimensional associative $\mathbb{C}$-algebra and $R(B)$ be the Jacobson radical of B. Assume that $\{\mathbf{e}_0, \ldots, \mathbf{e}_{n-1}\}$ is a fixed complete set of pairwise orthogonal primitive idempotents of B. Then for $i \in \{0, 1, \ldots, n-1\}$ we have the simple left B-module $L_i = B\mathbf{e}_i / R(B)\mathbf{e}_i$.

For $i \in \{0, \ldots, n-1\}$ let I_i denote the two-sided ideal of B, generated by $R(B)$ and all $\mathbf{e}_j$, $j \neq i$. Then $B/I_i \cong \mathbb{C}\mathbf{e}_i \cong \mathbb{C}$ is a simple algebra. For $i > 0$ consider the $B - B$-bimodules E_i and F_i defined as follows:

$$\mathrm{E}_i = \bigoplus_{s=1}^{a_i} \mathbf{e}_i B, \qquad \mathrm{F}_i = \bigoplus_{s=1}^{b_i} \mathbf{e}_{i-1} B. \qquad (7.2)$$

In both cases, we just take a_i and b_i copies of the corresponding right projective B-module, which defines on E_i and F_i the right B-module structure. To define the left B-module structure, we let the simple quotient algebras B/I_{i-1} and B/I_i act on E_i and F_i from the left via the identity maps, respectively. Set

$$\mathrm{E} = \bigoplus_{i=1}^{n} \mathrm{E}_i \otimes_B - : B\text{-mod} \to B\text{-mod}, \qquad (7.3)$$

$$\mathrm{F} = \bigoplus_{i=1}^{n} \mathrm{F}_i \otimes_B - : B\text{-mod} \to B\text{-mod}. \qquad (7.4)$$

Theorem 7.14. *Let* $\mathbf{x}$ *be as above. Then* $(B\text{-mod}, \mathrm{E}, \mathrm{F})$ *is a naïve categorification of* $(\mathbf{V}^{(n)}, \mathbf{x})$.

Proof. The functors E and F are exact as the bimodules E_i and F_i are right projective for all i by definition. For $i \in \{1, 2, \ldots, n-1\}$ we have

$$\mathrm{E}\, L_i \overset{(7.3)}{\cong} \bigoplus_{j=1}^{n} \mathrm{E}_j \otimes_B L_i$$

$$\overset{(7.2)}{\cong} \bigoplus_{j=1}^{n} \bigoplus_{s=1}^{a_j} \mathbf{e}_j B \otimes_B B\mathbf{e}_i / R(B)\mathbf{e}_i$$

$$\cong \bigoplus_{s=1}^{a_i} \mathbf{e}_i B \otimes_B B\mathbf{e}_i / R(B)\mathbf{e}_i$$

$$\cong \bigoplus_{s=1}^{a_i} \mathbb{C}\mathbf{e}_i \otimes \mathbf{e}_i.$$

Here, the two last isomorphisms follow from the fact that both $R(B)$ and $\mathbf{e}_j$, $j \neq i$, annihilate the module $L_i = B\mathbf{e}_i/R(B)\mathbf{e}_i$. The quotient B/I_{i-1} acts on the space $\bigoplus_{s=1}^{a_i} \mathbb{C}\mathbf{e}_i \otimes \mathbf{e}_i$ from the left via the identity map by construction. This yields $\mathrm{E}\, L_i \cong \bigoplus_{s=1}^{a_i} L_{i-1}$ and hence $[\mathrm{E}][L_i] = a_i[L_{i-1}]$. Analogous arguments also give $[\mathrm{E}][L_0] = 0$. Similarly, one can show that $[\mathrm{F}][L_{i-1}] = b_i[L_i]$ and $[\mathrm{F}][L_{n-1}] = 0$. The claim of the theorem follows. $\square$

Example 7.15. A special case of Theorem 7.14 is when $R(B) = 0$; that is when B is a semi-simple algebra with simple components given by $\mathbb{C}\mathbf{e}_i$, $i = 0, \ldots, n-1$. In this case we have $\mathbb{C}\mathbf{e}_i$-mod $\cong \mathbb{C}$-mod for every i. Under this identification, the functors $\mathrm{E}_i \otimes_B \_$ and $\mathrm{F}_i \otimes_B \_$ are isomorphic to the direct sum of a_i and b_i copies of the identity functor on $\mathbb{C}$-mod, respectively.

Exercise 7.16. Let $(B\text{-mod}, \mathrm{E}, \mathrm{F})$ be a naïve categorification of $(\mathbf{V}^{(n)}, \mathbf{x})$. Show that for every $i, j \in \{0, 1, \ldots, n-1\}$ we have $\mathrm{E}^{i+1}L_j = 0$ if, and only if, $j \leq i$; and $\mathrm{F}^{n-i}L_j = 0$ if, and only if, $j \geq i$.

As the name suggests, the above is a highly naïve way to categorify the module $\mathbf{V}^{(n)}$. It is hard to believe that all naïve categorifications of $\mathbf{V}^{(n)}$ may be described (classified). To be able to reduce the number of possible categorifications, we will need to introduce further restrictions, motivated by additional symmetries of $\mathbf{V}^{(n)}$. This is done in the next section.

7.3   Weak categorification of $\mathbf{V}^{(n)}$

Let $n \in \mathbb{N}$, $\mathbf{x}$ be an admissible basis of the module $\mathbf{V}^{(n)}$ and $(\{a_i\}, \{b_i\})$ be the corresponding tuple of structure constants. A naïve categorification $(B\text{-mod}, \mathrm{E}, \mathrm{F})$ of $(\mathbf{V}^{(n)}, \mathbf{x})$ is called a *weak categorification* provided that the functor E is both left and right adjoint to the functor F. This condition is motivated by the existence of the anti-involution $\star$ on $U(\mathfrak{g})$, for which we have $e^\star = f$ and $f^\star = e$.

Example 7.17. Any naïve categorification of $\mathbf{V}^{(1)}$ is a weak categorification.

Exercise 7.18. Show that any naïve categorification of $\mathbf{V}^{(3)}$, given by Example 7.15, is not a weak categorification.

Our main goal in this section is to prove the following result, which, when compared with Theorem 7.14, show that the class of weak categorifications is much smaller than the class of naïve categorifications. For a vector space V and $k \in \mathbb{N}$, we denote by $V^{\oplus k}$ the vector space $\underbrace{V \oplus V \oplus \cdots \oplus V}_{k \text{ summands}}$.

Theorem 7.19. *Let* $(B\text{-mod}, \mathrm{E}, \mathrm{F})$ *be a weak categorification of* $(\mathbf{V}^{(n)}, \mathbf{x})$. B *is then a direct sum of local algebras.*

Proof. We assume that B is basic, L_i, $i = 0, \ldots, n-1$, is a fixed complete set of pairwise non-isomorphic simple B-modules and P_i is the indecomposable projective cover of L_i. Since both E and F are left adjoint to exact functors, both E and F map projective modules to projective modules (see proof of Corollary 5.21). The images of E and F on indecomposable projective modules can be computed using the following:

Lemma 7.20. *Under the assumptions of Theorem 7.19 for all elements* $i \in \{0, 1, \ldots, n-1\}$ *we have*

$$\mathrm{E}\,P_i = \begin{cases} 0, & i = 0; \\ P_{i-1}^{\oplus b_i}, & i \neq 0; \end{cases} \qquad \mathrm{F}\,P_i = \begin{cases} 0, & i = n-1; \\ P_{i+1}^{\oplus a_{i+1}}, & i \neq n-1. \end{cases} \tag{7.5}$$

Proof. We prove the first formula; the second formula is proved similarly. As we already know that the module $\mathrm{E}P_i$ is projective, so to determine its decomposition into indecomposable projectives we should compute the dimension of the homomorphism space to all simple modules. As (E, F) is an adjoint pair of functors, for $j \in \{0, 1, \ldots, n-1\}$ we have:

$$\dim \operatorname{Hom}_B(\mathrm{E}\, P_i, L_j) = \dim \operatorname{Hom}_B(P_i, \mathrm{F}\, L_j)$$
$$= \begin{cases} 0, & j \neq i-1; \\ b_i, & j = i-1. \end{cases}$$

Here the last equality follows from the definition of the structure constants and the fact that $(B\text{-mod}, \mathrm{E}, \mathrm{F})$ is a naïve categorification of $(\mathbf{V}^{(n)}, \mathbf{x})$. The first formula in (7.5) follows. The second one is proved similarly. This completes the proof. $\qquad \square$

Our next step is the following:

Lemma 7.21. *For all $i, j \in \{0, 1, \ldots, n-1\}$ the inequality $[P_i : L_j] \neq 0$ implies $j \leq i$.*

Proof. From (7.5) it follows that $\mathrm{E}^{i+1} P_i = 0$. The functor E^{i+1} is exact and annihilates only the simple modules L_j, $j \leq i$ (Exercise 7.16). Hence $[P_i : L_j] \neq 0$ implies $j \leq i$. $\qquad \square$

Using the functor F instead of E in the proof of Lemma 7.21, we obtain the following:

Lemma 7.22. *For all $i, j \in \{0, 1, \ldots, n-1\}$ the inequality $[P_i : L_j] \neq 0$ implies $i \leq j$.*

From Lemmas 7.21 and 7.22 it follows that for all $i, j \in \{0, 1, \ldots, n-1\}$ the inequality $[P_i : L_j] \neq 0$ implies $j = i$. Hence any projective B-module is local and thus B is a direct sum of local algebras, as asserted. $\qquad \square$

We can use Theorem 7.19 to classify all weak categorifications of $\mathbf{V}^{(2)}$. The classification below shows how the difficulty of our problem (to classify all categorifications of all $\mathbf{V}^{(n)}$) grows for the first step from $\mathbf{V}^{(1)}$ to $\mathbf{V}^{(2)}$, even under the additional upgrade of naïve categorification to weak categorification.

Exercise 7.23. Show that, up to equivalence, the basis $\mathbf{v} = \{v_0, v_1\}$, given by (1.9), is the unique admissible basis of $\mathbf{V}^{(2)}$.

Let B' be a local algebra and E' be any auto-equivalence of B'-mod with inverse F'. Consider the algebra $B = B' \oplus B'$. Then

$$B\text{-mod} \cong B'\text{-mod} \oplus B'\text{-mod}.$$

Let E be the endofunctor of B-mod, given by $E(X, Y) = (0, E'X)$; and F be the endofunctor of B-mod, given by $F(X, Y) = (F'Y, 0)$.

Exercise 7.24. Show that the functor E from the previous paragraph is both left and right adjoint to the functor F.

For a category $\mathcal{C}$ we denote by $\text{Aut}(\mathcal{C})$ the group of all auto-equivalences on $\mathcal{C}$ (up to isomorphism of functors).

Corollary 7.25.

  (i) *The tuple $(B\text{-mod}, E, F)$, given by the above, is a weak categorification of the pair $(\mathbf{V}^{(2)}, \mathbf{v})$.*

 (ii) *Every weak categorification of the pair $(\mathbf{V}^{(2)}, \mathbf{v})$ is isomorphic to $(B\text{-mod}, E, F)$ for some B' and E' as above.*

(iii) *If two weak categorifications $(B\text{-mod}, E, F)$ and $(\tilde{B}\text{-mod}, \tilde{E}, \tilde{F})$, given by the above, are isomorphic, then B' and $\tilde{B}'$ are Morita equivalent.*

 (iv) *Two weak categorifications $(B\text{-mod}, E, F)$ and $(B\text{-mod}, \tilde{E}, \tilde{F})$, given by the above, are isomorphic if and only if E' and $\tilde{E}'$ are conjugate in $\text{Aut}(B\text{-mod})$.*

Proof. From the definition of E and F it follows that these functors send simple modules to simple modules, which yields that $(B\text{-mod}, E, F)$ is a naïve categorification of $(\mathbf{V}^{(2)}, \mathbf{v})$. Now the claim (i) follows from Exercise 7.24.

Let now $(B\text{-mod}, E, F)$ be a weak categorification of $(\mathbf{V}^{(2)}, \mathbf{v})$. From Theorem 7.19 we have that B decomposes into a direct sum $B_0 \oplus B_1$ of local algebras. The functors E and F induce exact functors

$$E' : B_1\text{-mod} \to B_0\text{-mod} \quad \text{and} \quad F' : B_0\text{-mod} \to B_1\text{-mod}.$$

From the adjointness of E and F we have that the functor E' is both left and right adjoint to the functor F'. From (1.9) it follows that both E' and F' send simple modules to simple modules. Hence the functors E' and F' are mutually inverse equivalences of categories between B_1-mod and B_0-mod (see proof of Theorem 3.58). The claim (ii) follows.

The claims (iii) and (iv) follow directly from the definitions. This completes the proof. $\qquad\square$

Corollary 7.26. *Let $n \in \mathbb{N}$ and $\mathbf{x}$ be an admissible basis of $\mathbf{V}^{(n)}$. Assume that $(B\text{-mod}, E, F)$ is a weak categorification of $(\mathbf{V}^{(n)}, \mathbf{x})$. Then the sets $\{[P]\}$ and $\{[I]\}$, where P and I run through the sets of representatives of* *isoclasses of indecomposable projective and injective modules, respectively, form bases in* $[B\text{-mod}]^{\mathbb{C}}$.

Proof. This follows from Theorem 7.19 and the simple fact that for a local algebra B' any B'-module M gives a non-zero element $[M]$, which automatically forms a basis in the complexified Grothendieck group $[B'\text{-mod}]^{\mathbb{C}}$. $\square$

Exercise 7.27. Let $(B\text{-mod}, E, F)$ be a weak categorification of $(\mathbf{V}^{(n)}, \mathbf{x})$. Show that the bases of $\mathbf{V}^{(n)}$, given by Corollary 7.26, coincide. Show further that this basis is admissible.

Exercise 7.28. Let $(B\text{-mod}, E, F)$ be a weak categorification of $(\mathbf{V}^{(n)}, \mathbf{x})$. Show that for $n = 1, 2$, the basis of $\mathbf{V}^{(n)}$, given by Corollary 7.26, is equivalent to $\mathbf{x}$.

We finish this section with a construction of a weak categorification of $(\mathbf{V}^{(n)}, \mathbf{v})$ with respect to the basis $\mathbf{v} = \{v_i\}$ from (1.9). By (1.9) we have the following structure constants with respect to $\mathbf{v}$: $b_i = 1$, $a_i = i(n - i)$, $i = 1, \ldots, n - 1$. Set $B_{n-1} = \mathbb{C}$ and for $i = 0, \ldots, n - 2$ set

$$B_i = \mathbb{C}[x]/[x^{a_{i+1}}] \bigotimes_{\mathbb{C}} \mathbb{C}[x]/[x^{a_{i+2}}] \bigotimes_{\mathbb{C}} \cdots \bigotimes_{\mathbb{C}} \mathbb{C}[x]/[x^{a_{n-1}}] \bigotimes_{\mathbb{C}} \mathbb{C}.$$

Then for every $i = 1, 2 \ldots, n - 1$, we can regard B_i as the subalgebra $\mathbb{C} \otimes B_i$ of B_{i-1}. Consider functors

$$F_i = \operatorname{Res}_{B_i}^{B_{i-1}} : B_{i-1}\text{-mod} \to B_i\text{-mod};$$

$$E_i = \operatorname{Ind}_{B_i}^{B_{i-1}} : B_i\text{-mod} \to B_{i-1}\text{-mod}.$$

Finally, set

$$B = \bigoplus_{i=0}^{n-1} B_i, \qquad F = \bigoplus_{i=1}^{n-1} F_i, \qquad E = \bigoplus_{i=1}^{n-1} E_i$$

(we use the convention that $FB_{n-1}\text{-mod} = 0$ and $EB_0\text{-mod} = 0$). For $i = 0, 1, \ldots, n - 1$ let L_i denote the simple B_i-module.

Theorem 7.29. *The tuple* $(B\text{-mod}, E, F)$ *is a weak categorification of* $(\mathbf{V}^{(n)}, \mathbf{v})$.

Proof. From the definitions, we have that (E, F) is an adjoint pair of functors, that the functor F is exact and that

$$[F][L_i] = \begin{cases} [L_{i+1}], & i \neq n - 1; \\ 0, & i = n - 1. \end{cases}$$

By construction, for every $i = 1, 2, \ldots, n-1$ the algebra B_{i-1} is a free B_i-module with the basis $\{x^j \otimes 1 \otimes \cdots \otimes 1 : j = 0, \ldots, a_i - 1\}$. This means that the functor E_i (and hence also the functor E) is exact and that

$$[E][L_i] = \begin{cases} a_i[L_{i-1}], & i \neq 0; \\ 0, & i = 0. \end{cases}$$

To complete the proof we are left to show that (F, E) is an adjoint pair of functors. To prove this we will need the following lemma:

Lemma 7.30. *For every* $i = 0, 1, \ldots, n-1$ *there is an isomorphism of* B_i-B_i-*bimodules as follows:* $\varphi_i : B_i \to \mathrm{Hom}_{\mathbb{C}}(B_i, \mathbb{C})$.

Proof. For $k \in \mathbb{N}_0$ consider the algebra $P_k = \mathbb{C}[x]/(x^k)$ and the linear map $\mathfrak{p} : P_k \to \mathbb{C}$, defined as follows:

$$\mathfrak{p}(x^s) = \begin{cases} 1, & s = k-1; \\ 0, & \text{otherwise.} \end{cases}$$

Consider the bilinear form $(\cdot, \cdot)_k$ on P_k defined as follows: $(x, y)_k = \mathfrak{p}(xy)$, $x, y \in P_k$. It is easy to see that the form $(\cdot, \cdot)_k$ is non-degenerate. The form $(\cdot, \cdot)_k$ is symmetric as P_k is commutative. Because of the associativity of multiplication in P_k, for $x, y, z \in P_k$ we have $(xz, y)_k = \mathfrak{p}((xz)y) = \mathfrak{p}(x(zy)) = (x, zy)_k$ and thus

$$(xz, y)_k = (x, zy)_k. \tag{7.6}$$

Using the bilinear form $(\cdot, \cdot)_k$, $k \in \mathbb{N}_0$, and formula (1.26), for every $i \in 0, 1, \ldots, n-1$ we define the bilinear form $(\cdot, \cdot)_{B_i}$ on B_i. From Exercise 1.47 it follows that $(\cdot, \cdot)_{B_i}$ is symmetric and non-degenerate. From (7.6) and the definitions it also follows that $(\cdot, \cdot)_{B_i}$ has the property (7.6).

Define the $\mathbb{C}$-linear map $\psi : B_i \to B_i^* = \mathrm{Hom}_{\mathbb{C}}(B_i, \mathbb{C})$, $x \mapsto \psi_x$, as follows: $\psi_x(y) = (x, y)_{B_i}$, $x, y \in B_i$. This map is bijective as B_i is finite-dimensional and $(\cdot, \cdot)_{B_i}$ is non-degenerate. For any $b_1, b_2 \in B_i$, using the definitions and the properties of $(\cdot, \cdot)_{B_i}$, we have:

$$b_1 \cdot \psi_x \cdot b_2(y) \overset{\text{def.}}{=} \psi_x(b_2 y b_1)$$
$$\text{(by definition)} = (x, b_2 y b_1)_{B_i}$$
$$\text{(by (7.6))} = (x b_2, y b_1)_{B_i}$$
$$((\cdot, \cdot)_{B_i} \text{ is symmetric}) = (y b_1, x b_2)_{B_i}$$
$$\text{(by (7.6))} = (y, b_1 x b_2)_{B_i}$$
$$((\cdot, \cdot)_{B_i} \text{ is symmetric}) = (b_1 x b_2, y)_{B_i}$$
$$\text{(by definition)} = \psi_{b_1 x b_2}(y).$$

Hence ψ is an isomorphism of B_i-B_i-bimodules, as required. $\square$

Fix now $i \in \{1, 2, \ldots, n-1\}$. The identity functor on B_{i-1}-mod can be described in two different ways, as $B_{i-1} \otimes_{B_{i-1}} -$ and as $\text{Hom}_{B_{i-1}}(B_{i-1}, -)$. Restricting the action of B_{i-1} to that of B_i (the left action in the first case and the right action in the second case), we will get two alternative descriptions of the restriction functor:

$$\text{Res}_{B_i}^{B_{i-1}} \cong B_{i-1} \otimes_{B_{i-1}} - \cong \text{Hom}_{B_{i-1}}(B_{i-1}, -). \tag{7.7}$$

Denote by $\mathbf{d} = \text{Hom}_{\mathbb{C}}(\_, \mathbb{C})$ the usual duality between left and right modules and note that it obviously commutes with $\text{Res}_{B_i}^{B_{i-1}}$. To distinguish left and right actions of an algebra X we denote the left action by X- and the right action by $-X$.

For $M \in B_{i-1}$-mod and $N \in B_i$-mod we have the following sequence of natural isomorphisms:

$$
\begin{aligned}
\text{Hom}_{B_{i-1}\text{-}}(M, \text{Ind}_{B_i}^{B_{i-1}} N) &\overset{\text{def.}}{=} \text{Hom}_{B_{i-1}\text{-}}(M, B_{i-1} \otimes_{B_i} N) \\
\text{(applying } \mathbf{d}) \quad &= \text{Hom}_{\text{-}B_{i-1}}(\mathbf{d}(B_{i-1} \otimes_{B_i} N), \mathbf{d}(M)) \\
\text{(}\mathbf{d} \text{ and } \otimes \text{ commute)} \quad &= \text{Hom}_{\text{-}B_{i-1}}(\mathbf{d}(N) \otimes_{B_i} \mathbf{d}(B_{i-1}), \mathbf{d}(M)) \\
\text{(by adjunction)} \quad &= \text{Hom}_{\text{-}B_i}(\mathbf{d}(N), \text{Hom}_{\text{-}B_{i-1}}(\mathbf{d}(B_{i-1}), \mathbf{d}(M))) \\
\text{(Lemma 7.30)} \quad &= \text{Hom}_{\text{-}B_i}(\mathbf{d}(N), \text{Hom}_{\text{-}B_{i-1}}(B_{i-1}, \mathbf{d}(M))) \\
\text{((7.7) for right modules)} \quad &= \text{Hom}_{\text{-}B_i}(\mathbf{d}(N), \text{Res}_{B_i}^{B_{i-1}} \mathbf{d}(M)) \\
\text{(applying } \mathbf{d}) \quad &= \text{Hom}_{B_i\text{-}}(\mathbf{d}(\text{Res}_{B_i}^{B_{i-1}} \mathbf{d}(M)), N) \\
\text{(Res and } \mathbf{d} \text{ commute)} \quad &= \text{Hom}_{B_i\text{-}}(\text{Res}_{B_i}^{B_{i-1}} M, N).
\end{aligned}
$$

This yields that $(\mathbf{F}, \mathbf{E})$ is an adjoint pair of functors and completes the proof of the theorem. □

Exercise 7.31. Prove that the bilinear form $(\cdot, \cdot)_k$, defined in the proof of Lemma 7.30, is non-degenerate.

Exercise 7.32. Prove the property (7.6) for the bilinear form $(\cdot, \cdot)_{B_i}$, $i = 0, 1, \ldots, n-1$.

A finite-dimensional associative $\mathbb{C}$-algebra A is called *symmetric* provided that the A–A-bimodules A and $A^* = \text{Hom}_{\mathbb{C}}(A, \mathbb{C})$ are isomorphic. Thus Lemma 7.30 says that all algebras B_i above are symmetric.

Exercise 7.33. Show that A is symmetric if, and only if, there is a symmetric non-degenerate bilinear form on A, which has the property (7.6).

Exercise 7.34. Show that both the direct sum and the tensor product of symmetric algebras are symmetric.

7.4   Categorification of $V^{(n)}$ via coinvariant algebras

The aim of this section is to present a weak categorification of $(V^{(n)}, \hat{\mathbf{w}})$ with respect to the basis $\hat{\mathbf{w}} = \{\hat{w}_i\}$ from (1.11). This categorification is the "correct" categorification of $V^{(n)}$ and can be proved to be even unique under some additional assumptions. The latter is, however, a rather extensive and complicated technical work, which we will not delve too far into. Moreover, to be able to formulate the main result we will have to recall several results from the classical invariant theory of reflection groups.

Let $n \in \mathbb{N}_0$. Consider the polynomial algebra $\mathbf{P}_n = \mathbb{C}[x_1, \ldots, x_n]$. The symmetric group $\mathbf{S}_n$ acts on $\mathbf{P}_n$ permuting the indices of the variables. For $\sigma \in \mathbf{S}_n$ we have:

$$\sigma(x_1^{s_1} x_2^{s_2} \cdots x_n^{s_n}) = x_{\sigma(1)}^{s_1} x_{\sigma(2)}^{s_2} \cdots x_{\sigma(n)}^{s_n}.$$

Let $\mathbf{J}_n^+$ denote the ideal of $\mathbf{P}_n$, generated by all $\mathbf{S}_n$-invariant (that is, symmetric) homogeneous polynomials of non-zero degree. The quotient algebra $\mathbf{C}_n = \mathbf{P}_n / \mathbf{J}_n^+$ is called the *coinvariant* algebra. As $\mathbf{J}_n^+$ is generated by homogeneous elements, the algebra $\mathbf{C}_n$ becomes automatically graded by the degree of monomials. The algebra $\mathbf{C}_n$ is commutative.

For $i \in \{1, 2, \ldots, n-1\}$ we denote by s_i the elementary transposition $(i, i+1)$ of $\mathbf{S}_n$, and by $\mathbf{S}_n^i$ the subgroup of $\mathbf{S}_n$, generated by all s_j, $j \neq i$. Since the ideal $\mathbf{J}_n^+$ is generated by symmetric polynomials, it is invariant with respect to the action of $\mathbf{S}_n$, in particular, with respect to the action of $\mathbf{S}_n^i$. Hence the algebra $\mathbf{C}_n$ inherits an action of $\mathbf{S}_n^i$ and we denote by $\mathbf{C}_n^i$ the subalgebra of $\mathbf{C}_n$, consisting of all elements, invariant with respect to the $\mathbf{S}_n^i$-action. Similarly, for $i = 1, \ldots, n-2$ we denote by $\mathbf{S}_n^{i,i+1}$ the subgroup of $\mathbf{S}_n$, generated by all s_j, $j \neq i, i+1$; and by $\mathbf{C}_n^{i,i+1}$ the subalgebra of $\mathbf{C}_n$, consisting of all elements, invariant with respect to the action of $\mathbf{S}_n^{i,i+1}$. As $\mathbf{S}_n^{i,i+1} \subset \mathbf{S}_n^i$, the algebra $\mathbf{C}_n^i$ is a subalgebra of $\mathbf{C}_n^{i,i+1}$ in the natural way.

Theorem 7.35 below presents a collection of classical facts from the invariant theory, which can be found, for example, in [56, 69].

Theorem 7.35.

(i)   The algebra $\mathbf{C}_n$ is a symmetric local algebra of dimension $n!$.

(ii) The algebra $\mathbf{C}_n^i$ is a symmetric local algebra of dimension $\binom{n}{i}$.

(iii) The algebra $\mathbf{C}_n^{i,i+1}$ is a symmetric local algebra of dimension $\frac{n!}{i!(n-i-1)!}$.

(iv) The algebra $\mathbf{C}_n^{i,i+1}$ is a free module over the subalgebra $\mathbf{C}_n^i$.

(v)   The algebra $\mathbf{C}_n^{i,i+1}$ is a free module over the subalgebra $\mathbf{C}_n^{i+1}$.

Set

$$\mathbf{C}_n^0 = \mathbf{C}_n^n = \mathbb{C}, \quad \mathbf{C}_n^{0,1} = \mathbf{C}_n^1, \quad \mathbf{C}_n^{n-1,n} = \mathbf{C}_n^{n-1}.$$

For $i = 0, \ldots, n-1$ consider the functors

$$\operatorname{Ind}_{\mathbf{C}_n^i}^{\mathbf{C}_n^{i,i+1}} : \mathbf{C}_n^i\text{-mod} \quad \to \mathbf{C}_n^{i,i+1}\text{-mod},$$

$$\operatorname{Ind}_{\mathbf{C}_n^{i+1}}^{\mathbf{C}_n^{i,i+1}} : \mathbf{C}_n^{i+1}\text{-mod} \quad \to \mathbf{C}_n^{i,i+1}\text{-mod},$$

$$\operatorname{Res}_{\mathbf{C}_n^i}^{\mathbf{C}_n^{i,i+1}} : \mathbf{C}_n^{i,i+1}\text{-mod} \to \mathbf{C}_n^i\text{-mod},$$

$$\operatorname{Res}_{\mathbf{C}_n^{i+1}}^{\mathbf{C}_n^{i,i+1}} : \mathbf{C}_n^{i,i+1}\text{-mod} \to \mathbf{C}_n^{i+1}\text{-mod}.$$

The usual adjunction between induction and restriction yields that the pairs $(\operatorname{Ind}_{\mathbf{C}_n^i}^{\mathbf{C}_n^{i,i+1}}, \operatorname{Res}_{\mathbf{C}_n^i}^{\mathbf{C}_n^{i,i+1}})$ and $(\operatorname{Ind}_{\mathbf{C}_n^{i+1}}^{\mathbf{C}_n^{i,i+1}}, \operatorname{Res}_{\mathbf{C}_n^{i+1}}^{\mathbf{C}_n^{i,i+1}})$ are adjoint pairs of functors.

Corollary 7.36. *The pairs* $(\operatorname{Res}_{\mathbf{C}_n^i}^{\mathbf{C}_n^{i,i+1}}, \operatorname{Ind}_{\mathbf{C}_n^i}^{\mathbf{C}_n^{i,i+1}})$ *and* $(\operatorname{Res}_{\mathbf{C}_n^{i+1}}^{\mathbf{C}_n^{i,i+1}}, \operatorname{Ind}_{\mathbf{C}_n^{i+1}}^{\mathbf{C}_n^{i,i+1}})$ *are adjoint pairs of functors.*

Proof. Both algebras $\mathbf{C}_n^i$ and $\mathbf{C}_n^{i,i+1}$ are symmetric by Theorem 7.35. Taking this instead of Lemma 7.30, the claim is proved identically to the proof of the second part of Theorem 7.29. $\square$

For $i = 0, 1, \ldots, n$ set $B_i = \mathbf{C}_n^i$ and for $i \neq 0$ consider the following functors:

$$\mathrm{F}_i = \operatorname{Res}_{\mathbf{C}_n^i}^{\mathbf{C}_n^{i-1,i}} \circ \operatorname{Ind}_{\mathbf{C}_n^{i-1}}^{\mathbf{C}_n^{i-1,i}} : B_{i-1}\text{-mod} \to B_i\text{-mod},$$

$$\mathrm{E}_i = \operatorname{Res}_{\mathbf{C}_n^{i-1}}^{\mathbf{C}_n^{i-1,i}} \circ \operatorname{Ind}_{\mathbf{C}_n^i}^{\mathbf{C}_n^{i-1,i}} : B_i\text{-mod} \quad \to B_{i-1}\text{-mod}.$$

Finally, set

$$B = \bigoplus_{i=0}^{n} B_i, \qquad \mathrm{F} = \bigoplus_{i=1}^{n} \mathrm{F}_i, \qquad \mathrm{E} = \bigoplus_{i=1}^{n} \mathrm{E}_i$$

(we again use the convention that $\mathrm{F}\,B_n\text{-mod} = 0$ and $\mathrm{E}\,B_0\text{-mod} = 0$). For $i = 0, 1, \ldots, n-1$ let L_i denote the simple B_i-module (recall that B_i is local by Theorem 7.35(ii)).

Theorem 7.37. *The tuple* $(B\text{-mod}, \mathrm{E}, \mathrm{F})$, *given by the above, is a weak categorification of* $(\mathbf{V}^{(n+1)}, \hat{\mathbf{w}})$.

Proof. From Corollary 7.36 it follows that the functor E is both left and right adjoint to the functor F. Hence we only need to show that the tuple $(B\text{-mod}, \mathrm{E}, \mathrm{F})$ is a naïve categorification of $(\mathbf{V}^{(n+1)}, \hat{\mathbf{w}})$.

By Theorem 7.35, the algebra $\mathbf{C}_n^{i-1,i}$ is a free module over $\mathbf{C}_n^{i-1}$ of rank $n + 1 - i$. Therefore, inducing up the one-dimensional $\mathbf{C}_n^{i-1}$-module

L_{i-1} from $\mathbf{C}_n^{i-1}$ to $\mathbf{C}_n^{i-1,i}$, we get a $\mathbf{C}_n^{i-1,i}$-module of dimension $n+1-i$. Restricting the latter to $\mathbf{C}_n^i$, we get a $\mathbf{C}_n^i$-module of dimension $n+1-i$. This and the definitions yield that

$$[\mathrm{F}][L_{i-1}] = \begin{cases} (n+1-i)[L_i], & i \neq n+1; \\ 0, & i = n+1. \end{cases}$$

Similarly, by Theorem 7.35, the algebra $\mathbf{C}_n^{i-1,i}$ is a free $\mathbf{C}_n^i$-module of rank i. Analogously to the above, this and the definitions yield that

$$[\mathrm{E}][L_i] = \begin{cases} i[L_{i-1}], & i \neq 0; \\ 0, & i = 0. \end{cases}$$

The claim of the theorem follows. $\qquad\square$

Exercise 7.38. Let $(B\text{-mod}, \mathrm{E}, \mathrm{F})$ be the weak categorification of the pair $(\mathbf{V}^{(n+1)}, \hat{\mathbf{w}})$, given by Theorem 7.37. Show that the classes of indecomposable projective B-modules in $[B\text{-mod}]^{\mathbb{C}}$ correspond to the basis $\mathbf{w}$ of $\mathbf{V}^{(n+1)}$, given by (1.10).

The basis $\mathbf{w}$ of $\mathbf{V}^{(n)}$, given by (1.10), is called *canonical* and the basis $\hat{\mathbf{w}}$ of $\mathbf{V}^{(n)}$, given by (1.11), is called *dual canonical*.

7.5 Addenda and comments

7.5.1

Exposition in Section 7.1 closely follows [91]. The idea of a weak categorification, presented in Section 7.3, is taken from [30]. The idea of a naïve categorification, presented in Section 7.2, is just a further simplification of weak categorification. The categorification via coinvariant algebras, given in Section 7.4, appears in a different guise in [30], and in both this and some other versions in [48].

7.5.2

The term *categorification* was introduced by L. Crane in [33] based on the ideas developed earlier in the joint work [34] with I. Frenkel. *Categorification* usually refers to a process of replacing set-theoretic concepts by category-theoretic analogs; for example, sets by categories, functions by functors, equations between functions by natural isomorphisms between functors etc. On the one hand, this makes the objects of study more complicated. On the other, it might reveal additional interesting structures.

The most famous application of categorification to date is categorification of the Jones polynomial, proposed by M. Khovanov in [75]. This led to discovery of more powerful knot and link invariants, generalizing the Jones polynomial, which are now called *Khovanov homology*. The papers [19, 27, 28, 115, 116] related this categorification to the category $\mathcal{O}$.

J. Chuang and R. Rouquier developed in [30] a general framework for (rather strong) categorification of finite-dimensional $\mathfrak{sl}_2$-modules and apply it to construct certain derived equivalences for blocks of the symmetric group $\mathbf{S}_n$, thus proving *Broué's abelian defect group conjecture* for symmetric groups.

7.5.3

The papers [30, 91] suggest that there are two further natural levels of categorification for (simple) finite-dimensional $\mathfrak{sl}_2$-modules. The first level will be simply called *categorification* and is defined in terms of defining relations of the algebra $U(\mathfrak{g})$. Let $n \in \mathbb{N}$ and $\mathbf{x}$ be an admissible basis of $\mathbf{V}^{(n)}$. A weak categorification $(B\text{-mod}, E, F)$ of $(\mathbf{V}^{(n)}, \mathbf{x})$ will be called a *categorification* provided that there is an isomorphism of functors as follows:

$$EFE \oplus EFE \cong E \oplus E \oplus FEE \oplus EEF. \qquad (7.8)$$

For example, any weak categorification of $\mathbf{V}^{(1)}$ is, in fact, a categorification.

In [30, 48] it is shown that the categorification $(B\text{-mod}, E, F)$ of $(\mathbf{V}^{(n)}, \hat{\mathbf{w}})$, given by Theorem 7.37, has the following property: For $i = 0, 1, \ldots, n$ let ID_i denote the identity functor on B_i-mod, then there are isomorphisms of functors as follows:

$$EF \cong FE \oplus \underbrace{\mathrm{ID}_i \oplus \cdots \oplus \mathrm{ID}_i}_{n - 2i \text{ summands}}, i \leq \tfrac{n}{2};$$

$$FE \cong EF \oplus \underbrace{\mathrm{ID}_i \oplus \cdots \oplus \mathrm{ID}_i}_{2i - n \text{ summands}}, i > \tfrac{n}{2}. \qquad (7.9)$$

It is easy to see that (7.9) implies (7.8) and hence $(B\text{-mod}, E, F)$ is a categorification of $(\mathbf{V}^{(n)}, \hat{\mathbf{w}})$.

The ultimate level of categorification is the one, suggested in [30]. We call it *strong categorification*. The definition is rather long, technical and requires the introduction of several new objects, so we will not formulate it here. Roughly speaking, a strong categorification $(B\text{-mod}, \mathrm{E}, \mathrm{F})$ of $(\mathbf{V}^{(n)}, \mathbf{x})$ is a weak categorification with a certain fixed action of some affine Hecke algebra on powers of the functors E and F by natural transformations. It is shown in [30] that the categorification $(B\text{-mod}, \mathrm{E}, \mathrm{F})$ of $(\mathbf{V}^{(n)}, \hat{\mathbf{w}})$, given by Theorem 7.37, is a strong categorification. A huge advantage of this notion is that the minimal model of such categorification (that is, the one for which the simple module, representing the highest weight, is also projective) is unique. In particular, the categorification given by Theorem 7.37 is the unique minimal strong categorification of $\mathbf{V}^{(n)}$. This also means that no strong categorification exists for any other choice of a basis in $\mathbf{V}^{(n)}$.

In [30], the authors also propose a more advanced notion for morphism of categorifications, which additionally requires existence of some natural transformations and commutativity of certain diagrams.

7.5.4

In [48] the categorification of $(\mathbf{V}^{(n)}, \hat{\mathbf{w}})$, given by Theorem 7.37, is called *geometric* categorification. The reason for this is that the algebras appearing in this categorification have geometric interpretation. The coinvariant algebra $\mathbf{C}_n$ is the cohomology algebra of the flag variety of all complete flags in $\mathbb{C}^n$. The subalgebra $\mathbf{C}_n^i$ is the cohomology algebra of the Grassmannian, corresponding to dimension i. The subalgebra $\mathbf{C}_n^{i,i+1}$ is the cohomology algebra of the partial flag variety corresponding to flags with components of dimensions i and $i+1$. Many properties of these algebras (for example, those mentioned in Theorem 7.35) can be explained and proved geometrically. We refer the reader to [56] for details.

7.5.5

The paper [48] also gives a categorification of simple finite-dimensional modules over the quantum universal enveloping algebra $U_q(\mathfrak{sl}_2)$.

7.5.6

The problem of classification of all possible naïve categorifications, weak categorifications and categorifications of $(\mathbf{V}^{(n)}, \mathbf{x})$ is open.

7.5.7

One might observe that our definition of categorification of a module M over an associative algebra A, formulated in Section 7.1, does not really reflect the algebra structure of A. There is a natural way to take the algebra structure of A into account using the following definition: A *proper categorification* of the A-module M is a tuple $(\mathcal{C}, \varphi, \mathcal{F}, \psi)$, where $\mathcal{C}$ is an abelian category, $\varphi : M \to [\mathcal{C}]^{\Bbbk}$ is a fixed isomorphism, $\mathcal{F}$ is an additive monoidal category of exact functors on $\mathcal{C}$ (where the monoidal structure is given by composition of functors) and $\psi : A \to [\mathcal{F}]^{\Bbbk}$ is a fixed isomorphism from A to the split Grothendieck ring of $\mathcal{F}$; such that $(\mathcal{C}, \varphi, \{\psi(a_i) : i \in I\})$ is a naïve categorification of the A-module M. One easily defines in this context also the notion of homomorphisms of categorifications.

There are some natural examples of proper categorification; for instance, the categorification of the right regular representation of the symmetric group $\mathbf{S}_n$ via the action of projective functors on the regular block of the category $\mathcal{O}$ for $\mathfrak{sl}_n$ (see [91]). Proper categorifications of certain modules over some universal enveloping algebras were recently constructed by M. Khovanov and A. Lauda in [76]. However, there is still no reasonably general theory for any kind of categorification.

7.6 Additional exercises

Exercise 7.39. Let $\Bbbk$, A and M be as in Section 7.1 and $(\mathcal{C}, \varphi, \{F_i : i \in I\})$ be naïve categorifications of M. Let $\mathcal{D}$ be an abelian subcategory of $\mathcal{C}$ such that the inclusion of $\mathcal{D}$ to $\mathcal{C}$ is exact. Assume that $\mathcal{D}$ is invariant with respect to all F_i, $i \in I$. Show that the action of F_i, $i \in I$, on $\mathcal{D}$ defines a naïve categorification of some submodule of the module M.

Exercise 7.40. Let $\Bbbk$, A and M be as in Section 7.1. Assume further that $(\mathcal{C}, \varphi, \{F_i : i \in I\})$ and $(\mathcal{D}, \psi, \{G_i : i \in I\})$ are two naïve categorifications of M. Finally, let $\Phi : \mathcal{C} \to \mathcal{D}$ be a homomorphism of naïve categorifications. Denote by $\mathcal{C}'$ the full subcategory of $\mathcal{C}$, which consists of all objects X such that $\Phi(X) = 0$.

(a) Show that the category $\mathcal{C}'$ is an abelian subcategory of $\mathcal{C}$ and that the inclusion of $\mathcal{C}'$ to $\mathcal{C}$ is exact.
(b) Show that the category $\mathcal{C}'$ is stable under the action of all F_i, $i \in I$.

(c) Show that by restricting the functors F_i, $i \in I$, to $\mathcal{C}'$, one obtains a naïve categorification of some submodule of the module M.

Exercise 7.41. Let $\mathcal{C}$ be an abelian category. A nonempty full subcategory $\mathcal{D}$ of $\mathcal{C}$ is called a *Serre subcategory* provided that for any short exact sequence

$$0 \to X \to Y \to Z \to 0$$

in $\mathcal{C}$ we have $Y \in \mathcal{D}$ if, and only if, $X, Z \in \mathcal{D}$. Let $\mathcal{D}$ be a Serre subcategory of $\mathcal{C}$.

(a) Show that $\mathcal{D}$ is abelian and that the inclusion from $\mathcal{D}$ to $\mathcal{C}$ is exact.
(b) Consider the category $\mathcal{C}/\mathcal{D}$, which has the same objects as $\mathcal{C}$, and for $X, Y \in \mathcal{C}$ we have

$$\mathcal{C}/\mathcal{D}(X, Y) = \varinjlim \mathcal{C}(X', Y/Y'),$$

where $X' \subset X$ and $Y' \subset Y$ are such that $X/X' \in \mathcal{D}$ and $Y' \in \mathcal{D}$. Show that $\mathcal{C}/\mathcal{D}$ is an abelian category.
(c) Show that the identity functor on $\mathcal{C}$ gives rise to the canonical exact functor $Q : \mathcal{C} \to \mathcal{C}/\mathcal{D}$.
(d) Let $\mathcal{X}$ be any abelian category and $F : \mathcal{C} \to \mathcal{X}$ be any exact functor such that $F M = 0$ for any $M \in \mathcal{D}$. Show that there exists a unique exact functor $\overline{F} : \mathcal{C}/\mathcal{D} \to \mathcal{X}$ such that $F = \overline{F}Q$.

Exercise 7.42. Let B be a finite-dimensional associative $\mathbb{C}$-algebra.

(a) Show that for every Serre subcategory $\mathcal{D}$ of B-mod there exists an idempotent $e \in B$ such that $\mathcal{D}$ coincides with the full subcategory of B-mod, consisting of all modules M such that $eM = 0$.
(b) Show that B-mod has only finitely many Serre subcategories.
(c) Let $\mathcal{D}$ be a Serre subcategory of B-mod and e be the corresponding idempotent, given by (a). Show that $\mathcal{D}$ is equivalent to $B/(BeB)$-mod.
(d) Let $\mathcal{D}$ be a Serre subcategory of B-mod and e be the corresponding idempotent, given by (a). Show that B-mod$/\mathcal{D}$ is equivalent to eBe-mod.

Exercise 7.43. Let $\Bbbk$, A and M be as in Section 7.1. Assume that $(\mathcal{C}, \varphi, \{F_i : i \in I\})$ is a naïve categorification of M. Let $\mathcal{D}$ be a Serre subcategory of $\mathcal{C}$, stable with respect to the action of all F_i, $i \in I$.

(a) Show that the functors F_i, $i \in I$, induce well-defined exact endofunctors on $\mathcal{C}/\mathcal{D}$.

(b) Show that, using (a), the category $\mathcal{C}/\mathcal{D}$ gives rise to a naïve categorification of some quotient of the module M.

Exercise 7.44. Let $\Bbbk$ and A be as in Section 7.1 and M_1, M_2 be two A-modules. Assume that $(\mathcal{C}, \varphi, \{F_i : i \in I\})$ is a naïve (weak) categorification of M_1 and $(\mathcal{D}, \psi, \{G_i : i \in I\})$ is a naïve (weak) categorification of M_2. Show that $(\mathcal{C} \oplus \mathcal{D}, \varphi \oplus \psi, \{F_i \oplus G_i : i \in I\})$ is a naïve (weak) categorification of $M_1 \oplus M_2$.

Exercise 7.45. Let $\Bbbk$, A and M be as in Section 7.1. Assume that $(B\text{-mod}, \varphi, \{F_i : i \in I\})$ is a naïve categorification of M, where B is a finite-dimensional associative $\mathbb{C}$-algebra. Show that for any local $\mathbb{C}$-algebra D there is an isomorphism $\varphi' : M \to [B \otimes D\text{-mod}]^{\Bbbk}$ such that the tuple

$$(B \otimes D\text{-mod}, \varphi', \{F_i \otimes \mathrm{ID}_{D\text{-mod}} : i \in I\})$$

is a naïve categorification of M.

Exercise 7.46. Let $n \in \mathbb{N}$, $\mathbf{x}$ be an admissible basis of $\mathbf{V}^{(n)}$, and $(B\text{-mod}, \mathrm{E}, \mathrm{F})$ be a weak categorification of $(\mathbf{V}^{(n)}, \mathbf{x})$. Show that for any local $\mathbb{C}$-algebra D the construction from Exercise 7.45 gives rise to a weak categorification of the pair $(\mathbf{V}^{(n)}, \mathbf{x})$.

Exercise 7.47. For $k \in \mathbb{N}$ set $D_k = \mathbb{C}[x]/(x^k)$. If $k|m$, we consider D_k as a subalgebra of D_m via the embedding $x \mapsto x^{m/k}$. Set $B_i = D_{\binom{n}{i}}$, $B_i' = D_{\frac{n!}{i!(n-i-1)!}}$ and for $i \neq 0$ consider the following functors:

$$F_i = \mathrm{Res}_{B_i}^{B_{i-1}'} \circ \mathrm{Ind}_{B_{i-1}}^{B_{i-1}'} : B_{i-1}\text{-mod} \to B_i\text{-mod},$$
$$E_i = \mathrm{Res}_{B_{i-1}}^{B_{i-1}'} \circ \mathrm{Ind}_{B_i}^{B_{i-1}'} : B_i\text{-mod} \to B_{i-1}\text{-mod}.$$

Set

$$B = \bigoplus_{i=0}^{n} B_i, \qquad F = \bigoplus_{i=1}^{n} F_i, \qquad E = \bigoplus_{i=1}^{n} E_i$$

and use the convention that $FB_n\text{-mod} = 0$ and $EB_0\text{-mod} = 0$.

Show that the tuple $(B\text{-mod}, E, F)$, given by the above, is a weak categorification of $(\mathbf{V}^{(n+1)}, \hat{\mathbf{w}})$.

Exercise 7.48. Generalize Exercise 7.47 to cover the case of an arbitrary admissible basis $\mathbf{x}$ of every $\mathbf{V}^{(n)}$.

Exercise 7.49. Show that the weak categorification of $(\mathbf{V}^{(n+1)}, \hat{\mathbf{w}})$, given by Exercise 7.47, is not equivalent to the weak categorification of $(\mathbf{V}^{(n+1)}, \hat{\mathbf{w}})$, given by Theorem 7.37, in the general case.

Exercise 7.50 ([30]). Let $D = \mathbb{C}[x]/(x^2)$. Show that the following picture defines a weak categorification of $(\mathbf{V}^{(3)}, \hat{\mathbf{w}})$:

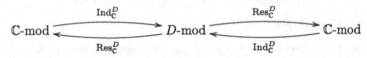

Exercise 7.51.

(a) Define the notion of a weak categorification for any finite-dimensional $U(\mathfrak{sl}_2)$-module.

(b) Let $D = \mathbb{C} \oplus \mathbb{C}$. We consider $\mathbb{C}$ as the unital subalgebra of D. Show that the following picture defines a weak categorification of $\mathbf{V}^{(3)} \oplus \mathbf{V}^{(1)}$:

$$\mathbb{C}\text{-mod} \underset{\mathrm{Res}_\mathbb{C}^D}{\overset{\mathrm{Ind}_\mathbb{C}^D}{\rightleftarrows}} D\text{-mod} \underset{\mathrm{Ind}_\mathbb{C}^D}{\overset{\mathrm{Res}_\mathbb{C}^D}{\rightleftarrows}} \mathbb{C}\text{-mod}$$

Exercise 7.52. Let $n \in \mathbb{N}$, $\mathbf{x}$ be an admissible basis of $\mathbf{V}^{(n)}$, and $(B\text{-mod}, \mathrm{E}, \mathrm{F})$ be a naïve categorification of $(\mathbf{V}^{(n)}, \mathbf{x})$ such that B is semi-simple. Show that $(B\text{-mod}, \mathrm{E}, \mathrm{F})$ is isomorphic to the naïve categorification given by Example 7.15.

Appendix A

Answers and hints to exercises

1.12. Hint: Any subspace of this module is a submodule.

1.16. Hint: Use the relations (1.3).

1.25. Hint: Use Theorem 1.22.

1.42. Hint: The self-adjoint operator E is not diagonalizable.

1.55. Answer: $\mathrm{Mat}_{3\times 3}(\mathbb{C}) \cong \mathbf{V}^{(1)} \oplus \mathbf{V}^{(1)} \oplus \mathbf{V}^{(2)} \oplus \mathbf{V}^{(2)} \oplus \mathbf{V}^{(3)}$ as $\mathfrak{g}$-module.

1.59. Hint: Use 1.7.3.

1.60. Hint: Consider the two-dimensional $\mathfrak{gl}_2$-module $\mathbb{C}^2$ on which $\mathfrak{sl}_2$ acts trivially and the identity matrix acts via $\begin{pmatrix} 1 & 1 \\ 0 & 1 \end{pmatrix}$.

1.61. (a) Hint: Use the Jordan decomposition for B. (b) Hint: Use the Jordan decomposition for A. Let λ be an eigenvalue of A. Consider $V(\lambda) = \{v \in V : (A - \lambda)^k(v) = 0 \text{ for some } k\}$. For $v \in V(\lambda)$ use $AB - BA = A^2$ to show by induction on the minimal k such that $(A - \lambda)^k(v) = 0$ that $B(v) \in V(\lambda)$ and hence that $BV(\lambda) \subset V(\lambda)$. For example, if $Av = \lambda v$, then $(AB - BA)(v) = A^2(v)$ implies $(A - \lambda)B(v) = A^2(v) \in V(\lambda)$. As $A - \lambda$ acts bijectively on $V(\mu)$, $\mu \neq \lambda$, it follows that $B(v) \in V(\lambda)$. Then use that the trace of any commutator is always zero to deduce that $\lambda = 0$. (c) Hint: This is a special case of *Kleineke–Shirokov's Theorem* (see [54]).

1.63. Hint: Use Exercise 1.54.

1.67. (b) Hint: Use the basis $\{v_i\}$ from (1.9) and write the matrix of the form in this basis with unknown coefficients. Then use the definition of $\natural$ to obtain linear relations for these unknown coefficients. The resulting matrix of the form will have a constant that alternates in sign on the second diagonal and all other entries equal to zero.

1.70. Hint: Use Theorem 1.50.

1.71. (a) See Proposition 3.4.2 in [20]. (c) Hint: Use induction on the length of α.

1.72. Hint: Use Sections 3.4 and 7.1 from [20].

1.73. Hint: Use Theorem 1.22 and Corollary 1.34.

2.1 Hint: Use Lemma 1.30.

2.2 Hint: Use Lemma 1.30.

2.11. Hint: Use Exercise 2.1 and definitions.

2.16 Hint: Use induction on the degree of the monomial.

2.18 Hint: Observe that we have only finitely many monomials for each degree. Moreover, the number of the "old" and "new" standard monomials of each degree coincide. Now show that for every n, every "old" standard monomial of degree at most n can be written as a linear combination of "new" standard monomials of degree at most n, and vice versa.

2.27 Answer: For example $x = \mathbf{e} + \mathbf{h}$.

2.45 Hint: Show that this defines the structure of a $U(\mathfrak{g})$-module and use Proposition 2.7.

2.46 Hint: Show that the linear span of all standard monomials of positive degree in $U(\mathfrak{g})$ is a submodule, which does not have any complement.

2.48 Answer: $\dim U(\mathfrak{g})^{(i)}/U(\mathfrak{g})^{(i-1)} = \binom{i+2}{2}$.

2.49 Hint: Use Lemma 2.22.

2.51 Hint: Consider the actions of u, v and 1 on the one-dimensional $U(\mathfrak{g})$-module.

2.54 Hint: Use the PBW Theorem.

2.56 Hint: For example, take as I_i the left ideal generated by e^i and use that $U(\mathfrak{g})$ is a domain to show that $e^i \notin I_{i+1}$.

2.57 Hint: Consider the image of $U(\mathfrak{g})$ in the algebra of all linear operators on $\mathbf{V}^{(n)}$.

2.58 Hint: To show that $V \cong \bigoplus_{i \in \mathbb{N}} \mathbf{V}^{(2i-1)}$ consider the intersections of V with every $U(\mathfrak{g})^{(k)}$.

3.2 Hint: Use Example 3.1 and Weyl's Theorem.

3.4 Hint: Use that φ commutes with the action of H.

3.9 Hint: Use computation from the proof of Proposition 3.8.

3.10 Hint: Use Exercise 3.9.

3.17 Hint: Show that any homomorphism between Verma modules is uniquely determined by the image of the generator.

3.31 Hint: Check that the linear operator E does not act injectively on the quotient $\mathbf{V}(n-1+2\mathbb{Z}, n^2)/M(-n-1)$.

3.38 Hint: Use Theorem 3.32.

3.39 Hint: Consider the action of the endomorphism on some weight generator.

3.40 Hint: Multiply the last and the second formulae from (3.9) with f^{-1} from both sides.

3.41 Hint: Use induction on i.

3.50 Hint: Show that $B_0 V = 0$ for any finite-dimensional $\mathfrak{g}$-module V.

3.51 Hint: Use that the adjoint action of f on $U(\mathfrak{g})$ is locally nilpotent.

3.56 Hint: Use Theorem 2.33.

3.57 Hint: Show that the functor of taking a weight subspace (of a fixed weight) is exact on the category of weight modules.

3.72 Hint: Compare the dimensions of V and V^ω.

3.74 Hint: Use Exercise 1.38.

3.78 Hint: Construct an injective homomorphism from $\mathbf{V}(\xi, \tau)$ to V.

3.79 Hint: Use Proposition 3.77(ii).

3.88 Hint: Argue similarly to the proof of Theorem 3.81(i).

3.89 Hint: Use that $U(\mathfrak{g})$ is a domain.

3.90 Hint: Use Lemma 3.5.

3.91 Hint: For example $U(\mathfrak{g}) \otimes_{\mathbb{C}[h,c]} V$, where V is a two-dimensional $\mathbb{C}[h, c]$-module on which c acts as a scalar and h acts as a nontrivial Jordan cell.

3.92 Hint: Take the module, constructed in Exercise 3.91, and show that it has a filtration whose subquotients are weight modules.

3.93 Hint: For example $\oplus_{i \in \mathbb{N}} \overline{M}(2i)$.

3.94 Hint: Use Theorem 2.33.

3.95 Hint: Use the universal property of Verma modules.

3.97 Hint: Use the universal property of Verma modules.

3.98 Hint: Use that E acts injectively on $\overline{M}(\mu)$, while it acts locally nilpotent on $M(\lambda)$.

3.99 Hint: The module $\tilde{\mathbf{V}}(\xi, \tau)$ is generated by $\tilde{\mathbf{V}}(\xi, \tau)_\lambda$ by definition.

3.101 Hint: Look at the kernel of F on $\overline{M}(-n+1)^\circledast$.

3.102 Hint: The assumption is satisfied for any $\mathbf{V}(\xi, \tau)$.

3.105 Hint: The category $\overline{\mathfrak{W}}^\xi$ contains simple modules of three or four possible kinds: those whose support is ξ, those whose support is a "one half" of ξ, that is a ray in one of two directions, or, sometimes, those whose support is a segment from ξ (finite-dimensional modules). Assume that our module has infinite length. At least one type of simple modules should occur infinitely many times in the composition series. Adding up their supports one can show that the dimensions of weight spaces are not uniformly bounded.

3.106 Hint: Use that every module $V \in \mathfrak{X}$ is generated by V_λ for any λ such that E^i acts injectively on V_λ for all $i \in \mathbb{N}$. Then use arguments similar to those used in the proof of Theorem 3.58.

3.109 Answer: The category from Exercise 3.108(a) has one indecomposable object; the category from Exercise 3.108(b) has four indecomposable objects; the category from Exercise 3.108(c) has nine indecomposable objects. All these objects are multiplicity-free.

3.110 Hint: Use Theorem 3.81(i) and exactness of $\mathbf{V}^{(n)} \otimes \_$.

3.111 Hint: Use Propositions 3.77 and 3.80, and exactness of $\mathbf{V}^{(n)} \otimes \_$.

3.112 Hint: Realize $N(-n-1)$ as a submodule of $\mathbf{W}(-n-1+2\mathbb{Z}, n^2)$.

3.113 Hint: Use Theorem 3.81.

3.115 Hint: Use that the adjoint action of $\mathfrak{g}$ on $U(\mathfrak{g})$ is locally nilpotent.

3.116 Hint: Use Theorem 3.81.

3.118 Hint: Use that $W^\circledast \cong W$.

3.119 Hint: First prove this for simple modules, then use Weyl's theorem.

3.120 Hint: Use Exercise 3.28 and the definition of $\circledast$.

3.121 Hint: Use Exercise 3.28.

4.5 Hint: Argue similarly to the proof of Theorem 4.2.

4.6 Hint: Argue similarly to the proof of Theorem 4.2.

4.10 Hint: Show that every endomorphism of L is algebraic over $\mathbb{C}$. To show this use the same arguments as in the proof of Theorem 4.7(i).

4.14 Hint: Consider powers of the ideal $\mathcal{I}_\lambda$, $\lambda \in \mathbb{C}$.

4.29 Hint: Show first that $c - (\lambda+1)^2$ annihilates $M(\lambda)^\circledast$ and then use Theorem 4.15.

4.30 Hint: Use Proposition 3.77(iv).

4.31 Hint: Follow the proof of Theorem 4.28.

4.32 Hint: Follow the proof of Theorem 4.28.

4.33 Hint: Follow the proof of Theorem 4.7.

4.34 Hint: Use Theorem 4.28.

4.35 Hint: Use Theorem 4.19.

4.36 Hint: Use Theorem 4.19.

4.37 Hint: Use Theorem 4.19.

4.40 Hint: Use Theorem 4.19.

4.42 Hint: Use Exercise 4.38.

5.6 Hint: Use Exercise 3.92.

5.7 Hint: First reduce the problem to the case where v is a weight vector. Then use the arguments from the proof of Proposition 5.5.

5.9 Hint: Use the proof of Proposition 5.8.

5.10 Hint: Use the proof of Proposition 5.8.

5.15 Hint: Use Proposition 3.55.

5.18 Hint: Use that the duality ⊛ is a contravariant self-equivalence and hence interchanges projective and injective objects.

5.22 Hint: As in the proof of Corollary 5.21 show that for every injective $I \in \mathcal{O}$ the functor $\mathrm{Hom}_{\mathfrak{g}}(\_, V \otimes I)$ is exact.

5.24 Hint: Use the same arguments as in the proof of Theorem 5.16.

5.25 Hint: Use that the duality ⊛ is a contravariant self-equivalence and hence swaps projective and injective objects.

5.26 Hint: Use ⊛.

5.35 (d) Hint: Use the usual adjunction between the functors Hom and $\otimes$.

5.42 Hint: Construct an isomorphism from D to this matrix algebra mapping the generator **a** of D to the matrix $\begin{pmatrix} 0\,0\,0 \\ 0\,0\,1 \\ 0\,0\,0 \end{pmatrix}$ and the generator **b** of D to the matrix $\begin{pmatrix} 0\,1\,0 \\ 0\,0\,0 \\ 0\,0\,0 \end{pmatrix}$.

5.45 Hint: Apply ⊛ to the claim of Proposition 5.43.

5.48 Hint: Apply ⊛ to the claim of Proposition 5.47.

5.50 Hint: Argue similarly to the proof of Lemma 5.32.

5.57 Hint: The sum of non-zero homogeneous maps of different degrees is not homogeneous.

5.62 Hint: Use arguments, similar to those used in the proof of Theorem 5.31(iii).

5.68 Hint: Apply ⊛ to the claim of Proposition 5.65.

5.72 Hint: Use Proposition 5.65.

5.96 Hint: Compute the value of both $\vartheta^i_i \circ \vartheta^i_i$ and $\vartheta^i_i \oplus \vartheta^i_i$ on $M(i)$ and use Corollary 5.95.

5.97 (a) Hint: Use Figure 5.2. (b) Hint: Use Exercise 3.110 or Exercise 5.92.

5.104 (a) Hint: Use the functor ⊚ ∘ ⊛.

5.105 (b) Hint: Use the fact that $\mathrm{supp}(N) \subset i - 2\mathrm{N}$ for all $N \in \mathcal{O}_i$.

5.107 Hint: Use that this property is additive and check it on all indecomposable modules. Alternatively, use that every module is a quotient of a projective module and check the property on all projective modules.

5.108 Hint: Argue similarly to Sections 5.1–5.3.

5.109 Hint: Use that all these categories are fully additive and check everything for indecomposable modules.

5.111 Answer:

$$
\begin{array}{lccc}
\text{position:} & -1 & 0 & 1 \\
L(\mathrm{p}): & 0 & \to T(\mathrm{p}) \to & 0 \\
L(\mathrm{q}): & 0 \to T(\mathrm{p})\langle -1\rangle \to & T(\mathrm{q}) \to T(\mathrm{p})\langle 1\rangle & \to 0 \\
P(\mathrm{p}): & 0 & \to T(\mathrm{q}) \to & 0 \\
P(\mathrm{q}): & 0 & \to T(\mathrm{q}) \to T(\mathrm{p})\langle 1\rangle & \to 0 \\
I(\mathrm{q}): & 0 \to T(\mathrm{p})\langle -1\rangle \to & T(\mathrm{q}) \to & 0
\end{array}
$$

5.114 Hint: Prove that the center of $\mathcal{O}_0$ is isomorphic to the center of D and then use 5.113.

5.118 Hint: Prove this for indecomposable projective modules first.

5.119 Hint: Apply both sides to $L(0)$ and show that the cohomologies of the resulting complexes are concentrated in different positions.

5.120 Hint: Apply both sides to $L(0)$ and show that the cohomologies of the resulting complexes are concentrated in different positions.

5.121 (a) Hint: Use that $V^{\circledast} \cong V$.

5.122 See, for example, [73, 91].

5.123 Hint: Use 5.122 and the fact that $\mathrm{Z} \cong \circledast \circ \hat{\mathrm{Z}} \circ \circledast$.

5.124 Hint: Use that Z commutes with both ID and ϑ_0^0.

5.125 Hint: Use 5.124 and the facts that $\mathrm{Z} \cong \circledast \circ \hat{\mathrm{Z}} \circ \circledast$ and $\mathrm{C} \cong \circledast \circ \mathrm{K} \circ \circledast$.

5.126 Hint: Use that T commutes with both ID and ϑ_0^0.

5.131 (c) Hint: This is not obvious only in the case $\lambda \in \mathbb{N}_0$. In this case use induction on n.

6.6 Hint: Use the fact that $\mathbb{C}[h]$ is preserved by $\square$.

6.12 Hint: Use that any $\alpha \in \mathrm{A}$ such that $\mathrm{n}(\alpha) = 0$ is invertible.

6.16 Hint: Use Proposition 6.15 and the observation that the condition $L_\alpha \cong L_\beta$ is symmetric.

6.17 Hint: See Chapter 3 in [61].

6.19 Hint: Take $g(h)$ to be the product of all denominators in all non-zero coefficients of α.

6.28 Hint: Let $\overline{L}$ be the simple A-module such that L is the simple $U(\mathcal{I}_\lambda)$-socle of $\overline{L}$. Take any non-zero $v \in L \subset \overline{L}$ and consider the epimorphism $\psi : \mathrm{A} \twoheadrightarrow \overline{L}$, which sends 1 to v. Then pick $\alpha \in U(\mathcal{I}_\lambda)$ such that α generates the kernel of ψ in A.

6.38 Hint: One can even choose $x, y \in \mathbf{B}_1$.

6.45 Hint: Show that $M(\alpha)$ surjects onto each of these Verma modules and hence on their direct sum and then use Lemma 6.35.

6.51 Hint: The left regular $U(\mathfrak{g})$-module, which has the trivial module as a quotient.

6.52 Hint: Use Exercise 6.5 and the fact that $N \otimes \_$ is a self-adjoint functor preserving the category of weight modules.

6.54 Hint: Argue similarly to the proof of Theorem 6.18.

6.57 Hint: Show that the dimension of any simple A-module over $\mathbb{C}$ is uncountable.

6.58 Hint: Show that the dimension of any simple A-module over $\mathbb{C}$ is uncountable.

6.60 Hint: Use that $\mathbb{C}[h]$ does not intersect the image of e.

6.61 Hint: Take an infinite direct sum of one-dimensional $\mathfrak{g}$-modules.

6.62 Hint: Use the resolution (6.14).

6.63 Hint: Use 6.62 and argue similarly to the proof of Theorem 6.40.

7.3 Hint: The element $a \in A$ corresponds to the element $\mathrm{id} + (1,2) \in \mathbb{C}S_2$.

7.9 Hint: Every column of X has a non-zero entry if and only if $Xv \neq 0$ for any non-zero vector v with non-negative coordinates.

7.12 Hint: From (1.9) it follows that if $v \in \mathbf{V}^{(n)}$ is a non-zero weight vector of weight λ, then $EF(v) = k_\lambda v$ and $FE(v) = m_\lambda v$ for some non-negative integers k_λ and m_λ, which depend only on the weight of v. If $\mathbf{x}$ and $\mathbf{y}$ are admissible bases of $\mathbf{V}^{(n)}$, then, up to a global scalar, the transformation from $\mathbf{x}$ to $\mathbf{y}$ is given in terms of (products of) divisors of k_λ's and m_λ's.

7.18 Hint: The identity functor is not adjoint to the direct sum of two identity functors.

7.27 Hint: Use that for a local algebra the dimensions of indecomposable projective and injective modules coincide, and that a functor, which is both left and right adjoint to an exact functor, preserves both the additive category of projective modules and the additive category of injective modules.

7.28 Hint: Use Exercise 7.23.

7.38 Hint: Use arguments similar to those used in the proof of Lemma 7.20.

7.42 (a) Hint: Let $L_1, \ldots, L_k$ be a complete list of pairwise non-isomorphic simple B-modules and $\mathbf{e}_1, \ldots, \mathbf{e}_k$ be the corresponding complete list of pairwise orthogonal primitive idempotents. Then $\mathbf{e}$ is the sum of all those $\mathbf{e}_i$'s, such that $\mathbf{e}_i M = 0$ for all $M \in \mathcal{D}$. (d) Hint: Show that $B\text{-mod}/\mathcal{D}$ is equivalent to $\mathrm{End}_B(B\mathbf{e})$-mod.

7.46 Hint: Use that the identity functor is both left and right adjoint to itself.

7.47 Hint: Use the same arguments as in the proof of Theorem 7.37.

7.49 Hint: Show that the corresponding associative algebras are not isomorphic.

7.52 Hint: Describe first all exact functors between semi-simple categories.

Bibliography

[1] Andersen, H. and Stroppel, C. (2003). Twisting functors on $\mathcal{O}$, *Representation Theory* **7**, pp. 681–699.

[2] Amitsur, S. (1958). Commutative linear differential operators, *Pacific J. Math.* **8**, pp. 1–10.

[3] Arkhipov, S. (2004). Algebraic construction of contragradient quasi-Verma modules in positive characteristic, *Adv. Stud. Pure Math.* **40**, pp. 27–68.

[4] Arnal, D. and Pinczon, G. (1973). Idéaux à gauche dans les quotients simples de l'algèbre enveloppante de $sl(2)$, *Bull. Soc. Math. France* **101**, pp. 381–395.

[5] Arnal, D. and Pinczon, G. (1974). On algebraically irreducible representations of the Lie algebra $sl(2)$, *J. Mathematical Phys.* **15**, pp. 350–359.

[6] Barthel, G., Brasselet, J.-P., Fieseler, K.-H. and Kaup, L. (2007). Hodge-Riemann relations for polytopes: a geometric approach, *Singularity theory*, (World Sci. Publ., Hackensack, NJ), pp. 379–410.

[7] Bass, H. (1968). *Algebraic K-theory*, (W. A. Benjamin, Inc.)

[8] Bavula, V. (1990). Classification of simple $sl(2)$-modules and the finite-dimensionality of the module of extensions of simple $sl(2)$-modules, *Ukrain. Mat. Zh.* **42**, 9, pp. 1174–1180.

[9] Bavula, V. (1991). Finite-dimensionality of Ext^n and Tor_n of simple modules over a class of algebras, *Funkt. Anal. i Prilozhen.* **25**, 3, pp. 80–82.

[10] Bavula, V. (1992). Generalized Weyl algebras and their representations, *Algebra i Analiz* **4**, 1, pp. 75–97.

[11] Bavula, V. and Bekkert, V. (2000). Indecomposable representations of generalized Weyl algebras, *Comm. Algebra* **28**, 11, pp. 5067–5100.

[12] Beilinson, A. and Bernstein, J. (1981). Localisation de $\mathfrak{g}$-modules, *C. R. Acad. Sci. Paris Ser. I Math.* **292**, 1, pp. 15–18.

[13] Beilinson, A., Ginzburg, V. and Soergel, W. (1996). Koszul duality patterns in representation theory. *J. Amer. Math. Soc.* **9**, 2, pp. 473–527.

[14] Benkart, G. and Roby, T. (1998). Down-up algebras, *J. Algebra* **209**, 1, pp. 305–344.

[15] Bergman, G. (1978). The diamond lemma for ring theory, *Adv. Math.* **29**, 2, pp. 178–218.

[16] Bernstein, J. and Gelfand, S. (1980). Tensor products of finite and infinite-dimensional representations of semisimple Lie algebras, *Compositio Math.* **41**, 2, pp. 245–285.

[17] Bernstein, I., Gelfand, I. and Gelfand, S. (1971). Structure of representations that are generated by vectors of higher weight, *Funckcional. Anal. i Prilozen.* **5**, 1, pp. 1–9.

[18] Bernstein, I., Gelfand, I. and Gelfand, S. (1976). A certain category of $\mathfrak{g}$-modules, *Funckcional. Anal. i Prilozen.* **10**, 2, pp. 1–8.

[19] Bernstein, J., Frenkel, I. and Khovanov, M. (1999). A categorification of the Temperley–Lieb algebra and Schur quotients of $U(\mathfrak{sl}_2)$ via projective and Zuckerman functors, *Selecta Math. (N.S.)* **5**, 2, pp. 199–241.

[20] Björner, A. and Brenti, F. (2005). *Combinatorics of Coxeter groups, Graduate Texts in Mathematics*, Vol. **231** (Springer).

[21] Block, R. (1979). Classification of the irreducible representations of $\mathfrak{sl}(2, C)$, *Bull. Amer. Math. Soc. (N.S.)* **1**, 1, pp. 247–250.

[22] Block, R. (1981). The irreducible representations of the Lie algebra $\mathfrak{sl}(2)$ and of the Weyl algebra, *Adv. in Math.* **39**, 1, pp. 69–110.

[23] Bourbaki, N. (1998). *Lie groups and Lie algebras. Chapters 1–3, Elements of Mathematics* (Springer).

[24] Bouwer, I. (1968). Standard representations of simple Lie algebras, *Canad. J. Math.* **20**, pp. 344–361.

[25] Britten, D. and Lemire, F. (1987). A classification of simple Lie modules having a 1-dimensional weight space, *Trans. AMS* **299**, 2, pp. 683–697.

[26] Brylinski, J.-L. and Kashiwara, M. (1981). Kazhdan–Lusztig conjecture and holonomic systems, *Invent. Math.* **64**, 3, pp. 387–410.

[27] Brundan, J. (2008). Symmetric functions, parabolic category $\mathcal{O}$, and the Springer fiber, *Duke Math. J.* **143**, 1, pp. 41–79.

[28] Brundan, J. (2008). Centers of degenerate cyclotomic Hecke algebras and parabolic category $\mathcal{O}$, *Represent. Theory* **12**, pp. 236–259.

[29] Carlin, K. (1986). Extensions of Verma modules, *Trans. Amer. Math. Soc.* **294**, 1, pp. 29–43.

[30] Chuang, J. and Rouquier, R. (2008). Derived equivalences for symmetric groups and $\mathfrak{sl}_2$-categorification, *Ann. of Math. (2)* **167**, 1, pp. 245–298.

[31] Cline, E., Parshall, B. and Scott, L. (1988). Finite-dimensional algebras and highest weight categories, *J. Reine Angew. Math.* **391**, pp. 85–99.

[32] Collingwood, D. and Irving, R. (1989). A decomposition theorem for certain self-dual modules in the category $\mathcal{O}$, *Duke Math. J.* **58** (1989), 1, pp. 89–102.

[33] Crane, L. (1995). Clock and category: is quantum gravity algebraic? *J. Math. Phys.* **36**, 11, pp. 6180–6193.

[34] Crane, L. and Frenkel, I. (1994). Four-dimensional topological quantum field theory, Hopf categories, and the canonical bases. Topology and physics, *J. Math. Phys.* **35**, 10, pp. 5136–5154.

[35] Dlab, V. (2000). Properly stratified algebras, *C. R. Acad. Sci. Paris Sér. I Math.* **331**, 3, pp. 191–196.

[36] Dlab, V. and Ringel, C.M. (1989). Quasi-hereditary algebras. *Illinois J. Math.* **33**, 2, pp. 280–291.

[37] Dixmier, J. (1996). *Enveloping algebras* (American Mathematical Society)

[38] Dixmier, J. (1970). Sur les algèbres de Weyl. II. *Bull. Sci. Math. (2)* **94**, pp. 289–301.

[39] Drozd, Y. (1983). Representations of Lie algebras $\mathfrak{sl}(2)$, *Visnik Kyiv. Univ. Ser. Mat. Mekh.* **25**, pp. 70–77.

[40] Drozd, Y., Futorny, V. and Ovsienko, S. (1991). On Gelfand–Zetlin modules, *Rend. Circ. Mat. Palermo (2) Suppl.* **26**, pp. 143–147.

[41] Drozd, Y., Guzner, B. and Ovsienko, S. (1996). Weight modules over generalized Weyl algebras, *J. Algebra* **184**, 2, pp. 491–504.

[42] Drozd, Y. and Kirichenko, V. (1994) *Finite-dimensional algebras*, (Springer).

[43] Duflo, M. (1975). Construction of primitive ideals in an enveloping algebra, *Lie groups and their representations*, pp. 77–93, (Halsted).

[44] Enright, T. (1979). On the fundamental series of a real semisimple Lie algebra: their irreducibility, resolutions and multiplicity formulae, *Ann. of Math. (2)* **110**, 1, pp. 1–82.

[45] Enright, T. and Wallach, N. (1980). Notes on homological algebra and representations of Lie algebras, *Duke Math. J.* **47**, 1, pp. 1–15.

[46] Erdmann, K. and Wildon, M. (2006). *Introduction to Lie algebras*, (Springer).

[47] Fernando, S. (1990). Lie algebra modules with finite-dimensional weight spaces, *I. Trans. Amer. Math. Soc.* **322**, 2, pp. 757–781.

[48] Frenkel, I., Khovanov, M. and Stroppel, C. (2006). A categorification of finite-dimensional irreducible representations of quantum $\mathfrak{sl}_2$ and their tensor products, *Selecta Math. (N.S.)* **12**, 3-4, pp. 379–431.

[49] Fulton, W. and Harris, J. (1991). *Representation theory. A first course*, *Graduate Texts in Mathematics*, **129** (Springer).

[50] Gabriel, P. (1959). Lectures at the Séminaire Godement, unpublished notes, (Paris).

[51] Gelfand, I. and Ponomarev, V. (1968). Indecomposable representations of the Lorentz group, *Uspehi Mat. Nauk* **23**, 2, pp. 3–60.

[52] Gelfand, I. and Zetlin, M. (1950). Finite-dimensional representations of the group of unimodular matrices, *Doklady Akad. Nauk SSSR (N.S.)* **71**, pp. 825–828.

[53] Griffiths, P. and Harris, J. (1994). *Principles of algebraic geometry*, (John Wiley & Sons).

[54] Halmos, P. (1982). *A Hilbert space problem book, Graduate Texts in Mathematics*, **19**, (Springer).

[55] Harish-Chandra, (1949). On representations of Lie algebras, *Ann. of Math. (2)* **50**, pp. 900–915.

[56] Hiller, H. (1982) *Geometry of Coxeter groups, Research Notes in Mathematics*, **54**, (Pitman).

[57] Humphreys, J. (1978). *Introduction to Lie algebras and representation theory, Graduate Texts in Mathematics*, **9**, (Springer).

[58] Humphreys, J. (2008). *Representations of semisimple Lie algebras in the BGG category* $\mathcal{O}$, *Graduate Studies in Mathematics*, **94**, (American Mathematical Society).

[59] Irving, R. (1993). Shuffled Verma modules and principal series modules over complex semisimple Lie algebras, *J. London Math. Soc. (2)* **48**, 2, pp. 263–277.

[60] Ito, T., Terwilliger, P. and Weng, C. (2006). The quantum algebra $U_q(\mathfrak{sl}_2)$ and its equitable presentation, *J. Algebra* **298**, 1, pp. 284–301.

[61] Jacobson, N. (1943). *The Theory of Rings, American Mathematical Society Mathematical Surveys* I, (American Mathematical Society).

[62] Jantzen, J.-C. (1974). Zur Charakterformel gewisser Darstellungen halbeinfacher Gruppen und Lie-Algebren, *Math. Z.* **140**, pp. 127–149.

[63] Jantzen, J.-C. (1979). *Moduln mit einem höchsten Gewicht, Lecture Notes in Mathematics* **750**, (Springer).

[64] Jantzen, J.-C. (1983). *Einhüllende Algebren halbeinfacher Lie-Algebren. Ergebnisse der Mathematik und ihrer Grenzgebiete (3)*, **3**, (Springer).

[65] Jantzen, J.-C. (1996). *Lectures on quantum groups, Graduate Studies in Mathematics*, **6**, (American Mathematical Society).

[66] Joseph, A. (1995). *Quantum groups and their primitive ideals, Ergebnisse der Mathematik und ihrer Grenzgebiete (3)*, **29**, (Springer).

[67] Joseph, A. (1994) Enveloping algebras: problems old and new, *Lie theory and geometry*, (Birkhäuser) *Progr. Math.* **123**, pp. 385–413.

[68] Joseph, A. (1982). The Enright functor on the Bernstein-Gelfand-Gelfand category $\mathcal{O}$, *Invent. Math.* **67**, 3, pp. 423–445.

[69] Kane, R. (2001). *Reflection groups and invariant theory, CMS Books in Mathematics*, **5**, (Springer).

[70] Kazhdan, D. and Lusztig, G. (1979). Representations of Coxeter groups and Hecke algebras, *Invent. Math.* **53**, 2, pp. 165–184.

[71] Kåhrström, J. (2007). Bilinear forms on $\mathfrak{sl}_2$-modules and a hypergeometric identity, Preprint arXiv:math/0702137.

[72] Khomenko, O. (2005). Categories with projective functors, *Proc. London Math. Soc. (3)* **90**, 3, pp. 711–737.

[73] Khomenko, O. and Mazorchuk, V. (2005). On Arkhipov's and Enright's functors, *Math. Z.* **249**, 2, pp. 357–386.

[74] Khomenko, O., Koenig, S. and Mazorchuk, V. (2005). Finitistic dimension and tilting modules for stratified algebras, *J. Algebra* **286**, 2, pp. 456–475.

[75] Khovanov, M. (2000). A categorification of the Jones polynomial, *Duke Math. J.* **101**, 3, pp. 359–426.

[76] Khovanov, M. and Lauda, A. (2008). A diagrammatic approach to categorification of quantum groups III, Preprint arXiv:0807.3250.

[77] Kostant, B. (1975). On the tensor product of a finite and an infinite dimensional representation, *J. Functional Analysis* **20**, 4, pp. 257–285.

[78] Kostant, B. (1978). On Whittaker vectors and representation theory, *Invent. Math.* **48**, 2, pp. 101–184.

[79] Martínez-Villa, R. and Saorín, M. (2004). Koszul equivalences and dualities, *Pacific J. Math.* **214**, 2, pp. 359–378.

[80] Mathieu, O. (2000). Classification of irreducible weight modules, *Ann. Inst. Fourier (Grenoble)* **50**, 2, pp. 537–592.

[81] Mazorchuk, V. (1999). Orthogonal Gelfand–Zetlin algebras. I, *Beiträge Algebra Geom.* **40**, 2, pp. 399–415.

[82] Mazorchuk, V. (2000). *Generalized Verma modules, Mathematical Studies Monograph Series*, **8**, (VNTL Publishers).

[83] Mazorchuk, V. (2001). A note on centralizers in q-deformed Heisenberg algebras, *AMA Algebra Montp. Announc.* Paper 2, 6 pp.

[84] Mazorchuk, V. (2004). Stratified algebras arising in Lie theory, *Fields Inst. Commun.* **40**, pp. 245–260.

[85] Mazorchuk, V. (2007). Some homological properties of the category $\mathcal{O}$, *Pacific J. Math.* **232**, 2, pp. 313–341.

[86] Mazorchuk, V. (2005). Applications of the category of linear complexes of tilting modules associated with the category $\mathcal{O}$, Preprint arXiv:math/0501220, to appear in *Alg. Rep. Theory*.

[87] Mazorchuk, V. and Ovsienko, S. (2005). A pairing in homology and the category of linear complexes of tilting modules for a quasi-hereditary algebra. With an appendix by Stroppel, C., *J. Math. Kyoto Univ.* **45**, 4, pp. 711–741.

[88] Mazorchuk, V., Ovsienko, S. and Stroppel, C. (2009). Quadratic duals, Koszul dual functors, and applications, *Trans. Amer. Math. Soc.* **361**, 3, pp. 1129–1172.

[89] Mazorchuk, V. and Stroppel, C. (2005). Translation and shuffling of projectively presentable modules and a categorification of a parabolic Hecke module, *Trans. Amer. Math. Soc.* **357**, 7, pp. 2939–2973.

[90] Mazorchuk, V. and Stroppel, C. (2007). On functors associated to a simple root, *J. Algebra* **314**, 1, pp. 97–128.

[91] Mazorchuk, V. and Stroppel, C. (2008). Categorification of (induced) cell modules and the rough structure of generalised Verma modules, *Adv. Math.* **219**, 4, pp. 1363–1426.

[92] Mazorchuk, V. and Stroppel, C. (2008). Projective-injective modules, Serre functors and symmetric algebras, *J. Reine Angew. Math.* **616**, pp. 131–165.

[93] Mazorchuk, V. and Turowska, L. (2001). Existence and uniqueness of σ-forms on finite-dimensional modules, *Methods Funct. Anal. Topology* **7**, 1, pp. 53–62.

[94] McConnell, J. and Robson, J. (1973). Homomorphisms and extensions of modules over certain differential polynomial rings, *J. Algebra* **26**, pp. 319–342.

[95] Miller, W. (1964). On Lie algebras and some special functions of mathematical physics, *Mem. Amer. Math. Soc.* **50**.

[96] Moody, R. and Pianzola, A. (1995). *Lie algebras with triangular decompositions*, (John Wiley & Sons, Inc.).

[97] Nazarova, L. and Roiter, A. (1973). A certain problem of I. M. Gelfand, *Funkcional. Anal. i Prilozen.* **7**, 4, pp. 54–69.

[98] Newman, M. (1942). On theories with a combinatorial definition of "equivalence", *Ann. of Math. (2)* **43**, pp. 223–243.

[99]   Puninskii, G. (1999). Left almost split morphisms and generalized Weyl algebras, *Mat. Zametki* **66**, 5, pp. 734–740.

[100] Quillen, D. (1969). On the endomorphism ring of a simple module over an enveloping algebra, *Proc. Amer. Math. Soc* **21**, pp. 171–172.

[101] Ringel, C.M. (1991). The category of modules with good filtrations over a quasi-hereditary algebra has almost split sequences, *Math. Z.* **208**, 2, pp. 209–223.

[102] Rocha-Caridi, A. (1980). Splitting criteria for $\mathfrak{g}$-modules induced from a parabolic and the Bernstein–Gelfand–Gelfand resolution of a finite-dimensional, irreducible $\mathfrak{g}$-module, *Trans. AMS* **262**, 2, pp. 335–366.

[103] Rueda, S. (2002). Some algebras similar to the enveloping algebra of $\mathfrak{sl}(2)$, *Comm. Algebra* **30**, 3, pp. 1127–1152.

[104] Rueda, S. (2002). Some representations of algebras similar to the enveloping algebra of $\mathfrak{sl}(2)$, *Colecc. Congr.* **35**, pp. 163–170.

[105] Ryom-Hansen, S. (2004). Koszul duality of translation- and Zuckerman functors, *J. Lie Theory* **14**, 1, pp. 151–163.

[106] Serre, J.-P. (1992). *Lie algebras and Lie groups, Lecture Notes in Mathematics*, **1500**, (Springer).

[107] Smith, S. (1990). A class of algebras similar to the enveloping algebra of $\mathfrak{sl}(2)$, *Trans. Amer. Math. Soc.* **322**, 1, pp. 285–314.

[108] Soergel, W. (1986). Équivalences de certaines catégories de $\mathfrak{g}$-modules, *C. R. Acad. Sci. Paris Sér. I Math.* **303**, 15, pp. 725–728.

[109] Soergel, W. (1990). Kategorie $\mathcal{O}$, perverse Garben und Moduln über den Koinvarianten zur Weylgruppe, *J. Amer. Math. Soc.* **3**, 2, pp. 421–445.

[110] Soergel, W. (1992). The combinatorics of Harish-Chandra bimodules, *J. Reine Angew. Math.* **429**, pp. 49–74.

[111] Soergel, W. (1997). Charakterformeln für Kipp-Moduln über Kac-Moody-Algebren, *Represent. Theory* **1**, pp. 115–132.

[112] Stroppel, C. (2004). Composition factors of quotients of the universal enveloping algebra by primitive ideals, *J. London Math. Soc. (2)* **70**, 3, pp. 643–658.

[113] Stroppel, C. (2003). Category $\mathcal{O}$: quivers and endomorphism rings of projectives, *Represent. Theory* **7**, pp. 322–345.

[114] Stroppel, C. (2003). Category $\mathcal{O}$: gradings and translation functors, *J. Algebra* **268**, 1, pp. 301–326.

[115] Stroppel, C. (2005). Categorification of the Temperley-Lieb category, tangles, and cobordisms via projective functors, *Duke Math. J.* **126**, 3, pp. 547–596.

[116] Stroppel, C. (2006). Perverse Sheaves on Grassmannians, Springer Fibres and Khovanov cohomology, Preprint math.RT/0608234, to appear in *Compositio Math.*

[117] Vybornov, M. (2007). Perverse sheaves, Koszul IC-modules, and the quiver for the category $\mathcal{O}$, *Invent. Math.* **167**, 1, pp. 19–46.

[118] Zuckerman, G. (1977). Tensor products of finite and infinite dimensional representations of semisimple Lie groups, *Ann. Math. (2)*, **106**, 2, pp. 295–308.

Index of Notation

θ_i^{i+1}, 172
θ_{i+1}^i, 172
θ_{-1}^i, 176
θ_i^{-1}, 176
θ_i^j, 176
$M_n(\lambda)$, 194
$\tilde{\mathcal{O}}$, 179

$\underline{H}_x$, 188
ε, 34
ϑ_i^i, 176
$\{\mathbf{e}, \mathbf{f}, \mathbf{h}\}$, 2
$\{\mathbf{h}\}$, 44
c, 45
u_M, 204

Index